CAMBRIDGE STUDIES IN
ADVANCED MATHEMATICS 76

HODGE THEORY AND COMPLEX ALGEBRAIC GEOMETRY I

HODGE THEORY AND COMPLEX ALGEBRAIC GEOMETRY I

CLAIRE VOISIN
CNRS, Institut de Mathématiques de Jussieu

Translated by Leila Schneps

CAMBRIDGE
UNIVERSITY PRESS

CAMBRIDGE UNIVERSITY PRESS
Cambridge, New York, Melbourne, Madrid, Cape Town, Singapore,
São Paulo, Delhi, Dubai, Tokyo, Mexico City

Cambridge University Press
The Edinburgh Building, Cambridge CB2 8RU, UK

Published in the United States of America by Cambridge University Press, New York

www.cambridge.org
Information on this title: www.cambridge.org/9780521718011

First published 2002
Reprinted 2004

A catalogue record for this publication is available from the British Library

Library of Congress Cataloguing in Publication data
Voisin, Claire.
Hodge theory and complex algebraic geometry / Claire Voisin.
p. cm. – (Cambridge studies in advanced mathematics)
Includes bibliographical references and index.
ISBN 0 521 80260 1
1. Hodge theory. 2. Geometry, Algebraic. I. Title. II. Series.
QA564 .V65 2002
516.3'5 – dc21 2002017389

ISBN 978-0-521-80260-4 Hardback
ISBN 978-0-521-71801-1 Paperback

Contents

0
Introduction

Kähler manifolds and projective manifolds. The goal of this first volume is to explain the existence of special structures on the cohomology of Kähler manifolds, namely, the Hodge decomposition and the Lefschetz decomposition, and to discuss their basic properties and consequences. The second volume will be devoted to the systematic application of these results in different directions, relating Hodge theory, topology and the study of algebraic cycles on smooth projective complex manifolds.

Indeed, smooth projective complex manifolds are special cases of compact Kähler manifolds. A Kähler manifold is a complex manifold equipped with a Hermitian metric whose imaginary part, which is a 2-form of type $(1, 1)$ relative to the complex structure, is closed. This 2-form is called the Kähler form of the Kähler metric. As complex projective space (equipped, for example, with the Fubini–Study metric) is a Kähler manifold, the complex submanifolds of projective space equipped with the induced metric are also Kähler. We can indicate precisely which members of the set of Kähler manifolds are complex projective, thanks to Kodaira's theorem:

Theorem 0.1 *A compact complex manifold admits a holomorphic embedding into complex projective space if and only if it admits a Kähler metric whose Kähler form is of integral class.*

In this volume, we are essentially interested in the class of Kähler manifolds, without particularly emphasising projective manifolds. The reason is that our goal here is to establish the existence of the Hodge decomposition and the Lefschetz decomposition on the cohomology of such a manifold, and for this, there is no need to assume that the Kähler class is integral. However, the Lefschetz decomposition will be defined on the rational cohomology only in the projective case, and this is already an important reason to restrict ourselves, later, to the case of projective manifolds. Indeed, in this text, we will introduce the notions of polarised Hodge structure and the polarised period domain parametrising polarised Hodge structures. These polarised period domains have curvature properties which the non-polarised period domains do not possess. The Lefschetz decomposition, when it is defined on the rational or integral cohomology, splits the cohomology of a Kähler manifold into a direct sum of polarised Hodge structures.

In studying the applications of Hodge theory, another reason to restrict ourselves to projective manifolds is the fact that a Kähler manifold does not, in general, have complex submanifolds, whereas projective manifolds have many, so many that in fact it is currently conjectured, as a vast generalisation of the Hodge conjecture, that the Hodge structures on a projective manifold X are governed by, and determine in a sense to be explained later, the geometry of the algebraic subvarieties of X, and more precisely the Chow groups of X.

The Hodge decomposition. If X is a complex manifold, the tangent space to X at each point x is equipped with a complex structure J_x. The data consisting of this complex structure at each point is what is known as the underlying almost complex structure. The J_x provide a decomposition

$$T_{X,x} \otimes \mathbb{C} = T^{1,0}_{X,x} \oplus T^{0,1}_{X,x}, \tag{0.1}$$

where $T^{0,1}_{X,x}$ is the vector space of complexified tangent vectors $u \in T_{X,x}$ such that $J_x u = -iu$ and $T^{1,0}_{X,x}$ in the complex conjugate of $T^{0,1}_{X,x}$. From the point of view of the complex structure, i.e. of the local data of holomorphic coordinates, the vector fields of type (0, 1) are those which kill the holomorphic functions.

The decomposition (0.1) induces a similar decomposition on the bundles of complex differential forms

$$\Omega^k_{X,\mathbb{C}} := \Omega^k_{X,\mathbb{R}} \otimes \mathbb{C} = \bigoplus\nolimits_{p+q=k} \Omega^{p,q}_X, \tag{0.2}$$

where

$$\Omega^{p,q}_X \cong \bigwedge^p \Omega^{1,0}_X \otimes \bigwedge^q \Omega^{0,1}_X$$

and

$$\Omega_{X,\mathbb{R}} \otimes \mathbb{C} = \Omega^{1,0}_X \oplus \Omega^{0,1}_X$$

is the dual decomposition of (0.1). The decomposition (0.2) has the property of Hodge symmetry

$$\overline{\Omega^{p,q}_X} = \Omega^{q,p}_X,$$

where complex conjugation acts naturally on $\Omega^k_{X,\mathbb{C}} = \Omega^k_{X,\mathbb{R}} \otimes \mathbb{C}$.

If we let $A^k_{\mathbb{C}}(X)$ denote the space of complex differential forms of degree k on X, i.e. the $\mathcal{C}^\infty$ sections of the vector bundle $\Omega^k_{X,\mathbb{C}}$, then we also have the exterior differential

$$d : A^k_{\mathbb{C}}(X) \to A^{k+1}_{\mathbb{C}}(X),$$

which satisfies $d \circ d = 0$. We then define the kth de Rham cohomology group of X by

$$H^k(X, \mathbb{C}) = \frac{\operatorname{Ker}\left(d : A^k_{\mathbb{C}}(X) \to A^{k+1}_{\mathbb{C}}(X)\right)}{\operatorname{Im}\left(d : A^{k-1}_{\mathbb{C}}(X) \to A^k_{\mathbb{C}}(X)\right)}.$$

The main theorem proved in this book is the following.

Theorem 0.2 *Let $H^{p,q}(X) \subset H^k(X, \mathbb{C})$ be the set of classes which are representable by a closed form α which is of type (p, q) at every point x in the decomposition (0.2). Then we have a decomposition*

$$H^k(X, \mathbb{C}) = \bigoplus_{p+q=k} H^{p,q}(X). \tag{0.3}$$

Note that by definition, we have the Hodge symmetry

$$H^{p,q}(X) = \overline{H^{q,p}(X)},$$

where complex conjugation acts naturally on $H^k(X, \mathbb{C}) = H^k(X, \mathbb{R}) \otimes \mathbb{C}$. Here $H^k(X, \mathbb{R})$ is defined by replacing the complex differential forms by real differential forms in the above definition.

This theorem immediately gives constraints on the cohomology of a Kähler manifold, which reveal the existence of compact complex manifolds which are not Kähler. For example, the decomposition (0.3) and the Hodge symmetry imply that the dimensions $\dim_{\mathbb{C}} H^k(X, \mathbb{C})$ (called the Betti numbers) are even for odd k, a property not satisfied by Hopf surfaces. These surfaces are the quotients of $\mathbb{C}^2 - \{0\}$ by the fixed-point-free action of a group isomorphic to $\mathbb{Z}$, where a generator g acts via

$$g(z_1, z_2) = (\lambda_1 z_1, \lambda_2 z_2),$$

where the λ_i are non-zero complex numbers of modulus strictly less than 1. These surfaces are compact, equipped with the quotient complex structures, and their π_1 is isomorphic to $\mathbb{Z}$ since $\mathbb{C}^2 - \{0\}$ is simply connected. Thus, their first Betti number is equal to 1, which implies that they are not Kähler.

The Lefschetz decomposition. The Lefschetz decomposition is another decomposition of the cohomology of a compact Kähler manifold X, this time of a topological nature. It depends only on the cohomology class of the Kähler form

$$[\omega] \in H^2(X, \mathbb{R}).$$

The exterior product on differential forms satisfies Leibniz' rule

$$d(\alpha \wedge \beta) = d\alpha \wedge \beta + (-1)^{d^0\alpha} \alpha \wedge d\beta,$$

so the exterior product with ω sends closed forms (i.e. forms killed by d) to closed forms and exact forms (i.e. forms in the image of d) to exact forms. Thus it induces an operator, called the Lefschetz operator,

$$L : H^k(X, \mathbb{R}) \to H^{k+2}(X, \mathbb{R}).$$

The following theorem is sometimes called the hard Lefschetz theorem.

Theorem 0.3 *For every $k \leq n = \dim X$, the map*

$$L^{n-k} : H^k(X, \mathbb{R}) \to H^{2n-k}(X, \mathbb{R}) \tag{0.4}$$

is an isomorphism.

(Note that the spaces on the right and on the left are of the same dimension by Poincaré duality, which is valid for all compact oriented manifolds.)

A very simple consequence of the above isomorphism is the following result, which is an additional topological constraint satisfied by Kähler manifolds.

Corollary 0.4 *The morphism*

$$L : H^k(X, \mathbb{R}) \to H^{k+2}(X, \mathbb{R})$$

is injective for $k < n = \dim X$. Thus, the odd Betti numbers $b_{2k-1}(X)$ increase with k for $2k - 1 \leq n$, and similarly, the even Betti numbers $b_{2k}(X)$ increase for $2k \leq n$.

An algebraic consequence of Lefschetz' theorem is the Lefschetz decomposition, which as we noted earlier is particularly important in the case of projective manifolds. Let us define the primitive cohomology of a compact Kähler manifold X by

$$H^k(X, \mathbb{R})_{\text{prim}} := \text{Ker}\,(L^{n-k+1} : H^k(X, \mathbb{R}) \to H^{2n-k+2}(X, \mathbb{R}))$$

for $k \leq n$. (One can extend this definition to the cohomology of degree $> n$ by using the isomorphism (0.4).)

Theorem 0.5 *The natural map*

$$i : \bigoplus_{k-2r\geq 0} H^{k-2r}(X, \mathbb{R})_{\text{prim}} \to H^k(X, \mathbb{R})$$

$$(\alpha_r) \mapsto \sum_r L^r \alpha_r$$

is an isomorphism for $k \leq n$.

Once again, we can extend this decomposition to the cohomology of degree $> n$ by using the isomorphism (0.4).

Harmonic forms and cohomology. Let us now express the main principle of Hodge theory, which has immense applications. The study of the cohomology of Kähler manifolds and the proof of the theorems 0.2 and 0.3, which are the main content of this book, are among the most important applications, but the principle applies in various other situations. The vanishing theorems for the cohomology of line bundles equipped with a Chern connection with positive curvature, whose proofs will only be sketched here, provide another example of possible applications. The applications to the topology of compact Riemannian manifolds under certain curvature hypotheses are also very important, but they lie outside of the scope of this book.

Following Weil (1957), we restrict ourselves here to giving an explanation of the main idea, which is the notion of a harmonic form, and the application of the theory of elliptic operators which makes it possible to represent the cohomology classes by harmonic forms, but we will omit the proof of the fundamental theorem on elliptic operators, which uses estimations and notions from analysis (Sobolev spaces), which are in different directions from the aims of this book. The delicate point consists in passing from spaces of L^2 differential forms, in which the Hodge decomposition is algebraically obvious, to spaces of C^∞ differential forms. One of the problems we encounter is the fact that the operators considered here are differential operators, and thus do not define continuous operators on the spaces of L^2 forms. We refer to Demailly (1996) for a presentation of this analytic aspect of Hodge theory.

The idea that we want to explain here is the following: using the metric on X, we can define the L^2 metric on the spaces of differential forms

$$(\alpha, \beta)_{L^2} = \int_X \langle \alpha, \beta \rangle_x \text{Vol},$$

where α, β are differential forms of degree k and the scalar product $\langle \alpha, \beta \rangle_x$ at a point $x \in X$ is induced by the evaluation of the forms at the point x and by the metric at the point x.

The operator $d : A^k(X) \to A^{k+1}(X)$ is a differential operator, and we can construct its formal adjoint $d^* : A^k(X) \to A^{k-1}(X)$, which is also a differential operator, and satisfies the identity

$$(\alpha, d\beta)_{L^2} = (d^*\alpha, \beta)_{L^2}$$

for $\alpha \in A^k(X)$, $\beta \in A^{k-1}(X)$. This adjunction relation only makes d^* into a formal adjoint, since these operators are not defined on the Hilbert space $L^2(\Omega_X^*)$ of L^2 differential forms, which is the completion of $A^*(X)$ for the L^2 metric.

The idea of Hodge theory consists in using the adjoint d^* to write the decompositions

$$A^k(X) = \operatorname{Im} d \oplus \operatorname{Im} d^\perp = \operatorname{Im} d \oplus \operatorname{Ker} d^*,$$
$$A^k(X) = \operatorname{Ker} d \oplus \operatorname{Ker} d^\perp = \operatorname{Ker} d \oplus \operatorname{Im} d^*,$$

and finally, using the inclusion $\operatorname{Im} d \subset \operatorname{Ker} d$,

$$A^k(X) = \operatorname{Im} d \oplus \operatorname{Im} d^* \oplus \operatorname{Ker} d \cap \operatorname{Ker} d^*.$$

Of course, these identities, which would be valid on finite-dimensional spaces or Hilbert spaces since the operator d has closed image there, require the analysis mentioned above in order to justify them here.

Apart from this issue, if we accept these identities, we see that the space

$$\mathcal{H}^k := \operatorname{Ker} d \cap \operatorname{Ker} d^* \subset A^k(X)$$

of harmonic forms projects bijectively onto $H^k(X, \mathbb{R})$ (or $H^k(X, \mathbb{C})$ if we study the cohomology with complex coefficients), since it is a supplementary subspace of $\operatorname{Im} d$ inside $\operatorname{Ker} d$.

Another characterisation of harmonic forms uses the Laplacian

$$\Delta_d = dd^* + d^*d.$$

Indeed, it is very easy to see that we have

$$\mathcal{H}^k = \operatorname{Ker} \Delta_d.$$

The operator Δ_d is an elliptic operator. This property of a differential operator can be read directly from its symbol, which is essentially its homogeneous term of largest order (which is 2 for the Laplacian). The decompositions written above are special cases of the decomposition associated to an elliptic operator.

Kähler identities. The Hodge decomposition (0.3) is obtained by combining the Hodge theory sketched above and the study of the properties of the Laplacian of a Kähler manifold. We have already mentioned various operators acting on the spaces of differential forms of a Kähler manifold, namely d, L and their formal adjoints d^*, Λ for the L^2 metric. Moreover, the complex structure makes it possible to decompose d as

$$d = \partial + \overline{\partial},$$

where the Dolbeault operator $\overline{\partial}$ sends $\alpha \in A^{p,q}(X)$ to the component of bidegree $(p, q+1)$ of $d\alpha$. Here $A^{p,q}(X)$ is the space of differential forms of bidegree (p, q) at every point of x; it is also the space of sections of the bundle $\Omega_X^{p,q}$ which appears in the decomposition (0.2) given by the complex structure. The differential operators ∂ and $\overline{\partial}$ are differential operators of order 1, and have formal adjoint operators ∂^* and $\overline{\partial}^*$.

The Kähler identities establish commutation relations between these operators. For example, we have the identity

$$[\Lambda, \partial] = i\overline{\partial}^*,$$

and the other identities follow from this one via passage to the complex conjugate or to the adjoint.

From these identities, and from the fact that L commutes with d while ∂ and $\overline{\partial}$ anticommute, we deduce the following result.

Theorem 0.6 *The Laplacians Δ_d, Δ_∂ and $\Delta_{\overline{\partial}}$ associated to the operators d, ∂ and $\overline{\partial}$ respectively satisfy the equalities*

$$\Delta_d = 2\Delta_\partial = 2\Delta_{\overline{\partial}}. \tag{0.5}$$

We deduce that the harmonic forms for d are also harmonic for ∂ and $\overline{\partial}$, and in particular are also ∂- and $\overline{\partial}$-closed. Finally, as the operators ∂ and $\overline{\partial}$ are bihomogeneous (of bidegree $(1, 0)$ and $(0, 1)$ respectively) for the bigraduation of the spaces of differential forms given by the decomposition (0.2), it follows easily that each of the Laplacians Δ_∂ and $\Delta_{\overline{\partial}}$ is bihomogeneous of bidegree $(0, 0)$, i.e. preserves the forms of type (p, q) for every bidegree (p, q). The same then holds for Δ_d by the equality (0.5). The Hodge decomposition is then obtained simply by the decomposition of the harmonic forms as sums of forms of type (p, q):

Corollary 0.7 *Let X be a compact Kähler manifold. If ω is a harmonic form (for the Laplacian associated to the operator d and to the metric), its components of type (p, q) are harmonic. Thus, we have a decomposition*

$$\mathcal{H}^k(X) = \bigoplus \mathcal{H}^{p,q}, \tag{0.6}$$

where $\mathcal{H}^{p,q}$ is the space of harmonic forms of type (p, q) at every point of X.

The Hodge decomposition (0.3) is obtained by combining the theorem of representation of cohomology classes by harmonic forms with the decomposition (0.6).

The Lefschetz decomposition is also an easy consequence of the decomposition (0.6). Indeed, we first show that theorem 0.3 holds for the operator L acting on differential forms. Furthermore, the Kähler identities show that L commutes with the Laplacian, so that the operators L^r send harmonic forms to harmonic forms, and once the theorem is proved on the level of forms, it remains valid on the level of harmonic forms, and thus also on cohomology classes.

De Rham cohomology and Betti cohomology. The Hodge decomposition (0.3) gives an extremely interesting structure when it is combined with the integral structure on the cohomology

$$H^k(X, \mathbb{C}) = H^k(X, \mathbb{Z}) \otimes_{\mathbb{Z}} \mathbb{C}.$$

For this equality, which follows from the change of coefficients theorem, one must adopt a different definition of cohomology, which does not make use of differential forms.

For one possible definition, we can introduce the singular cohomology

$$H^k_{\mathrm{sing}}(X, \mathbb{Z}).$$

We start from the complex

$$C_*(X), \quad \partial : C_k(X) \to C_{k-1}(X)$$

of singular chains, where $C_k(X)$ is the free abelian group generated by the continuous maps from the simplex Δ_k of dimension k to X. The map ∂ is given by the restriction to the boundary

$$\partial\phi = \sum_i (-1)^i \phi_{|\partial\Delta_{k,i}},$$

where $\Delta_{k,i}$ is the ith face of Δ_k. The complex $(C^*_{\mathrm{sing}}(X), d)$ of singular cochains is then defined as the dual complex of $(C_*(X), \partial)$. Its cohomology is the singular cohomology $H^*_{\mathrm{sing}}(X, \mathbb{Z})$. We have the following theorem, due to de Rham.

Theorem 0.8 *For $K = \mathbb{R}$ or $K = \mathbb{C}$, we have*

$$H^k(X, K) = H^k_{\text{sing}}(X, \mathbb{Z}) \otimes_{\mathbb{Z}} K.$$

If we consider the complex of differentiable chains, we can prove this theorem by using the natural map from $A^k(X)$ to $C^k_{\text{sing}}(X)$ given by integration:

$$\alpha \mapsto \left(\phi \mapsto \int_{\Delta_k} \phi^* \alpha\right).$$

Sheaves and cohomology. A much more conceptual proof of de Rham's theorem can be given by using the language of sheaf theory and sheaf cohomology, which we present here, and whose usefulness will appear frequently throughout this book: it will be used, for example, in the Hodge decomposition, to describe the spaces $H^{p,q}$ as the Dolbeault cohomology groups $H^q(X, \Omega^p_X)$, which are defined for every complex manifold X as the qth cohomology group of X with values in the sheaf Ω^p_X of holomorphic differential forms of degree p. (Note, however, that this identification is valid only in the Kähler case. In general, without the Kähler hypothesis, we cannot identify $H^q(X, \Omega^p_X)$ with the space of cohomology classes of degree $p + q$ which are representable by a closed form of type (p, q) at every point.)

The notion of a sheaf $\mathcal{F}$ (of groups, for example) over a topological space X is a set-theoretic notion. It is given by the following data: the group $\mathcal{F}(U)$ of "sections of $\mathcal{F}$ over U" for every open subset U of X, and restriction maps

$$\mathcal{F}(U) \to \mathcal{F}(V)$$

for every inclusion $V \subset U$. These restrictions are compatible in an obvious way when we take three open sets $W \subset V \subset U$. We also require that a section from $\mathcal{F}$ to U is determined exactly by its restrictions to the open subsets of an open cover of U, which of course must coincide on the intersections of two of these open sets. The first typical example of a sheaf is the sheaf of local sections of a vector bundle over X. Another example is given by the constant sheaves of stalk G, where G is a fixed abelian group; to an open set U, we associate the locally constant maps defined on U with values in G.

The sheaves of abelian groups over X form an abelian category which has "sufficiently many injective objects" (see chapter 4). Thus, the theory of derived functors applies to this category. The main functors which interest us here are the functors of global sections Γ of the category of sheaves of abelian groups on X to the category of abelian groups, or the direct image functor from the category of sheaves of abelian groups on X to the category of sheaves of abelian groups on Y, for a continuous map $\phi : X \to Y$.

These functors are left-exact. Given a left-exact functor $F : \mathcal{A} \to \mathcal{B}$ of an abelian category $\mathcal{A}$ having sufficiently many injective objects to an abelian category $\mathcal{B}$, we define $R^i F(M)$, $M \in \mathcal{O}b(\mathcal{A})$ as the ith cohomology group of the complex $(F(M\cdot), d)$, where $(M\cdot, d)$ is an injective resolution of M. In fact, more generally, we can take resolutions by F-acyclic objects. The important point is that given two such resolutions, we have a canonical isomorphism between the objects $R^i F(M)$ calculated via the two resolutions.

Returning to the case of the functor of global sections Γ, we show using Poincaré's theorem that the sheaves of differential forms form a Γ-acyclic resolution of the constant sheaf $\mathbb{C}_X$ (often written $\mathbb{C}$) of stalk $\mathbb{C}$, so that the space $H^k(X, \mathbb{C})$ defined above must be understood as the kth cohomology group of X with values in $\mathbb{C}_X$, i.e. $R^k\Gamma(\mathbb{C}_X)$. Similarly, we can interpret the singular cohomology as the cohomology of the complex of global sections of a Γ-acyclic resolution of the constant sheaf of stalk $\mathbb{Z}$. Thus, we have $H^k_{\text{sing}}(X, \mathbb{Z}) = H^k(X, \mathbb{Z})$ canonically.

De Rham's theorem thus reduces to proving a change of coefficients theorem for the cohomology of the sheaves

$$H^k(X, \mathbb{C}) = H^k(X, \mathbb{Z}) \otimes \mathbb{C},$$

which is not difficult.

These different interpretations of the cohomology, corresponding to different resolutions, are all equally important, since they carry different types of information. For example, the Hodge decomposition of the cohomology of a Kähler manifold requires the de Rham version of the cohomology, while that of the integral structure requires another version, singular or Čech for example.

The Frölicher spectral sequence. With the exception of the statement concerning the Hodge symmetry, the theorem of Hodge decomposition can be reformulated as a theorem of degeneracy of a spectral sequence. The justification of this reformulation, particularly in the case of projective manifolds, is that it consists in a completely algebraic translation, where in fact we may even use Serre's "GAGA" principle of comparison of algebraic geometry and analytic geometry to replace the sheaves of holomorphic differential forms and their cohomology relative to the usual topology by sheaves of algebraic differential forms and their cohomology relative to the Zariski topology. Thus, we can almost give meaning to Hodge's theorem 0.2 for smooth projective manifolds defined over an arbitrary field. Under certain "lifting" hypotheses, Deligne and Illusie prove this statement for manifolds in non-zero characteristic (Illusie 1996).

The differentiable de Rham complex of a differentiable manifold, i.e. the complex of sheaves of differential forms equipped with the exterior differential, is by Poincaré's lemma a resolution of the constant sheaf (real or complex according to the context). When the manifold is complex, this complex is equipped with the decomposition (0.2), and the differential d decomposes as $d = \partial + \overline{\partial}$. Thus, the differentiable de Rham complex of a complex manifold is naturally the simple complex associated to the double complex $\mathcal{A}^{p,q}$ equipped with the two differentials ∂ and $\overline{\partial}$, which anticommute. Such a complex admits a filtration by the subcomplexes

$$F^p \mathcal{A}_X^k = \bigoplus_{l \geq p} \mathcal{A}_X^{l,k-l},$$

and the associated graded object $\mathrm{Gr}_F^p \mathcal{A}^*$ is the complex

$$0 \to \mathcal{A}_X^{p,0} \stackrel{\overline{\partial}}{\to} \mathcal{A}_X^{p,1} \stackrel{\overline{\partial}}{\to} \mathcal{A}_X^{p,2} \cdots.$$

This last complex, introduced by Dolbeault, is an acyclic resolution of the sheaf Ω_X^p of holomorphic differential forms of degree p on X, and the cohomology of the space of global sections of this complex in degree q is thus equal to $H^q(X, \Omega_X^p)$.

The theory of spectral sequences presented in chapter 8 now shows that given a filtered complex (K^*, D, F), where the filtration is decreasing and bounded, we have a spectral sequence $(E_r^{p,q}, d_r)$ which enables us to compute, via successive approximations, the graded object $\mathrm{Gr}_F^p H^n(K^*)$, where the filtration F on $H^n(K^*)$ is given by

$$F^p H^n(K^*) := \mathrm{Im}\,(H^n(F^p K^*) \to H^n(K^*)). \tag{0.7}$$

Specifically, each $(E_r^{*,*}, d_r)$ is a complex, where the differential d_r sends $E_r^{p,q}$ to $E_r^{p+r,q-r+1}$, and we have

$$E_{r+1}^{p,q} = H^{p,q}(E_r^{*,*}, d_r),$$

and for r sufficiently large,

$$E_r^{p,q} =: E_\infty^{p,q} = \mathrm{Gr}_F^p H^{p+q}(K^*).$$

Moreover, the first term of the spectral sequence is given by

$$E_1^{p,q} = H^{p+q}\left(\mathrm{Gr}_F^p K^*, \overline{d}\right),$$

where $\overline{d}$ is the differential induced by d on $\mathrm{Gr}_F^p K^*$.

Thus, in the case of the de Rham complex equipped with its filtration by the truncations (called either Hodge or naive), we see that for the corresponding spectral sequence, called the Frölicher spectral sequence, we have $E_1^{p,q} = H^q(X, \Omega_X^p)$, which enables us to realise $\mathrm{Gr}_F^p H^{p+q}(X, \mathbb{C})$ for a certain filtration on the cohomology of X as the quotient of a subspace of $H^q(X, \Omega_X^p)$, determined by the differentials d_r, $r \geq 1$. A statement very nearly equivalent to the Hodge theorem is then as follows.

Theorem 0.9 *The Frölicher spectral sequence of a compact Kähler manifold degenerates at E_1, i.e. the differentials d_r vanish for $r \geq 1$.*

The filtration F of (0.7) is then the Hodge filtration determined by the Hodge decomposition $F^p H^n(X, \mathbb{C}) = \bigoplus_{r \geq p} H^{r,n-r}(X)$; and by the degeneracy at E_1, we have

$$\mathrm{Gr}_F^p H^{p+q}(X, \mathbb{C}) = H^q\big(X, \Omega_X^p\big).$$

We must pay attention to the fact that the degeneracy at E_1 of the Frölicher spectral sequence does not imply the Hodge symmetry in the form

$$\dim H^{p,q}(X) = \dim H^{q,p}(X),$$

where for any complex manifold X, we set $H^{p,q}(X) = \mathrm{Gr}_F^p H^{p+q}(X, \mathbb{C})$, where F is the filtration of (0.7). In fact, all compact complex surfaces satisfy the degeneracy condition at E_1 of the Frölicher spectral sequence, while the Hopf surfaces mentioned above do not satisfy Hodge symmetry, since Hodge symmetry would imply that their Betti number b_1 is even.

Hodge structures. By definition, an integral Hodge structure of weight k is given by a abelian group of finite type $H_{\mathbb{Z}}$, and a Hodge decomposition

$$H_{\mathbb{C}} := H_{\mathbb{Z}} \otimes \mathbb{C} = \bigoplus_{p+q=k} H^{p,q},$$

with $\overline{H^{p,q}} = H^{q,p}$. Thus, this is the structure which exists on the degree k cohomology of a Kähler manifold X. This structure is very rich; it has an important moduli space, which parametrises the isomorphism classes of such objects. Indeed, it is determined by the position of the complex subspaces $H^{p,q}$ in the space $H_{\mathbb{C}}^k$. The moduli space we consider is thus essentially the quotient of a product of Grassmannians (parametrising each $H^{p,q}$) by the group of automorphisms $\mathrm{Aut}_{\mathbb{Z}} H_{\mathbb{Z}}^k$. Note that the decompositions (0.3) alone, without taking the integral structures into account, do not have moduli, and yield only the dimensions of the spaces $H^{p,q}$ as information, since the group of automorphisms $\mathrm{Aut}_{\mathbb{C}} H_{\mathbb{C}}^k$ acts transitively on these decompositions.

Hodge filtration and the period map. Hodge decompositions thus provide an important piece of qualitative information about the complex structure of X. Indeed, the Hodge structure on $H^k(X, \mathbb{Z})$ depends only on the complex structure of X, and not on the choice of a Kähler metric, although the proof of its existence uses such a metric in a crucial way. The second part of this book will be devoted to the study of the dependence of the Hodge structure on the complex structure. Suppose that we let the complex structure of X vary, i.e. we take a family of complex manifolds $(X_t)_{t\in B}$ which are all differentiably equivalent. As the cohomology group $H^k(X_t, \mathbb{C})$ is a differentiable and even a topological invariant, it does not depend on t, and we can see the Hodge decomposition (0.3) as a varying decomposition on a fixed vector space. This gives rise to the period map, defined on the set of small deformations up to isomorphism of the complex structure of a Kähler manifold. Indeed, we will show that sufficiently small deformations of the complex structure of a Kähler manifold are still Kähler, so that the Kähler deformations form an open subset of the space of all deformations.

We define the Hodge filtration associated to a Hodge structure by

$$F^p H^k_{\mathbb{C}} = \bigoplus_{r\geq p} H^{r,k-r}.$$

For every p, the Hodge filtration satisfies the condition

$$H^k_{\mathbb{C}} = F^p H^k_{\mathbb{C}} \oplus \overline{F^{k-p+1} H^k_{\mathbb{C}}}. \tag{0.8}$$

The Hodge decomposition is then determined by $H^{p,q} = F^p H^k_{\mathbb{C}} \cap \overline{F^q H^k_{\mathbb{C}}}$, so that these data are actually equivalent.

We will consider the (local) period domain $\mathcal{D}$ as the space parametrising the filtrations $F^p H^k_{\mathbb{C}}$ by complex vector subspaces of fixed dimension satisfying the condition (0.8). The global period domain is essentially a quotient of the preceding one by the group $\mathrm{Aut}_{\mathbb{Z}} H^k_{\mathbb{Z}}$. One can show that $\mathcal{D}$ has the structure of a complex manifold.

Now let $\pi : \mathcal{X} \to B$ be a proper holomorphic submersive map, where B is a ball. The fibres X_t are differentiably equivalent complex manifolds by Ehresmann's theorem. Such a family gives rise to the notion of a holomorphic deformation of the complex structure of $X = X_0$. We then have the following result, due to Griffiths.

Theorem 0.10 *The period map $\mathcal{P} : B \to \mathcal{D}$, which to $t \in B$ associates the Hodge filtration on $H^k(X_t, \mathbb{C}) \cong H^k(X_0, \mathbb{C})$, is holomorphic.*

In this text, we will also prove Griffiths' transversality property, which will play a major role in the second volume, and which describes the way in which the Hodge filtration varies infinitesimally with the complex structure.

These considerations lead us to study small deformations of the complex structure of a manifold. We define a deformation of a manifold X parametrised by B as a family $\pi : \mathcal{X} \to B$ together with an isomorphism $X \cong \pi^{-1}(0)$, where this pair is given up to isomorphism. We can then speak of deformation to a finite order (B is then a scheme of finite length), and the study of the finite order deformations can be identified with the study of the first order infinitesimal neighbourhoods of 0 in the universal family of deformations of X, when it exists.

We restrict ourselves here to writing the first order deformations of a manifold, or the Zariski tangent space to the universal family of the deformations of X if it exists. We give various descriptions of the Kodaira–Spencer map, which classifies the first order deformations. Finally, following Griffiths, we compute the differential of the period map and give some applications of this computation to the Torelli problem.

Classes of cycles. The final part of this volume is devoted to the definition of invariants associated to analytic cycles of a complex or Kähler manifold.

In the second volume, this aspect of Hodge theory, especially in the case of projective manifolds, will be developed to the point of predicting a perfect interaction between the complexity or the size of the Chow groups of a smooth algebraic variety and the complexity or the level of its Hodge structures. The invariants we will describe, namely the Hodge class and the Abel–Jacobi invariant of a cycle, are only the first steps in this direction.

An analytic cycle of codimension k in a complex manifold is a combination with integral coefficients of irreductible analytic subsets of codimension k. Such a cycle Z has a cohomology class $[Z] \in H^{2k}(X, \mathbb{Z})$, which we can describe in various different ways, all of which make use of the existence of a stratification of an analytic subset, where the strata are complex manifolds of codimension $\geq k$, and the open stratum is the open set of smooth points.

When the manifold X is also compact and Kähler, it is easy to see that the image of the class $[Z] \in H^{2k}(X, \mathbb{Z})$ in $H^{2k}(X, \mathbb{C})$ lies in $H^{k,k}(X)$ relative to the Hodge decomposition. Such a class is called an integral Hodge class.

When X is a projective manifold, the Hodge conjecture predicts the following.

Conjecture 0.11 *If* $\alpha \in H^{2k}(X, \mathbb{Q}) \cap H^{k,k}(X) =: \mathrm{Hdg}^{2k}(X, \mathbb{Q})$ *is a rational Hodge class, there exists a cycle Z of codimension k in X, with rational coefficients, such that* $[Z] = \alpha$.

This conjecture holds for the classes of degree 2 (Lefschetz theorem on the classes of type $(1, 1)$). In the Kähler case, the Hodge conjecture in the form given above is false even for the classes of degree 2, since there exist Kähler manifolds having Hodge classes of degree 2 and not containing any complex hypersurface (consider, for example, complex tori). From this example, we see, however, that the classes we consider lie in the group generated by the Chern classes of holomorphic vector bundles (we can show that these are also Hodge classes). Voisin (2002) shows that the Hodge conjecture becomes false when generalised to Kähler manifolds, with the classes of algebraic cycles replaced by the Chern classes of coherent sheaves on X.

The class of an analytic cycle is defined without the Kähler hypothesis. However, the Abel–Jacobi invariant of a cycle homologous to 0 makes use of Hodge theory. If X is a compact Kähler manifold, then for each integer k, we can define an intermediate Jacobian which is a complex torus

$$J^{2k-1}(X) = \frac{H^{2k-1}(X, \mathbb{C})}{F^k H^{2k-1}(X) \oplus H^{2k-1}(X, \mathbb{Z})}.$$

The Abel–Jacobi map Φ_X^k is defined on the group of cycles of codimension k which are cohomologous to 0, and has values in $J^{2k-1}(X)$. If Z is cohomologous to 0, or homologous to 0, we can write $Z = \partial\Gamma$, where Γ is a real differentiable chain of dimension $2n - 2k + 1$, $n = \dim X$. Even though the integration contour Γ is not closed, we can then define a linear form

$$\int_\Gamma \in F^{n-k+1} H^{2n-2k+1}(X)^*.$$

Hodge theory plays an essential role here. As the chain Γ is defined up to a cycle T, $\int_\Gamma$ is also defined up to a period

$$\int_T \in \operatorname{Im}(H_{2n-2k+1}(X, \mathbb{Z}) \to F^{n-k+1} H^{2n-2k+1}(X)^*).$$

The Abel–Jacobi invariant of Z is then defined by

$$\Phi_X^k(Z) = \int_\Gamma \in F^{n-k+1} H^{2n-2k+1}(X)^* / H_{2n-2k+1}(X, \mathbb{Z}) = J^{2k-1}(X).$$

We will describe some of the first properties of the Abel–Jacobi map, the fact that it is holomorphic, and the relation between the Abel–Jacobi map for families of cycles parametrised by a curve and the Hodge classes on the product of X and this curve.

We conclude this part with an introduction to Deligne cohomology, a subtle object which combines Hodge classes and intermediate Jacobians. We will construct the Deligne class of a cycle, an invariant which combines the Hodge class and the Abel–Jacobi invariant.

The organisation of this text. Four chapters are devoted to preliminaries. The first chapter is quite rudimentary, and its principal goal, apart from recalling the main results of the theory of analytic functions, is the proof of two essential results: the Riemann and Hartogs extension theorems and the existence of local solutions to the equation $\partial f/\partial\overline{z} = g$, which enables us, in the following chapter, to prove the local exactness of the Dolbeault complex.

The next chapter is an introduction to complex manifolds and to holomorphic vector bundles over them. In the real analytic case, we prove the Newlander–Nirenberg theorem which determines the almost complex structures coming from a complex structure on a differentiable manifold. We introduce the operator $\overline{\partial}$ and the Dolbeault complex of a holomorphic vector bundle.

The third chapter is a introduction to Kähler geometry. We give various characterisations of the Kähler metrics, and introduce Chern connections of a holomorphic vector bundle equipped with a metric. One of of the characterisations of Kähler metrics is the equality of the Chern connection and the Levi-Civita connection on the tangent bundle. We also give some constructions of Kähler manifolds.

This part ends with an introduction to the theory of sheaves and their cohomology. Apart from the definition of the cohomology of a topological space with values in a sheaf, we give different types of acyclic resolutions, so as to prove the theorem of de Rham mentioned above.

The second part of this text is devoted to proving the Hodge and Lefschetz decomposition theorems. One of the chapters presents the ideas of Hodge theory, omitting, however, the proofs of the necessary results from analysis. The next chapter is centred around the application of Hodge theory to Kähler manifolds. We prove the Kähler identities, and develop the study of the Lefschetz operator, which as we explained above leads to the decomposition theorems.

The following two chapters give conceptual applications of these results. We first explain the notion of Hodge structure, and introduce the essential notion of a polarised Hodge structure. In general, given a compact differentiable manifold X, we have a perfect Poincaré duality between the groups $H^k(X)$ and $H^{m-k}(X)$, $m = \dim X$. If X is a Kähler manifold, the Lefschetz isomorphism (0.4) allows us to put an intersection form on each cohomology group. One can show that this intersection form has well-defined signs on each component of type (p, q) of the primitive cohomology. We simultaneously develop the notion of a polarised manifold, and we prove the Kodaira embedding theorem, which says that a complex manifold is projective if and only if it admits an integral polarisation. The chapter ends by exploring the relation between Hodge structures of weight 1 and complex tori.

The following chapter is devoted to the holomorphic de Rham complex and the interpretation of the Hodge theorem in terms of degeneracy of the Frölicher spectral sequence. A good part of this chapter is an introduction to spectral sequences. We conclude by introducing the holomorphic logarithmic de Rham complex on quasi-projective smooth varieties, and sketch the proof of the existence of a mixed Hodge structure on their cohomology.

The third part consists of two chapters devoted to studying variations of Hodge structures. We first introduce the notion of a family of complex manifolds, and construct the Kodaira–Spencer map. We also introduce the Gauss–Manin connection associated to the local system of cohomology of the fibres of a family. This is necessary in order to formulate the transversality property for Hodge bundles. Finally, we describe the differential of the period map, using the Kodaira–Spencer map and the cup-product in Dolbeault cohomology.

In the last part we define the different "cycle classes" mentioned above. We first study the basic properties of analytic subsets, so as to construct their cohomology class (the class of a complex submanifold is easy to construct; one has simply to see what happens with singularities). We also devote one section to the notion of a Hodge class which arises in this way, and particularly to its relation with correspondences or morphisms of Hodge structure.

The last chapter is devoted to Deligne cohomology and the Abel–Jacobi map. These objects will be studied much more deeply in the second volume of this book.

Part I

Preliminaries

1

Holomorphic Functions of Many Variables

In this chapter, we recall the main properties of holomorphic functions of several complex variables. These results will be used freely in the remainder of the text, and will enable us to introduce the notions of a complex manifold, and a holomorphic function defined locally on a complex manifold.

The $\mathbb{C}$-valued holomorphic functions of the complex variables $z_1, \ldots, z_n$ are those whose differential is $\mathbb{C}$-linear, or equivalently, those which are annihilated by the operators $\frac{\partial}{\partial \bar{z}_i}$. It follows immediately from this definition that the set of holomorphic functions forms a ring, and that the composition of two holomorphic functions is holomorphic. The following theorem, however, requires a certain amount of work.

Theorem 1.1 *The holomorphic functions of the complex variables $z_1, \ldots, z_n$ are complex analytic, i.e. they locally admit expansions as power series in the variables z_i.*

This result is an easy consequence of Cauchy's formula in several variables, which is a generalisation of the formula

$$f(z) = \frac{1}{2i\pi} \int_{|\zeta|=1} \frac{f(\zeta)}{\zeta - z} d\zeta,$$

where f is a holomorphic function defined in a disk of radius > 1, and $|z| < 1$.

Cauchy's formula can also be used to prove Riemann's theorem of analytic continuation:

Theorem 1.2 *Let f be a bounded holomorphic function on the pointed disk. Then f extends to a holomorphic function on the whole disk.*

And also Hartogs' theorem:

Theorem 1.3 *Let f be a holomorphic function defined on the complement of the subset F defined by the equations $z_1 = z_2 = 0$ in a ball B of $\mathbb{C}^n$, $n \geq 2$. Then f extends to a holomorphic function on B.*

(More generally, this theorem remains true if F is an analytic subset of codimension 2, but we only need the present version here.) Hartogs' theorem is used more in complex geometry than Riemann's theorem, because it does not impose any conditions on the function f. More generally, it enables us to show that a holomorphic section of a complex vector bundle over a complex manifold is defined everywhere if it is defined on the complement of an analytic subset of codimension 2. This is classically used to show the invariance of the "plurigenera" under birational transformations.

We conclude this chapter with a proof of an explicit formula for the local solution of the equation

$$\frac{\partial f}{\partial \bar{z}} = g,$$

where g is a differentiable function defined on an open set of $\mathbb{C}$. This will be used in the following chapter, to prove the local exactness of the Dolbeault complex. A good reference for the material in this chapter is Hörmander (1979).

1.1 Holomorphic functions of one variable

1.1.1 Definition and basic properties

Let $U \subset \mathbb{C} \cong \mathbb{R}^2$ be an open set, and $f : U \to \mathbb{C}$ a C^1 map. Let x, y be the linear coordinates on $\mathbb{R}^2$ such that $z = x + iy$ is the canonical linear complex coordinate on $\mathbb{C}$. Consider the complex-valued differential form

$$dz = dx + i dy \in \mathrm{Hom}_{\mathbb{R}}(T_U, \mathbb{C}) \cong \Omega_{U,\mathbb{R}} \otimes \mathbb{C}.$$

Clearly dz and its complex conjugate $d\bar{z}$ form a basis of $\Omega_{U,\mathbb{R}} \otimes \mathbb{C}$ over $\mathbb{C}$ at every point of U, since

$$2dx = dz + d\bar{z}, \quad 2i dy = dz - d\bar{z}. \tag{1.1}$$

The complex differential form $df \in \mathrm{Hom}_{\mathbb{R}}(T_U, \mathbb{C})$ can thus be uniquely written

$$df_u = f_z(u)dz + f_{\bar{z}}(u)d\bar{z}, \tag{1.2}$$

where the complex-valued functions $u \mapsto f_z(u)$, $u \mapsto f_{\bar{z}}(u)$ are continuous.

Definition 1.4 *We write $f_z = \frac{\partial f}{\partial z}$ and $f_{\bar{z}} = \frac{\partial f}{\partial \bar{z}}$.*

By (1.1) we obviously have

$$\frac{\partial f}{\partial z} = \frac{1}{2}\left(\frac{\partial f}{\partial x} - i\frac{\partial f}{\partial y}\right), \quad \frac{\partial f}{\partial \bar{z}} = \frac{1}{2}\left(\frac{\partial f}{\partial x} + i\frac{\partial f}{\partial y}\right). \tag{1.3}$$

We can also consider the decomposition (1.2) as the decomposition of $df \in \mathrm{Hom}_{\mathbb{R}}(\mathbb{C}, \mathbb{C})$ into $\mathbb{C}$-linear and $\mathbb{C}$-antilinear parts:

Lemma 1.5 *We have $\frac{\partial f}{\partial \bar{z}}(u) = 0$ if and only if the $\mathbb{R}$-linear map*

$$df_u : T_{U,u} \cong \mathbb{C} \to \mathbb{C}$$

is $\mathbb{C}$-linear, i.e. is equal to multiplication by a complex number, which is then equal to $\frac{\partial f}{\partial z}(u)$.

Proof Because $\frac{\partial}{\partial y} = i\frac{\partial}{\partial x}$ for the natural complex structure on $T_{U,u}$, the morphism $df_u : T_{U,u} \to \mathbb{C}$ is $\mathbb{C}$-linear if and only if we have

$$\frac{\partial f}{\partial y}(u) = i\frac{\partial f}{\partial x}(u),$$

and by (1.3), this is equivalent to $\frac{\partial f}{\partial \bar{z}}(u) = 0$. Moreover, we then have $df_u = \frac{\partial f}{\partial z}(u)dz$, i.e.

$$df_u\left(\frac{\partial}{\partial x}\right) = \frac{\partial f}{\partial z}(u), \quad df_u\left(\frac{\partial}{\partial y}\right) = i\frac{\partial f}{\partial z}(u),$$

which proves the second assertion, since the natural isomorphism $T_{U,u} \cong \mathbb{C}$ sends $\frac{\partial}{\partial x}$ to 1. □

Definition 1.6 *The function f is said to be holomorphic if it satisfies one of the equivalent conditions of lemma 1.5 at every point of U.*

Lemma 1.7 *If f is holomorphic and does not vanish on U, then $\frac{1}{f}$ is holomorphic. Similarly, if f, g are holomorphic, fg and $f + g$ and $g \circ f$ (when g is defined on the image of f) are all holomorphic.*

Proof The map $z \mapsto \frac{1}{z}$ is holomorphic on $\mathbb{C}^*$, so that the first assertion follows from the last one. Furthermore, if g and f are C^1 and g is defined on the image of f, then $g \circ f$ is C^1 and we have

$$d(g \circ f)_u = dg_{f(u)} \circ df_u.$$

If $dg_{f(u)}$ and df_u are both $\mathbb{C}$-linear for the natural identifications of $T_{\mathbb{C},u}$, $T_{\mathbb{C},f(u)}$ and $T_{\mathbb{C},g\circ f(u)}$ with $\mathbb{C}$, then $d(g \circ f)_u$ is also $\mathbb{C}$-linear, and the last assertion is proved. The other properties are proved similarly. □

In particular, we will use the following corollary.

Corollary 1.8 *If f is holomorphic on U, the map g defined by*

$$g(z) = \frac{f(z)}{z-a}$$

is holomorphic on $U - \{a\}$.

1.1.2 Background on Stokes' formula

Let α be a C^1 differential k-form on an n-dimensional manifold U (cf. definition 2.3 and section 2.1.2 in the following chapter). If $x_1, \ldots, x_n$ are local coordinates on U, we can write

$$\alpha = \sum_I \alpha_I dx_I,$$

where the indices I parametrise the totally ordered subsets $i_1 < \cdots < i_k$ of $\{1, \ldots, n\}$, with $dx_I = dx_{i_1} \wedge \cdots \wedge dx_{i_k}$. We can then define the continuous $(k+1)$-form

$$d\alpha = \sum_{I,i} \frac{\partial \alpha_I}{\partial x_i} dx_i \wedge dx_I; \tag{1.4}$$

we check that it is independent of the choice of coordinates. This follows from the more general fact that if V is an m-dimensional manifold and $\phi : V \to U$ is a C^1 map given in local coordinates by $\phi^* x_i := x_i \circ \phi = \phi_i(y_1, \ldots, y_m)$, then for every differential form $\alpha = \sum_I \alpha_I dx_{i_1} \wedge \cdots \wedge dx_{i_k}$, we can define its inverse image

$$\phi^*\alpha = \sum_I \alpha_I \circ \phi \; d\phi_{i_1} \wedge \cdots \wedge d\phi_{i_k}.$$

Moreover, if ϕ is C^2, this image inverse satisfies

$$d(\phi^*\alpha) = \phi^*(d\alpha),$$

where the coordinates y_i (and the formulae (1.4)) are used on the left, while the coordinates x_i are used on the right.

A C^0 differential k-form α can be integrated over the compact oriented k-dimensional submanifolds of U with boundary, or over the image of such manifolds under differentiable maps.

To begin with, let us recall that a k-dimensional manifold with boundary is a topological space covered by open sets U_i which are homeomorphic, via certain maps ϕ_i, to open subsets of $\mathbb{R}^k$ or to $]0,1]\times V$, where V is an open set of $\mathbb{R}^{k-1}$. We require the transition functions $\phi_i\circ\phi_j^{-1}$ to be differentiable on $\phi_j(U_i\cap U_j)$. When $\phi_j(U_i\cap U_j)$ contains points on the boundary of U_j, i.e. is locally isomorphic to $]0,1]\times V$, where V is an open set of $\mathbb{R}^{k-1}$, $\phi_i(U_i\cap U_j)$ must also be locally isomorphic to $]0,1]\times V'$, where V' is an open set of $\mathbb{R}^{k-1}$, and the differentiable map $\phi_i\circ\phi_j^{-1}$ must locally extend to a diffeomorphism of a neighbourhood in $\mathbb{R}^k$ of $]0,1]\times V$ to a neighbourhood of $]0,1]\times V'$, inducing a diffeomorphism from $1\times V$ to $1\times V'$. In particular, the boundary of M, which we denote by ∂M and which is defined, with the preceding notation, as the union of the $\phi_i^{-1}(1\times V)$, is a closed set of M which possesses an induced differentiable manifold structure.

The manifold with boundary M is said to be oriented if the diffeomorphisms $\phi_i\circ\phi_j^{-1}$ have positive Jacobian. The boundary of M is then also naturally oriented by the charts $1\times V$, where V is an open set of $\mathbb{R}^{k-1}$ as above, since the induced transition diffeomorphisms

$$\phi_i\circ\phi_j^{-1}{}_{|1\times V}:V\to V'$$

also have positive Jacobian.

If M is a k-dimensional manifold with boundary and $\phi: M\to U$ is a C^1 differentiable map (along the boundary of M, which is locally isomorphic to $]0,1]\times V$, we require ϕ to extend locally to a C^1 map on a neighbourhood $]0,1+\epsilon[\times V$ of $\{1\}\times V$), then for every continuous k-form α, we have the inverse image $\beta=\phi^*\alpha$ defined above, which is a continuous k-form on M. If moreover M is oriented and compact, such a form can be integrated over M as follows. Let f_i be a partition of unity subordinate to a covering of M by open sets U_i as above, which we may assume to be diffeomorphic to $]0,1]\times]0,1[^{k-1}$ or to $]0,1[^k$. Then $\beta=\sum_i f_i\beta$, and the form $f_i\beta$ on U_i extends to a continuous form on $[0,1]^k$. Setting $f_i\beta=g_i(x_1,\ldots,x_k)dx_1\wedge\cdots\wedge dx_k$, we then define

$$\int_M\beta=\sum_i\int_{U_i}f_i\beta$$

$$\int_{U_i}f_i\beta=\int_0^1\cdots\int_0^1 g_i(x_1,\ldots,x_k)dx_1\ldots dx_k.$$

The change of variables formula for multiple integrals and the fact that the authorised variable changes have positive Jacobians ensure that $\int_M \beta$ is well-defined independently of the choice of oriented charts, i.e. of local orientation-preserving coordinates.

Remark 1.9 *If we change the orientation of M, i.e. if we compose all the charts with local diffeomorphisms of $\mathbb{R}^k$ with negative Jacobians, the integrals $\int_M \phi^*\alpha$ change sign. This follows from the change of variables formula for multiple integrals, which uses only the absolute value of the Jacobian, whereas the change of variables formula for differential forms of maximal degree uses the Jacobian itself.*

Suppose now that α is a C^1 $(k-1)$-form on U. Then, as $\phi_{|\partial M}$ is differentiable and ∂M is a compact oriented manifold of dimension $k-1$, we can compute the integral $\int_{\partial M} \phi^*\alpha$. Moreover, we can integrate the differential $\phi^* d\alpha = d\phi^*\alpha$ over M. We then have

Theorem 1.10 *(Stokes' formula) The following equality holds:*

$$\int_M \phi^* d\alpha = \int_{\partial M} \phi^*\alpha. \tag{1.5}$$

In particular, if $d\alpha = 0$, we have $\int_{\partial M} \phi^*\alpha = 0$.

Proof Using a partition of unity, we are reduced to showing (1.5) when $\phi^*\alpha$ has compact support in an open set U_i of M as above. This follows immediately from the formula (1.4) for the differential, and the equality

$$\int_0^1 f'(t)dt = f(1) - f(0),$$

which holds for any C^1 function f. □

We will use Stokes' formula very frequently throughout this text. In particular, it will enable us to pair the de Rham cohomology with the singular homology. The following consequence will be particularly useful.

Corollary 1.11 *If α is a differential form of degree $n-1$ on a compact n-dimensional manifold without boundary, then $\int_M d\alpha = 0$.*

1.1.3 Cauchy's formula

We propose to apply Stokes' formula (1.5), using the following lemma.

Lemma 1.12 *Let $f : U \to \mathbb{C}$ be a holomorphic map. Then the complex differential form $f dz$ is closed.*

Proof We have $d(f dz) = d(f dx + i f dy) = \frac{\partial f}{\partial y} dy \wedge dx + i \frac{\partial f}{\partial x} dx \wedge dy$. Thus $\frac{\partial f}{\partial y} = i \frac{\partial f}{\partial x}$ implies that $d(f dz) = 0$. □

By corollary 1.8, we thus also have the following.

Corollary 1.13 *If f is holomorphic on U, the differential form $\frac{f}{z-z_0} dz$ is closed on $U - \{z_0\}$.*

Suppose now that U contains a closed disk D. For every $z_0 \in D$, let D_ϵ be the open disk of radius ϵ centred at z_0 which is contained in D for sufficiently small ϵ. Then $D - D_\epsilon$ is a manifold with boundary, whose boundary is the union of the circle ∂D and the circle of centre z_0 and radius ϵ, the first with its natural orientation, the second with the opposite orientation. For holomorphic f, Stokes' formula and corollary 1.1.3 then give the equality

$$\frac{1}{2i\pi} \int_{\partial D} \frac{f(z)}{z - z_0} dz = \frac{1}{2i\pi} \int_{|z-z_0|=\epsilon} \frac{f(z)}{z - z_0} dz. \tag{1.6}$$

Furthermore, we have the following.

Lemma 1.14 *If f is a function which is continuous at z_0, then*

$$\lim_{\epsilon \to 0} \frac{1}{2i\pi} \int_{|z-z_0|=\epsilon} \frac{f(z)}{z - z_0} dz = f(z_0).$$

Proof The circle of radius ϵ and centre z_0 is parametrised by the map $\gamma : t \mapsto z_0 + \epsilon e^{2i\pi t}$ on the segment $[0, 1]$. We have $\gamma^* \left(\frac{1}{2i\pi} \frac{f(z)}{z-z_0} dz \right) = f(z_0 + \epsilon e^{2i\pi t}) dt$. Thus,

$$\frac{1}{2i\pi} \int_{|z-z_0|=\epsilon} \frac{f(z)}{z - z_0} dz = \int_0^1 f(z_0 + \epsilon e^{2i\pi t}) dt. \tag{1.7}$$

But as f is continuous at z_0, the functions $f_\epsilon(t) = f(z_0 + \epsilon e^{2i\pi t})$ converge uniformly, as ϵ tends to 0, to the constant function equal to $f(z_0)$. We thus have

$$\lim_{\epsilon\to 0}\int_0^1 f(z_0+\epsilon e^{2i\pi t})dt = \int_0^1 f(z_0)dt = f(z_0).$$

□

Combining lemma 1.14 and equality (1.6), we now have

Theorem 1.15 *(Cauchy's formula) Let f be a holomorphic function on U and D a closed disk contained in U. Then for every point z_0 in the interior of D, we have the equality*

$$f(z_0) = \frac{1}{2i\pi}\int_{\partial D}\frac{f(z)}{z-z_0}dz. \tag{1.8}$$

1.2 Holomorphic functions of several variables

1.2.1 Cauchy's formula and analyticity

Let U be an open set of $\mathbb{C}^n$, and let $f : U \to \mathbb{C}$ be a C^1 map. For $u \in U$, we have a canonical identification $T_{U,u} \cong \mathbb{C}^n$. We can thus generalise the notion of a holomorphic function to higher dimensions.

Definition 1.16 *The function f is said to be holomorphic if for every $u \in U$, the differential*

$$df_u \in \mathrm{Hom}(T_{U,u}, \mathbb{C}) \cong \mathrm{Hom}(\mathbb{C}^n, \mathbb{C})$$

is $\mathbb{C}$-linear.

It is easy to prove that lemma 1.7 remains true in higher dimensions. Furthermore, we have the three following characterisations of holomorphic functions.

Theorem 1.17 *The following three properties are equivalent for a C^1 function f:*
(i) f is holomorphic.
(ii) In the neighbourhood of each point $z_0 \in U$, f admits an expansion as a power series of the form

$$f(z_0+z) = \sum_I \alpha_I z^I, \tag{1.9}$$

where I runs through the set of the n-tuples of integers $(i_1, \ldots, i_n)$ with $i_k \geq 0$, and $z^I := z_1^{i_1}\cdots z_n^{i_n}$. The coefficients of the series (1.9) satisfy the following *property: there exist $R_1 > 0, \ldots, R_n > 0$ such that the power series*

$$\sum_I |\alpha_I| r^I$$

converges for every $r_1 < R_1, \ldots, r_n < R_n$.
(iii) If $D = \{(\zeta_1, \ldots, \zeta_n) | \, |\zeta_i - a_i| \leq \alpha_i\}$ is a polydisk contained in U, then for every $z = (z_1, \ldots, z_n) \in D^0$, we have the equality

$$f(z) = \left(\frac{1}{2i\pi}\right)^n \int_{|\zeta_i - a_i| = \alpha_i} f(\zeta) \frac{d\zeta_1}{\zeta_1 - z_1} \wedge \cdots \wedge \frac{d\zeta_n}{\zeta_n - z_n}. \tag{1.10}$$

In the preceding formula, the integral is taken over a product of circles, equipped with the orientation which is the product of the natural orientations.

Remark 1.18 *Because of property (ii), holomorphic functions are also known as complex analytic functions.*

Proof The implications (iii)⇒(ii)⇒(i) are obvious: indeed, (iii)⇒(ii) is obtained, for ζ in the product of circles $\{\zeta | \, |\zeta_i - a_i| = \alpha_i\}$ and z in the interior of D, by expanding the functions

$$\frac{f(\zeta)}{(\zeta_1 - z_1) \cdots (\zeta_n - z_n)} = \frac{f(\zeta)}{((\zeta_1 - a_1) - (z_1 - a_1)) \cdots ((\zeta_n - a_n) - (z_n - a_n))}$$

as a power series in $z_1 - a_1, \ldots, z_n - a_n$ whose coefficients are continuous functions of ζ. The uniform convergence in ζ of this expansion then shows that the integral (1.10) admits the corresponding power series expansion in $z_1 - a_1, \ldots, z_n - a_n$, so that (ii) holds for the function defined by (1.10).

(ii)⇒(i) follows from the fact that every polynomial function of $z_1, \ldots, z_n$ is holomorphic, and that as the series (1.9) is a uniform limit of polynomials whose derivatives also converge uniformly, its differential is the limit of the differentials of these polynomials. As the differential of each polynomial is $\mathbb{C}$-linear, the same holds for the series, which is thus also holomorphic.

It remains to see that (i)⇒(iii), which is Cauchy's formula in several variables. We can prove it by induction on the dimension, using Cauchy's formula (1.8). We can also directly apply Stokes' formula, using the following analogue of lemma 1.12.

Lemma 1.19 *If f is holomorphic, then the differential form $f(z)dz_1 \wedge \cdots \wedge dz_n$ is closed.*

The product of circles $\prod_i \{|\zeta_i - z_i| = \epsilon\}$ is contained in D for sufficiently small ϵ, and homotopic in $D - \bigcup_i \{\zeta \mid \zeta_i = z_i\}$ to the product of circles $\prod_i \{|\zeta_i - a_i| = \alpha_i\}$, which means that there exists an oriented compact manifold M of dimension n and a differentiable map

$$\phi : [0,1] \times M \to D - \cup_i \{\zeta \mid \zeta_i = z_i\}$$

such that $\phi_{|0\times M}$ is a diffeomorphism from M to the first product of circles, and $\phi_{|1\times M}$ is a diffeomorphism from M to the second product of circles, the first isomorphism being compatible with the orientations, and the second changing the orientation. We then deduce from lemma 1.19 that if $\beta = f(\zeta)\frac{d\zeta_1}{\zeta_1 - z_1} \wedge \cdots \wedge \frac{d\zeta_n}{\zeta_n - z_n}$, the differential form $\phi^*\beta$ is closed on $[0,1] \times M$, and thus, by Stokes' formula, satisfies

$$\int_{\partial[0,1]\times M} \phi^*\beta = 0.$$

For ϵ sufficiently small, this gives the equality

$$\left(\frac{1}{2i\pi}\right)^n \int_{|\zeta_i - a_i| = \alpha_i} f(\zeta)\frac{d\zeta_1}{\zeta_1 - z_1} \wedge \cdots \wedge \frac{d\zeta_n}{\zeta_n - z_n}$$
$$= \left(\frac{1}{2i\pi}\right)^n \int_{|\zeta_i - z_i| = \epsilon} f(\zeta)\frac{d\zeta_1}{\zeta_1 - z_1} \wedge \cdots \wedge \frac{d\zeta_n}{\zeta_n - z_n}.$$

But the limit of the right-hand term as ϵ tends to 0 is equal to $f(z)$ by the same argument as above.

□

Remark 1.20 *The homotopy must have values in $D - \bigcup_i \{\zeta \mid \zeta_i = z_i\}$ and not only in D, in order to guarantee that the form $\phi^* f(\zeta)\frac{d\zeta_1}{\zeta_1 - z_1} \wedge \cdots \wedge \frac{d\zeta_n}{\zeta_n - z_n}$ is C^1 in $[0,1] \times M$ and to be able to apply Stokes' formula.*

1.2.2 Applications of Cauchy's formula

Let us give some applications of theorem 1.17. To begin with, we have

Theorem 1.21 *(The maximum principle) Let f be a holomorphic function on an open subset U of $\mathbb{C}^n$. If $|f|$ admits a local maximum at a point $u \in U$, then f is constant in the neighbourhood of this point.*

Proof Let $R_1, \ldots, R_n$ be positive real numbers such that for every $\epsilon_i \le R_i$, the polydisk $D_{\epsilon.} = \{\zeta \in \mathbb{C}^n \mid |\zeta_i - u_i| \le \epsilon_i\}$ is contained in U. Then we have

Cauchy's formula

$$f(u) = \left(\frac{1}{2i\pi}\right)^n \int_{|\zeta_i - u_i| = \epsilon_i} f(\zeta) \frac{d\zeta_1}{\zeta_1 - u_1} \wedge \cdots \wedge \frac{d\zeta_n}{\zeta_n - u_n}.$$

Parametrising the circles $|\zeta_i - u_i| = \epsilon_i$ by $\gamma_i(t) = u_i + \epsilon_i e^{2i\pi t}$, $t \in [0, 1]$, this can be written as

$$f(u) = \int_0^1 \cdots \int_0^1 f(u_1 + \epsilon_1 e^{2i\pi t_1}, \ldots, u_n + \epsilon_n e^{2i\pi t_n}) dt_1 \cdots dt_n. \quad (1.11)$$

But we have the inequality

$$\left| \int_0^1 \cdots \int_0^1 f(u_1 + \epsilon_1 e^{2i\pi t_1}, \ldots, u_n + \epsilon_n e^{2i\pi t_n}) dt_1 \cdots dt_n \right|$$
$$\leq \int_0^1 \cdots \int_0^1 |f(u_1 + \epsilon_1 e^{2i\pi t_1}, \ldots, u_n + \epsilon_n e^{2i\pi t_n})| \, dt_1 \cdots dt_n, \quad (1.12)$$

and equality holds if and only if the argument of $f(u_1 + \epsilon_1 e^{2i\pi t_1}, \ldots, u_n + \epsilon_n e^{2i\pi t_n})$ is constant, necessarily equal to that of $f(u)$ by (1.11).

Now, for sufficiently small ϵ_i, we have by hypothesis

$$|f(u)| \geq |f(u_1 + \epsilon_1 e^{2i\pi t_1}, \ldots, u_n + \epsilon_n e^{2i\pi t_n})| \, .$$

Combining this inequality with (1.11) and (1.12), we obtain

$$|f(u)| \leq \int_0^1 \cdots \int_0^1 |f(u_1 + \epsilon_1 e^{2i\pi t_1}, \ldots, u_n + \epsilon_n e^{2i\pi t_n})| \, dt_1 \cdots dt_n$$
$$\leq \int_0^1 \cdots \int_0^1 |f(u)| \, dt_1 \cdots dt_n = |f(u)| \, .$$

The equality of the two extreme terms then implies equality at every step; the first equality implies that the argument of f is constant, equal to that of $f(u)$ on each product of circles as above, and the second equality shows that the function f must have constant modulus equal to $|f(u)|$, for sufficiently small ϵ_i. Letting the multiradius of the polydisks $D_{\epsilon.}$ vary, we have thus shown that f is constant, equal to $f(u)$ on a neighbourhood of u possibly minus the hyperplanes $\{\zeta_i = u_i\}$, i.e. in fact constant in the neighbourhood of u by continuity. □

Another essential application is the principle of analytic continuation.

Theorem 1.22 *Let U be a connected open set of $\mathbb{C}^n$, and f a holomorphic function on U. If f vanishes on an open set of U, then f is identically zero.*

Proof This follows from the fact that by the characterisation (ii), f is in particular analytic (i.e. locally equal to the sum of its Taylor series). We can thus apply the principle of analytic continuation to f. We recall that the latter is shown by noting that if f is analytic, the open set consisting of the points in whose neighbourhood f vanishes is equal to the closed set consisting of the points where f and all its derivatives vanish. □

Let us now give some subtler applications of Cauchy's formula (1.10) or its generalisations. These theorems show that the possible singularities of a holomorphic function cannot exist unless the function is not bounded (Riemann), and is not defined on the complement of an analytic subset of codimension 2 (Hartogs).

Theorem 1.23 *(Riemann) Let f be a holomorphic function defined on $U - \{z \mid z_1 = 0\}$, where U is an open set of $\mathbb{C}^n$. Then if f is locally bounded on U, f extends to a holomorphic map on U.*

Proof Since this is a local statement, it suffices to show that if U contains a polydisk $D = \{(z_1, \ldots, z_n) \in \mathbb{C}^n \mid |z_i| \leq r_i\}$ on which f is bounded, then we can extend f to the points in the interior of D. We propose to show that for a point z in the interior of D such that $z_1 \neq 0$, Cauchy's formula

$$f(z) = \left(\frac{1}{2i\pi}\right)^n \int_{\partial D} f(\zeta) \frac{d\zeta_1}{\zeta_1 - z_1} \wedge \cdots \wedge \frac{d\zeta_n}{\zeta_n - z_n}, \tag{1.13}$$

where

$$\partial D := \{(\zeta_1, \ldots, \zeta_n) \mid |\zeta_i| = r_i, \forall i\},$$

holds. Note that the right-hand term in (1.13) is well-defined, since the integration locus is contained in the locus of definition of f.

Let $\epsilon_1 \in \mathbb{R}$, $0 < \epsilon_1 < |z_1|$ be such that the closed disk of radius ϵ_1 and centre z_1 is contained in the disk $\{\zeta \mid |\zeta| < r_1\}$. Then the polydisk

$$D_{\epsilon_1} := \{(\zeta_1, \ldots, \zeta_n) \mid |\zeta_1 - z_1| \leq \epsilon_1, \ |z_i| \leq r_i, \quad i \geq 2\}$$

is contained in $D - \{\zeta_1 = 0\}$, so that Cauchy's formula gives

$$f(z) = \left(\frac{1}{2i\pi}\right)^n \int_{\partial D_{\epsilon_1}} f(\zeta) \frac{d\zeta_1}{\zeta_1 - z_1} \wedge \cdots \wedge \frac{d\zeta_n}{\zeta_n - z_n}, \tag{1.14}$$

where

$$\partial D_{\epsilon_1} := \{(\zeta_1, \ldots, \zeta_n) \mid |\zeta_1 - z_1| = \epsilon_1, \ |\zeta_i| = r_i, \quad i \geq 2\}.$$

Consider, also, the product of circles

$$\partial D'_\epsilon := \{(\zeta_1, \dots, \zeta_n) |\ |\zeta_1| = \epsilon,\ |\zeta_i| = r_i, i \geq 2\}.$$

Then when ϵ is sufficiently small, $\partial D - \partial D_{\epsilon_1} - \partial D'_\epsilon$ is the boundary of the manifold

$$M = \{(\zeta_1, \dots, \zeta_n) |\ |\zeta_1 - z_1| \geq \epsilon_1,\ |\zeta_1| \geq \epsilon,\ |\zeta_i| = r_i,\ i \geq 2\},$$

which is contained in D and intersects neither the hypersurface $\{\zeta_1 = 0\}$ nor the hypersurfaces $\{\zeta_i = z_i\}$. Here, the signs given to the components of the boundary are positive when the orientation as part of the boundary of M coincides with the natural orientation, negative otherwise. Stokes' formula and (1.14) then give

$$f(z) = \left(\frac{1}{2i\pi}\right)^n \left[\int_{\partial D} f(\zeta) \frac{d\zeta_1}{\zeta_1 - z_1} \wedge \cdots \wedge \frac{d\zeta_n}{\zeta_n - z_n} \right.$$
$$\left. - \int_{\partial D'_\epsilon} f(\zeta) \frac{d\zeta_1}{\zeta_1 - z_1} \wedge \cdots \wedge \frac{d\zeta_n}{\zeta_n - z_n}\right].$$

The proof of formula (1.13) can then be finished using the following lemma.

Lemma 1.24 *When f is bounded, and for z such that $z_1 \neq 0, |z_i| < r_i$, we have*

$$\lim_{\epsilon \to 0} \int_{\partial D'_\epsilon} f(\zeta) \frac{d\zeta_1}{\zeta_1 - z_1} \wedge \cdots \wedge \frac{d\zeta_n}{\zeta_n - z_n} = 0. \tag{1.15}$$

Proof Let us parametrise the product of circles $\partial D'_\epsilon$ by $[0, 1]^n$, $(t_1, \dots, t_n) \mapsto (\epsilon e^{2i\pi t_1}, r_2 e^{2i\pi t_2}, \dots, r_n e^{2i\pi t_n})$. The integral (1.15) is thus equal to

$$(2i\pi)^n \int_0^1 \cdots \int_0^1 \epsilon r_2 \cdots r_n \prod_j e^{2i\pi t_j} \frac{f(\epsilon e^{2i\pi t_1}, r_2 e^{2i\pi t_2}, \dots, r_n e^{2i\pi t_n})}{(\epsilon e^{2i\pi t_1} - z_1) \cdots (r_n e^{2i\pi t_n} - z_n)} dt_1 \cdots dt_n.$$

As f is bounded, under the hypotheses on z, the integrand in this formula tends uniformly to 0 with ϵ, and thus the integral in the formula tends to 0 with ϵ.

□

As Cauchy's formula (1.13) is proved, Riemann's extension theorem follows immediately, since it is clear that the function defined by the right-hand term in (1.13) extends holomorphically to D. □

To conclude this section, we will prove the following version of Hartogs' extension theorem.

Theorem 1.25 *Let U be an open set of $\mathbb{C}^n$ and f a holomorphic function on $U - \{z \mid z_1 = z_2 = 0\}$. Then f extends to a holomorphic function on U.*

Remark 1.26 *This implies the more general theorem mentioned above, using theorem 11.11, which proves that an analytic subset of codimension 2 can be stratified into smooth analytic submanifolds of codimension at least 2; this theorem will be proved in section 11.1.1.*

Proof Let D be a closed polydisk contained in U:

$$D = \{(z_1, \dots, z_n) \mid |z_i| \le r_i\}.$$

Let $z \in D - \{\zeta \mid \zeta_1 = \zeta_2 = 0\}$. As in the preceding proof, we will show that Cauchy's formula

$$f(z) = \left(\frac{1}{2i\pi}\right)^n \int_{\partial D} f(\zeta) \frac{d\zeta_1}{\zeta_1 - z_1} \wedge \dots \wedge \frac{d\zeta_n}{\zeta_n - z_n} \tag{1.16}$$

is satisfied, and this will enable us to conclude, as above, that the function $f(z)$, given in the form of an integral as in (1.16), extends holomorphically to D. Let ϵ_1, ϵ_2 be two positive real numbers, sufficiently small for the polydisk

$$D_\epsilon = \{\zeta \mid |\zeta_i - z_i| \le \epsilon_i,\ i = 1, 2,\ |\zeta_i| \le r_i,\ i > 2\}$$

to be contained in $D - \{\zeta \mid \zeta_1 = \zeta_2 = 0\}$. Then we have Cauchy's formula

$$f(z) = \left(\frac{1}{2i\pi}\right)^n \int_{\partial D_\epsilon} f(\zeta) \frac{d\zeta_1}{\zeta_1 - z_1} \wedge \dots \wedge \frac{d\zeta_n}{\zeta_n - z_n}, \tag{1.17}$$

where

$$\partial D_\epsilon = \{\zeta \mid |\zeta_i - z_i| = \epsilon_i,\ i = 1, 2,\ |\zeta_i| = r_i,\ i > 2\}.$$

It thus suffices to show that $\partial D - \partial D_\epsilon$ is a boundary in

$$D - (\{\zeta \mid \zeta_1 = \zeta_2 = 0\} \cup \bigcup_i \{\zeta \mid \zeta_i = z_i\}),$$

in order to apply Stokes' formula and conclude that (1.16) holds.

Let $\alpha_1(t), \alpha_2(t)$, $t \in [0, 1]$, be two positive-valued differentiable functions such that $\alpha_i(1) = \epsilon_i$, $\alpha_i(0) = r_i$. For every $t \in [0, 1]$, let

$$\partial D_t = \{\zeta \mid |\zeta_i - t z_i| = \alpha_i(t),\ i = 1, 2,\ |\zeta_i| = r_i,\ i > 2\}.$$

Lemma 1.27 *For a suitable choice of functions α_1, α_2, ∂D_t is contained in*

$$D - (\{\zeta \mid \zeta_1 = \zeta_2 = 0\} \cup \bigcup_i \{\zeta \mid \zeta_i = z_i\})$$

for every $t \in [0, 1]$.

Proof Firstly, ∂D_t still lies in D if $\alpha_i(t) + t\,|z_i| \leq r_i$, $i = 1, 2$. Moreover, ∂D_t still lies in $D - \bigcup_i \{\zeta \mid \zeta_i = z_i\}$ if $\alpha_i(t) \neq (1-t)\,|z_i|$, $i = 1, 2$. Now, note that $(1-t)\,|z_i| < r_i - t\,|z_i|$, since $|z_i| < r_i$. Furthermore, the conditions $\alpha_i(t) \leq r_i - t\,|z_i|$ and $\alpha_i(t) > (1-t)\,|z_i|$ are both satisfied for $t = 1$ and $t = 0$. It thus suffices to take functions $\alpha_i(t)$ satisfying

$$(1-t)\,|z_i| < \alpha_i(t) \leq r_i - t\,|z_i|, \quad \alpha_i(0) = r_i, \quad \alpha_i(1) = \epsilon_i.$$

It remains to see that D_t does not meet $\{\zeta \mid \zeta_1 = \zeta_2 = 0\}$ for any $t \in [0, 1]$, for a suitable choice of the pair (α_1, α_2). But D_t meets $\{\zeta \mid \zeta_1 = \zeta_2 = 0\}$ if we have $\alpha_i(t) = t\,|z_i|$ for $i = 1$ and 2. For fixed t, this imposes two conditions on the pair (α_1, α_2), and t varies in a segment, so it is clear that this last condition is not satisfied by a pair of sufficiently general functions. □

Lemma 1.27 gives a differentiable homotopy in $D - (\{\zeta \mid \zeta_1 = \zeta_2 = 0\} \cup \{\zeta \mid \zeta_i = z_i\})$ from ∂D to ∂D_ϵ, so we can conclude by Stokes' formula that (1.16) holds. Thus theorem 1.25 is proved. □

1.3 The equation $\frac{\partial g}{\partial \bar{z}} = f$

The following theorem will play an essential role in the proof of the local exactness of the operator $\bar{\partial}$.

Theorem 1.28 *Let f be a C^k function (for $k \geq 1$) on an open set of $\mathbb{C}$. Then, locally on this open set, there exists a C^k function g (for $k \geq 1$), such that*

$$\frac{\partial g}{\partial \bar{z}} = f. \tag{1.18}$$

Remark 1.29 *Such a function g is defined up to the addition of a holomorphic function.*

Proof As the statement is local, we may assume that f has compact support, and thus is defined and C^k on $\mathbb{C}$. Now set

$$g = \frac{1}{2i\pi} \int_{\mathbb{C}} \frac{f(\zeta)}{\zeta - z} d\zeta \wedge d\bar{\zeta}.$$

□

Remark 1.30 *This is a singular integral. By definition, it is equal to the limit, as ϵ tends to 0, of the integrals*

$$\frac{1}{2i\pi} \int_{\mathbb{C}-D_\epsilon} \frac{f(\zeta)}{\zeta - z} d\zeta \wedge d\bar{\zeta},$$

where D_ϵ is a disk of radius ϵ centred at z. It is easy to see that this limit exists (the function $\frac{1}{\zeta - z}$ is L^1).

Making the change of variable $\zeta' = \zeta - z$, we also have

$$g(z) = \lim_{\epsilon \to 0} g_\epsilon(z), \quad g_\epsilon(z) = \frac{1}{2i\pi} \int_{\mathbb{C}_\epsilon} \frac{f(\zeta' + z)}{\zeta'} d\zeta' \wedge d\overline{\zeta'},$$

where

$$\mathbb{C}_\epsilon = \mathbb{C} - D'_\epsilon,$$

and D'_ϵ is a disk of radius ϵ centred at 0. The convergence of the g_ϵ when ϵ tends to 0 is uniform in z. Moreover, we can differentiate under the integral sign the (non-singular) integral defining g_ϵ

$$\frac{\partial g_\epsilon}{\partial \bar{z}} = \frac{1}{2i\pi} \int_{\mathbb{C}_\epsilon} \frac{\partial f(\zeta' + z)}{\partial \bar{z}} \frac{d\zeta' \wedge d\overline{\zeta'}}{\zeta'}.$$

As $\frac{\partial f(\zeta' + z)}{\partial \bar{z}}$ is C^{k-1}, with $k - 1 \geq 0$, the functions $\frac{\partial g_\epsilon}{\partial \bar{z}}$ converge uniformly, and we conclude that g is at least C^1 and satisfies

$$\frac{\partial g}{\partial \bar{z}} = \frac{1}{2i\pi} \int_{\mathbb{C}} \frac{\partial f(\zeta' + z)}{\partial \bar{z}} \frac{d\zeta' \wedge d\overline{\zeta'}}{\zeta'}.$$

By induction on k, the same argument actually shows that g is C^k. Thus, it remains to show the equality $\frac{\partial g}{\partial \bar{z}} = f$. Again making the change of variable $\zeta = \zeta' + z$, we have

$$\frac{\partial g}{\partial \bar{z}}(z) = \lim_{\epsilon \to 0} \frac{1}{2i\pi} \int_{\mathbb{C} - D_\epsilon} \frac{\partial f}{\partial \bar{\zeta}}(\zeta) \frac{d\zeta \wedge d\bar{\zeta}}{\zeta - z}. \tag{1.19}$$

Now, we have the equality on $\mathbb{C} - D_\epsilon$

$$\frac{\partial f}{\partial \bar{\zeta}}(\zeta) \frac{d\zeta \wedge d\bar{\zeta}}{\zeta - z} = -d\left(f \frac{d\zeta}{\zeta - z}\right);$$

indeed, for a differentiable function $\phi(\zeta)$ we know that $d\phi = \frac{\partial \phi}{\partial \zeta} d\zeta + \frac{\partial \phi}{\partial \bar{\zeta}} d\bar{\zeta}$, and thus

$$d(\phi d\zeta) = -\frac{\partial \phi}{\partial \bar{\zeta}} d\zeta \wedge d\bar{\zeta}.$$

Stokes' formula thus gives

$$\frac{1}{2i\pi} \int_{\mathbb{C} - D_\epsilon} \frac{\partial f}{\partial \bar{\zeta}}(\zeta) \frac{d\zeta \wedge d\bar{\zeta}}{\zeta - z} = \frac{1}{2i\pi} \int_{\partial D_\epsilon} f(\zeta) \frac{d\zeta}{\zeta - z}. \tag{1.20}$$

Using lemma 1.14 and the equalities (1.19), (1.20) we have thus proved the equality (1.18).

Exercises

1. Let $\phi : U \to V$ be a holomorphic map from an open subset of $\mathbb{C}^n$ to an open subset of $\mathbb{C}^n$. Show that the set

$$R = \{x \in U \mid d\phi_x \text{ is not an isomorphism}\}$$

is defined in U by exactly one holomorphic equation.
This set is called the ramification divisor of ϕ, when it is different from U.

2. Let f be a holomorphic function defined over an open subset U of $\mathbb{C}^n$. We assume that f does not vanish outside the set

$$\{z = (z_1, \ldots, z_n) \in U \mid z_1 = z_2 = 0\}.$$

Show that f does not vanish at any point of U.

3. Let f be a meromorphic function defined on an open subset U of $\mathbb{C}$. This means that f is locally the quotient of two holomorphic functions.
(a) Show that for any compact subset $K \subset U$, the number of zeros or poles of f in K is finite.
(b) Let $x \in U$. Show that there exists an integer $k_x \in \mathbb{Z}$ such that f can be written as $(z - x)^{k_x}\phi$ in a neighbourhood of x, with ϕ holomorphic and invertible (that is non-zero).
The divisor of f is defined as the locally finite sum

$$\sum_{x \in U} k_x x.$$

(c) Let $x \in U$ and $D \subset U$ be a disk centred in x, such that x is the only pole or zero of f in D. Show that

$$k_x = \int_{\partial D} \frac{1}{2i\pi} \frac{df}{f}.$$

2

Complex Manifolds

In this chapter, we introduce and study the notion of a complex structure on a differentiable or complex manifold. A complex manifold X of (complex) dimension n is a differentiable manifold locally equipped with complex-valued coordinates (called holomorphic coordinates) $z_1, \ldots, z_n$, such that the diffeomorphisms from an open set of $\mathbb{C}^n$ to an open set of $\mathbb{C}^n$ given by coordinate changes are holomorphic. By the definition of a holomorphic transformation, we then see that the structure of a complex vector space on the tangent space $T_{X,x}$ given by the identification $T_{X,x} \cong \mathbb{C}^n$ induced by the holomorphic coordinates $z_1, \ldots, z_n$ does not depend on the choice of holomorphic coordinates. The tangent bundle T_X of a complex manifold X is thus equipped with the structure of a complex vector bundle. Such a structure is called an almost complex structure.

After some preliminaries on manifolds and vector bundles, we turn to the proof of the Newlander–Nirenberg theorem, which characterises the almost complex structures induced as above by a complex structure. This 'integrability' criterion is extremely important in the study of the deformations of the complex structure of a manifold. Indeed, we could describe them as the deformations of the almost complex structure (which are essentially parametrised by a vector space of differentiable sections of a certain bundle over X) satisfying the integrability condition. This will enable us to put the structure of an infinite-dimensional manifold on this space of deformations, whose quotient by the group of diffeomorphisms of X describes the deformations of the complex structure of X up to isomorphism.

The Newlander–Nirenberg integrability criterion is as follows.

Theorem 2.1 *The almost complex structure $J \in \mathrm{End}\, T_X$, $J^2 = -1$ is integrable if and only if the bracket of two vector fields of type $(0, 1)$ for J is of type $(0, 1)$ for J.*

Following Weil, we prove this theorem only in the real analytic case, where it becomes a clever application of the Frobenius integrability theorem, which we prove beforehand. This theorem concerns distributions on a differentiable manifold X, i.e. a vector subbundle F of the tangent bundle T_X. We say that F is integrable if locally there exists a submersion $\phi : X \to \mathbb{R}^k$ such that F is identified with the bundle of tangent vectors to the fibres of ϕ.

Theorem 2.2 *(Frobenius) The distribution F is integrable if and only if the bracket of two vector fields which are tangent to the distribution F at every point also lies in F.*

This chapter concludes with the introduction of holomorphic vector bundles over a complex manifold. These vector bundles are those whose "transition matrices" are holomorphic. It turns out that we can define a differential operator $\overline{\partial}$ (the Dolbeault operator) on the space of sections of such a vector bundle E, and more generally, on the space of differential forms with values in such a bundle. The holomorphic sections σ of E are then characterised by the equation $\overline{\partial}\sigma = 0$.

One can show that the Dolbeault operator satisfies the condition $\overline{\partial} \circ \overline{\partial} = 0$, and that the complex defined in this way is locally exact. This will be used in the following chapters to compute, or rather to represent, the cohomology of X with values in the sheaf of holomorphic sections of E using $\overline{\partial}$-closed differential forms with coefficients in E.

2.1 Manifolds and vector bundles

2.1.1 Definitions

A topological manifold is a topological space X equipped with a covering by open sets U_i, which are homeomorphic, via maps ϕ_i called "local charts", to open sets of $\mathbb{R}^n$. One can show (Milnor 1965) that such an n is necessarily independent of i when X is connected; n is then called the dimension of X.

Definition 2.3 *A C^k differentiable manifold is a topological manifold equipped with a system of local charts $\phi_i : U_i \to \mathbb{R}^n$ such that the open sets U_i cover X, and the change of chart morphisms*

$$\phi_j \circ \phi_i^{-1} : \phi_i(U_i \cap U_j) \to \phi_j(U_i \cap U_j)$$

are differentiable of class C^k.

Definition 2.4 *A C^k differentiable function on such a manifold (or on an open set) is a function f such that for each U_i, $f \circ \phi_i^{-1}$ is differentiable of class C^k.*

Note that the fact that the change of chart maps are C^k implies that for a function f on $U_i \cap U_j$, $f \circ \phi_i^{-1}$ is C^k on $\phi_i(U_i \cap U_j)$ if and only if $f \circ \phi_j^{-1}$ is C^k on $\phi_j(U_i \cap U_j)$; this shows that it suffices to check the differentiability of f in any chart.

A real (resp. complex) topological vector bundle of rank m over a topological space X is a topological space E equipped with a map $\pi : E \to X$ such that for an open cover $\{U_i\}$ of X, we have "local trivialisation" homeomorphisms

$$\tau_i : \pi^{-1}(U_i) \cong U_i \times \mathbb{R}^m \text{ (resp. } U_i \times \mathbb{C}^m),$$

such that:

(i) $\mathrm{pr}_1 \circ \tau_i = \pi$.

(ii) The transition functions

$$\tau_j \circ \tau_i^{-1} : \tau_i(\pi^{-1}(U_i \cap U_j)) \to \tau_j(\pi^{-1}(U_i \cap U_j))$$

are $\mathbb{R}$-linear (resp. $\mathbb{C}$-linear) on each fibre $u \times \mathbb{R}^m$ (resp. $u \times \mathbb{C}^m$). Such a transformation

$$U_i \cap U_j \times \mathbb{R}^m \to U_i \cap U_j \times \mathbb{R}^m$$

must respect the first projection, by condition (i) above, and is thus described by a real matrix of type (m, m), whose coefficients, by continuity, are continuous functions of $u \in U_i \cap U_j$. (In the complex case, we must consider complex matrices.) These matrices are called transition matrices.

Definition 2.5 *If X is a C^k differentiable manifold, a vector bundle E over X is equipped with a C^k differentiable structure if we are given local trivialisations whose transition matrices are C^k.*

Remark 2.6 *The bundle E is then equipped with the structure of a C^k manifold for which π is C^k as well as the local trivialisations.*

A section of a vector bundle $E \stackrel{\pi}{\to} X$ is a map $\sigma : X \to E$ such that $\pi \circ \sigma = \mathrm{Id}_X$. This section is said to be continuous, resp. differentiable, resp. C^k differentiable, if σ is continuous, resp. differentiable, resp. C^k differentiable. If $\pi : E \to X$ is a vector bundle and $x \in X$, we write $E_x := \pi^{-1}(x)$. It is canonically a vector space, with structure given by any of the trivialisations of E in the neighbourhood of x. E_x is called the fibre of E at the point x.

A vector bundle $\pi : E \to X$ is said to be trivial if it admits a global trivialisation $\phi : E \cong X \times \mathbb{R}^n$. Equivalently, E must admit n global sections which provide a basis of the fibre E_x at each point. These sections are given by $\sigma_i = \phi^{-1} \circ \tilde{e}_i$, where $\tilde{e}_i : X \to X \times \mathbb{R}^n$ is given by $\tilde{e}_i(x) = (x, e_i)$, where the e_i form the standard basis of $\mathbb{R}^n$.

Let $\pi_E : E \to X$ and $\pi_F : F \to X$ be vector bundles over X. A morphism $\phi : E \to F$ of vector bundles is a continuous map such that $\pi_F \circ \phi = \pi_E$, and ϕ is linear on each fibre. This means that in local trivialisations, ϕ becomes linear ($\mathbb{C}$-linear in the case of complex bundles) on the fibres $u \times \mathbb{R}^k$; this definition is independent of the choice of the open set containing u, since the transition functions are also linear on the fibres. We have an analogous definition for differentiable bundles.

Given a vector bundle E, we can define its dual E^* and its exterior powers $\bigwedge^k E$, which are differentiable of the same class as E. The points of E^* are the linear forms on the fibres of $\pi_E : E \to X$. E^* admits a natural trivialisation when E is trivialised, and the transition matrices of E^* are the inverses of the transposes of the transition matrices of E. Similarly, the points of $\bigwedge^k E$ can be identified with the alternating k-linear forms on the fibres of $\pi_{E^*} : E^* \to X$.

2.1.2 The tangent bundle

If X is a C^k differentiable manifold, the tangent bundle T_X of X is a C^{k-1} differentiable bundle of rank $n = \dim X$ which we can define as follows. If X is covered by open sets U_i equipped with C^k diffeomorphisms ϕ_i to open sets of $\mathbb{R}^n$, then T_X is covered by open sets $U_i \times \mathbb{R}^n$, where the identifications (or transition morphisms) between $U_i \cap U_j \times \mathbb{R}^n \subset U_i \times \mathbb{R}^n$ and $U_i \cap U_j \times \mathbb{R}^n \subset U_j \times \mathbb{R}^n$ are given by

$$(u, v) \to (u, \phi_{ij_*}(v)).$$

Here $\phi_{ij} = \phi_j \circ \phi_i^{-1}$ is the transition diffeomorphism between the open sets $\phi_i(U_i \cap U_j)$ and $\phi_j(U_i \cap U_j)$ of $\mathbb{R}^n$, and ϕ_{ij_*} is its Jacobian matrix at the point u. A section of the tangent bundle of a differentiable manifold is called a vector field.

There exist two intrinsic ways of describing the elements of the tangent bundle. The points of the tangent bundle can be identified with equivalence classes of differentiable maps $\gamma : [-\epsilon, \epsilon] \to X$ (for an $\epsilon \in \mathbb{R}$, $\epsilon > 0$ varying with γ) for the equivalence relation

$$\gamma \equiv \gamma' \Leftrightarrow \gamma(0) = \gamma'(0), \quad \frac{d}{dt}\gamma\bigg|_{t=0} = \frac{d}{dt}\gamma'\bigg|_{t=0}.$$

The second equality in this definition makes sense in any local chart for X in the neighbourhood of $\gamma(0)$. We call these equivalence classes "jets of order 1". To check that the set defined in this way has the structure of a vector bundle introduced earlier, it suffices to note that the jets of order 1 of an open set U of $\mathbb{R}^n$ can be identified, via the map $\gamma \mapsto (\gamma(0), \dot{\gamma}(0))$, with $U \times \mathbb{R}^n$, and that a diffeomorphism $\psi : U \cong V$ between two open sets of $\mathbb{R}^n$ induces the isomorphism (ψ, ψ_*) between the spaces of jets of order 1 of U and V.

Another definition of the tangent vectors, i.e. of the elements of the tangent bundle, consists in identifying them with the derivations of the algebra of the real differentiable functions on X with values in $\mathbb{R}$ supported at a point $x \in X$. This means that we consider the linear maps

$$\psi : C^1(X) \to \mathbb{R}$$

satisfying Leibniz' rule

$$\psi(fg) = f(x)\psi(g) + g(x)\psi(f)$$

for a point $x \in X$. The equivalence between the two definitions is realised by the map which to a jet γ associates the derivation $\psi_\gamma(f) = \frac{d(f\circ\gamma)}{dt}\big|_{t=0}$.

Definition 2.7 *A differential form of degree k is a section of* $\bigwedge^k(T_X)^*$. *We write* $d^0\alpha := k$ *for the degree of such a form* α.

In general, we write $\Omega_{X,\mathbb{R}}$ for the bundle of real differential 1-forms, and $\Omega_{X,\mathbb{C}} = \mathrm{Hom}(T_X, \mathbb{C})$ for its complexification. Similarly, the bundle of real (resp. complex) k-forms is written $\Omega^k_{X,\mathbb{R}}$ (resp. $\Omega^k_{X,\mathbb{C}}$). We see immediately that if f is a real C^k differentiable function on X, then df is a C^{k-1} section of $\Omega_{X,\mathbb{R}}$. We also see that if $x_1, \ldots, x_n$ are local coordinates defined on an open set $U \subset X$, then the $dx_I = dx_{i_1} \wedge \cdots \wedge dx_{i_k}$, $1 \le i_1 < \cdots < i_k \le n$ provide a basis of the fibre of $\Omega^k_{X,\mathbb{R}}$ at each point of the open set U. Indeed, by the definition of T_X, the coordinates x_i provide a local trivialisation of T_X, where the corresponding local basis is given at each point $x \in U$ by the derivations $\frac{\partial}{\partial x_i}\big|_x$. The dx_i simply form the dual basis of $\Omega_{X,\mathbb{R}}$ at each point of U.

2.1.3 Complex manifolds

Let X be a differentiable manifold of dimension $2n$.

Definition 2.8 *We say that X is equipped with a complex structure if X is covered by open sets U_i which are diffeomorphic, via maps called ϕ_i, to open sets of $\mathbb{C}^n$, in such a way that the transition diffeomorphisms*

$$\phi_j \circ \phi_i^{-1} : \phi_i(U_i \cap U_j) \to \phi_j(U_i \cap U_j)$$

are holomorphic.

The (complex) dimension of X is by definition equal to n. On a complex manifold, a map f with values in $\mathbb{C}$ defined on an open set U is said to be holomorphic if $f \circ \phi_i^{-1}$ is holomorphic on $\phi_i(U \cap U_i)$. Once again, this definition does not depend on the choice of chart, since the change of chart morphisms is holomorphic and compositions of holomorphic functions are also holomorphic.

We can also define the notion of a holomorphic vector bundle.

Definition 2.9 *A differentiable complex vector bundle $\pi_E : E \to X$ over a complex manifold X is said to be equipped with a holomorphic structure if we have trivialisations*

$$\tau_i : \pi_i^{-1}(U_i) \cong U_i \times \mathbb{C}^n$$

such that the transition matrices $\tau_{ij} = \tau_j \circ \tau_i^{-1}$ have holomorphic coefficients.

The above trivialisations will be called "holomorphic trivialisations". If E is a holomorphic vector bundle, E is in particular a complex manifold such that π_E is holomorphic. Indeed, we can assume, in the definition above, that the U_i are charts, i.e. identified via ϕ_i with open sets of $\mathbb{C}^n$; then the $(\phi_i \times \mathrm{Id}_{\mathbb{C}^n}) \circ \tau_i$ give charts for E whose transition functions are clearly holomorphic.

A holomorphic section of a holomorphic vector bundle $\pi_E : E \to X$ over an open set U of X is a section $s : X \to E$ of π_E which is a holomorphic map. For example, a holomorphic local trivialisation τ_i of E as above is given by the choice of a family of holomorphic sections of E, whose values at each point u of U_i form a basis of the fibre E_u over $\mathbb{C}$.

Example 2.10 **The holomorphic tangent bundle.** *This bundle is defined exactly like the real tangent bundle of a differentiable manifold. Given a system of charts $\phi_i : U_i \cong V_i \subset \mathbb{C}^n$, we define T_X as the union of the $U_i \times \mathbb{C}^n$, glued* *by identifying $U_i \cap U_j \times \mathbb{C}^n \subset U_i \times \mathbb{C}^n$ and $U_i \cap U_j \times \mathbb{C}^n \subset U_j \times \mathbb{C}^n$ via*

$$(u, v) \mapsto (u, \phi_{ij*}(v)).$$

Here, the holomorphic Jacobian matrix ϕ_{ij*} is the matrix with holomorphic coefficients $\frac{\partial \phi_{ij}^k}{\partial z_l}(u)$, where $\phi_{ij} = \phi_j \circ \phi_i^{-1}$, and the operator $\frac{\partial}{\partial z_l}$ is defined in (1.3). The fact that $\frac{\partial \phi_{ij}^k}{\partial z_l}$ is holomorphic follows from theorem 1.17.

We can also, as in 2.1.2, define the holomorphic tangent bundle as the set of complex-valued derivations of the $\mathbb{C}$-algebra of holomorphic functions, or as the set of jets of order 1 *of holomorphic maps from the complex disk to X.*

2.2 Integrability of almost complex structures

2.2.1 Tangent bundle of a complex manifold

Let X be a complex manifold, and let $\phi_i : U_i \to \mathbb{C}^n$ be holomorphic local charts. Then the real tangent bundle $T_{U_i,\mathbb{R}}$ can be identified, via the differential ϕ_{i*}, with $U_i \times \mathbb{C}^n$. Moreover, the change of chart morphisms $\phi_j \circ \phi_i^{-1}$ are holomorphic by hypothesis, i.e. have $\mathbb{C}$-linear differentials, for the natural identifications:

$$T_{\mathbb{C}^n,x} \cong \mathbb{C}^n, \quad \forall x \in \mathbb{C}^n.$$

It follows that the $\mathbb{R}$-linear operators

$$I_i : T_{U_i,\mathbb{R}} \to T_{U_i,\mathbb{R}},$$

identified with $1 \times i$ acting on $U_i \times \mathbb{C}^n$, glue together on $U_i \cap U_j$ and define a global endomorphism, written I, of the bundle $T_{X,\mathbb{R}}$. Obviously I satisfies the identity $I^2 = -1$; thus I defines the structure of a $\mathbb{C}$-vector space of rank n on each fibre $T_{X,x}$. The differentiability of I even shows that $T_{X,\mathbb{R}}$ is thus equipped with the structure of a differentiable complex vector bundle. This leads us to introduce the following definition.

Definition 2.11 *An almost complex structure on a differentiable manifold is an endomorphism I of $T_{X,\mathbb{R}}$ such that $I^2 = -1$; equivalently, it is the structure of a complex vector bundle on $T_{X,\mathbb{R}}$.*

We saw that a complex structure on X naturally induces an almost complex structure.

Definition 2.12 *An almost complex structure I on a manifold X is said to be integrable if there exists a complex structure on X which induces I.*

In the case of a complex manifold, the relation between $T_{X,\mathbb{R}}$ seen as a complex vector bundle and the holomorphic tangent bundle T_X of X is as follows: the bundle T_X is generated, in the charts U_i, by the elements

$$\frac{\partial}{\partial z_j} = \frac{1}{2}\left(\frac{\partial}{\partial x_j} - i\frac{\partial}{\partial y_j}\right),$$

which are naturally elements of $T_{U_i} \otimes \mathbb{C}$. Thus, in fact, we have an inclusion of complex vector bundles

$$T_X \subset T_{X,\mathbb{R}} \otimes \mathbb{C}.$$

Moreover, for an almost complex manifold (X, I), the complexified tangent bundle $T_{X,\mathbb{R}} \otimes \mathbb{C}$ contains a complex vector subbundle, denoted by $T_X^{1,0}$ and defined as the bundle of eigenvectors of I for the eigenvalue i. As a real vector bundle, $T_X^{1,0}$ is naturally isomorphic to $T_{X,\mathbb{R}}$ via the application $\Re$ (real part) which to a complex field $u+iv$ associates its real part u. Moreover, this identification identifies the operators i on $T_X^{1,0}$ and I on $T_{X,\mathbb{R}}$. Clearly $T_X^{1,0}$ is generated by the $u - iIu,\ u \in T_{X,\mathbb{R}}$.

Furthermore, in the case where $X = \mathbb{C}^n$, consider the isomorphism $T_{\mathbb{C}^n,\mathbb{R}} \cong \mathbb{C}^n \times \mathbb{R}^{2n}$ given by the sections $\frac{\partial}{\partial x_j}, \frac{\partial}{\partial y_j}$ of the tangent bundle of $\mathbb{C}^n$, where $z_k = x_k + iy_k$ and the z_k are complex linear coordinates on $\mathbb{C}^n$. The induced complex structure operator I on $T_{\mathbb{C}^n,\mathbb{R}}$ sends $\frac{\partial}{\partial x_j}$ to $\frac{\partial}{\partial y_j}$. Thus, the tangent vectors of type $(1, 0)$ are generated over $\mathbb{C}$ at each point by the

$$\frac{\partial}{\partial x_j} - iI\frac{\partial}{\partial y_j} = 2\frac{\partial}{\partial z_j}.$$

In conclusion, we have shown the following.

Proposition 2.13 *If X is a complex manifold, then X admits an almost complex structure, and the subbundle $T_X^{1,0} \subset T_{X,\mathbb{R}} \otimes \mathbb{C}$ defined by I is equal, as a complex vector subbundle of $T_{X,\mathbb{R}} \otimes \mathbb{C}$, to the holomorphic tangent bundle T_X.*

Complex conjugation acts naturally on the complexified tangent bundle $T_{X,\mathbb{C}}$ of a differentiable manifold X. If I is an almost complex structure on X, we have the subbundle $T_X^{0,1}$ of $T_{X,\mathbb{C}}$, defined as the complex conjugate of $T_X^{1,0}$. We can also define it as the set of the complexified tangent vectors which are the eigenvectors of I associated to the eigenvalue $-i$. Thus, it is clear that we have a direct sum decomposition

$$T_{X,\mathbb{C}} = T_X^{1,0} \oplus T_X^{0,1}. \tag{2.1}$$

Remark 2.14 *When X is an almost complex manifold, the vector bundle $T_X^{1,0}$ does not a priori have the structure of a holomorphic bundle. In what follows, if X is a complex manifold, a section of T_X will be taken to mean a holomorphic section of T_X, while a section of $T_X^{1,0}$ will be a differentiable section.*

If $\phi : X \to Y$ is a holomorphic map between two complex manifolds, we define a morphism of holomorphic vector bundles

$$\phi_* : T_X \to \phi^*(T_Y)$$

in the obvious way. In holomorphic local charts which trivialise T_X and T_Y, the matrix of ϕ_* is given by the holomorphic Jacobian matrix $\left(\frac{\partial \phi_k}{\partial z_j}\right)$ of ϕ. This morphism can in fact be identified with the morphism of real vector bundles

$$\phi_{*,\mathbb{R}} : T_{X,\mathbb{R}} \to \phi^*(T_{Y,\mathbb{R}}),$$

via the identifications of real bundles $T_X \cong T_{X,\mathbb{R}}$, $T_Y \cong T_{Y,\mathbb{R}}$ given by the real part $\Re$. As a morphism of complex bundles, ϕ_* can be deduced from $\phi_{*,\mathbb{R}}$ by noting that $\phi_{*,\mathbb{R}}$ is compatible with the almost complex structures of X and Y, since ϕ is holomorphic, and thus induces a $\mathbb{C}$-linear morphism

$$\phi_*^{1,0} : T_X^{1,0} \to \phi^*\left(T_Y^{1,0}\right).$$

2.2.2 The Frobenius theorem

A C^l vector field χ over a manifold X defines a derivation

$$\chi : C^k(X) \to C^{k-1}(X), \quad k \leq l+1, \quad \chi(f) = df(\chi),$$

i.e. a linear map satisfying Leibniz' rule: $\chi(fg) = f\chi(g) + g\chi(f)$. Conversely, as explained in the preceding subsection, such a derivation gives a tangent vector at each point of X, and thus a vector field which is easily shown to be C^{k-1}. This enables us to define the Lie bracket of two C^l fields, thanks to the following elementary lemma.

Lemma 2.15 *Let χ, ψ be two derivations*

$$C^{l+1}(X) \to C^l(X), \quad l \geq 1.$$

Then the commutator

$$\chi \circ \psi - \psi \circ \chi : C^2(X) \to C^0(X)$$

is again a derivation.

Thus we can give the following definition.

Definition 2.16 *The bracket* $[\chi, \psi]$ *of the vector fields* χ, ψ *is the vector field corresponding to the derivation* $\chi \circ \psi - \psi \circ \chi$.

In local coordinates x_i on X, the vector field χ can be written uniquely as $\chi = \sum_i \chi_i \frac{\partial}{\partial x_i}$, and we have a similar expression for the vector field ψ.

Lemma 2.17 *We have the formula*

$$[\chi, \psi] = \sum_i (\chi(\psi_i) - \psi(\chi_i)) \frac{\partial}{\partial x_i}.$$

Proof We must check that for a C^2 function f, we have

$$[\chi, \psi](f) = \sum_i (\chi(\psi_i) - \psi(\chi_i)) \frac{\partial f}{\partial x_i}.$$

But $\psi(f) = \sum_i \psi_i \frac{\partial f}{\partial x_i}$ and $\chi(f) = \sum_i \chi_i \frac{\partial f}{\partial x_i}$, so we obtain

$$\chi \circ \psi(f) = \sum_{j,i} \chi_i \frac{\partial}{\partial x_i}\left(\psi_j \frac{\partial f}{\partial x_j}\right) = \sum_{j,i} \chi_i \left(\frac{\partial \psi_j}{\partial x_i}\frac{\partial f}{\partial x_j} + \psi_j \frac{\partial^2 f}{\partial x_i \partial x_j}\right),$$

and a similar expression for $\psi \circ \chi(f)$. The symmetry of the second derivatives then gives

$$[\chi, \psi](f) = \sum_{i,j} \left(\chi_i \frac{\partial \psi_j}{\partial x_i} - \psi_i \frac{\partial \chi_j}{\partial x_i}\right) \frac{\partial f}{\partial x_j}.$$

□

The following is an immediate consequence of lemma 2.17.

Corollary 2.18 *If* χ, ψ *are two* C^1 *vector fields and* f *is a* C^1 *function, all of which are differentiable, then*

$$[\chi, f\psi] = f[\chi, \psi] + \chi(f)\psi.$$

Now let X be an n-dimensional manifold, and let $E \subset T_X$ be a C^1 vector sub-bundle of rank k. Such an E is called a distribution on X.

Definition 2.19 *We say that the distribution* E *is integrable if* X *is covered by open sets* U *such that there exists a* C^1 *map*

$$\phi_U : U \to \mathbb{R}^{n-k}$$

such that for every $x \in U$, *the vector subspace* $E_x \subset T_{X,x}$ *is equal to* $\operatorname{Ker} d\phi_x$.

Clearly ϕ is then a submersion, and each fibre $\phi^{-1}(v)$ is a closed submanifold of U having the property that its tangent space at each point is equal to the fibre of E at that point. The following theorem characterises the integrable distributions.

Theorem 2.20 *(Frobenius) A distribution E is integrable if and only if for all C^1 vector fields χ, ψ contained in E, the bracket $[\chi, \psi]$ is also contained in E.*

Proof Obviously, if E is integrable, then E is stable under the bracket. Indeed, the sections of E are then exactly the fields χ (i.e. derivations) which annihilate the functions $f_i = x_i \circ \phi$, where ϕ is as in definition 2.19. But if we have $\chi(f_i) = 0$ and $\psi(f_i) = 0$, then we also have $[\chi, \psi](f_i) = 0$ by definition of the bracket. Thus the real content of this theorem is the converse result.

This can be shown by induction on k. When $k = 1$, E is generated by a single field. By corollary 2.18, there is then no integrability condition. Furthermore, the conclusion of theorem 2.20 is satisfied by the flow-box theorem (Arnold 1984).

Suppose now that the theorem is proved for $k-1$, and consider a distribution E of rank k satisfying the integrability condition $[E, E] \subset E$. Let U be an open set of X such that there exists a non-zero section χ of E over U, and a submersive map

$$\phi : U \to \mathbb{R}^{n-1}$$

whose fibres are the trajectories of χ. By the flow-box theorem, we may assume that U is diffeomorphic to $V \times (0, 1)$ where V is an open set of $\mathbb{R}^{n-1}$, that ϕ is the first projection and that χ is identified with $\frac{\partial}{\partial t}$. We will show the following result.

Lemma 2.21 *The integrability condition implies that there exists a distribution F of rank $k-1$ on V such that*

$$E = (\phi_*)^{-1}(F).$$

Moreover, E satisfies the integrability condition if and only if F does.

Here $\phi_* : T_{U,x} \to T_{V,\phi(x)}$ is the differential of ϕ. The notation means that $\forall x \in U$ we have $E_x = (\phi_*)^{-1}(F_{\phi(x)})$. Admitting the lemma, the conclusion is immediate, since we can then apply the induction hypothesis to the distribution F. This gives, at least locally,

$$\psi : V \to \mathbb{R}^{n-k},$$

whose fibres are integral manifolds of the distribution F. Then the fibres of $\psi \circ \phi : U \to \mathbb{R}^{n-k}$ are integral manifolds of the distribution E. □

Proof of lemma 2.21 At each point $x \in U$, the differential

$$\phi_* : T_{U,x} \to T_{V,\phi(x)}$$

enables us to define a vector subspace $F_x = \phi_*(E_x) \subset T_{V,\phi(x)}$ of rank $k-1$, since $\langle \chi \rangle = \operatorname{Ker} \phi_*$ is contained in E. At the point x, we obviously have

$$E_x = (\phi_*)^{-1}(F_x).$$

To show the first assertion, it suffices to see that $F_x \subset T_{V,\phi(x)}$ depends only on the point $\phi(x)$, and not on the choice of the point x in the fibre $\phi^{-1}(\phi(x))$. For this, we first note the following easy lemma.

Lemma 2.22 *Let $K \subset (0,1) \times \mathbb{R}^m$ be a differentiable vector subbundle of rank k of the trivial bundle of rank m over the segment* $(0, 1)$. *If for every differentiable section σ of K we have $\frac{d\sigma}{dt} \in K$, then K is a trivial subbundle, i.e. $K = (0,1) \times \mathbb{R}^k \subset (0,1) \times \mathbb{R}^m$.*

Proof This follows from the theorem of constant rank, applied to the composition map

$$K \hookrightarrow (0,\ 1) \times \mathbb{R}^m \stackrel{\mathrm{pr}_2}{\to} \mathbb{R}^m.$$

Indeed, the hypothesis means that this map is of rank k everywhere. Thus, the image is a k-dimensional submanifold which contains each rank k vector space K_t, so we see that K_t must be constant. □

To apply this lemma to our situation, consider the subbundle

$$K = \phi_* E \subset (0,\ 1) \times T_{V,v}$$

over the fibre $\phi^{-1}(v) = \{v\} \times (0, 1)$. We now check the following lemma.

Lemma 2.23 *Let σ be a differentiable section of the bundle E over* $(0, 1) \times v$. *Then $\phi_*(\sigma)$ is a differentiable section of the bundle K, and we have*

$$\left.\frac{d}{dt}(\phi_*(\sigma))\right|_{t=t_0} = \phi_{*,(v,t_0)}([\chi, \tilde{\sigma}])$$

in $T_{V,v}$, where $\tilde{\sigma}$ is a differentiable section of E over $V \times$ $(0, 1)$ *extending σ.*

We recall that $\chi \in E$ is the vector field $\frac{\partial}{\partial t}$. The proof of this lemma follows immediately from lemma 2.17. □

When the integrability condition is satisfied, lemma 2.23 shows that the hypothesis of lemma 2.22 is satisfied by K, since every differentiable section of K is of the form $\phi_*\sigma$, for a section σ of E as above. Thus, $K_t = \phi_* E_{v,t} \subset T_{V,v}$ is independent of t, and we can define a subbundle F of T_V by

$$F_v = \phi_* E_{v,t} \subset T_{V,v}.$$

To conclude the proof of lemma 2.21 and thus of theorem 2.20, it suffices to see that the distribution $F \subset T_V$ constructed in this way also satisfies the integrability condition. Now, let $\sigma,\ \rho$ be differentiable sections of F over V. By definition of F, there exist sections $\tilde{\sigma},\ \tilde{\rho}$ of E over $V \times (0, 1)$ such that

$$\phi_{*(v,t)}\big(\tilde{\sigma}_{(v,t)}\big) = \sigma_v, \quad \phi_{*(v,t)}\big(\tilde{\rho}_{(v,t)}\big) = \rho_v, \quad \forall (v, t).$$

We then show, using lemma 2.17, that

$$[\sigma, \rho]_{(v,t)} = \phi_{*(v,t)}\big([\tilde{\sigma}, \tilde{\rho}]_{(v,t)}\big),$$

so that $[\tilde{\sigma}, \tilde{\rho}] \in E$ implies $[\sigma, \rho] \in F$. □

2.2.3 The Newlander–Nirenberg theorem

Note first that the bracket of vector fields over a differentiable manifold X extends by $\mathbb{C}$-linearity to the complexified vector fields, i.e. to the differentiable sections of $T_{X,\mathbb{C}}$. Now, let (X, I) be an almost complex manifold. As mentioned above, the almost complex structure operator I splits the bundle $T_{X,\mathbb{C}}$ into elements of type $(1, 0)$, eigenvectors associated to the eigenvalue i of I, and elements of type $(0, 1)$, eigenvectors associated to the eigenvalue $-i$ of I. The bundle $T_X^{1,0}$ is the complex conjugate of the bundle $T_X^{0,1}$. The following theorem gives an exact description of the integrable almost complex structures.

Theorem 2.24 *(Newlander–Nirenberg) The almost complex structure I is integrable if and only if we have*

$$\big[T_X^{0,1}, T_X^{0,1}\big] \subset T_X^{0,1}.$$

Remark 2.25 *By passing to the conjugate, this is equivalent to the condition that the bracket of two vector fields of type* $(1, 0)$ *is of type* $(1, 0)$.

This theorem is a difficult theorem in analysis, for it implies, in particular, that the manifold X which was assumed to be only differentiable actually admits the structure of a real analytic manifold. Following Weil (1957), we will show that when (X, I) are assumed to be real analytic, this theorem follows easily from the following analytic version of the Frobenius theorem 2.20.

Theorem 2.26 *Let X be a complex manifold of dimension n, and let E be a holomorphic distribution of rank k over X, i.e. a holomorphic vector subbundle of rank k of the holomorphic tangent bundle T_X. Then E is integrable in the holomorphic sense if and only if we have the integrability condition*

$$[E, E] \subset E.$$

Here, the integrability in the holomorphic sense means that X is covered by open sets U such that there exists a holomorphic submersive map

$$\phi_U : U \to \mathbb{C}^{n-k}$$

satisfying

$$E_u = \mathrm{Ker}\big(\phi_* : T_{U,u} \to T_{\mathbb{C}^{n-k},\phi(u)}\big)$$

for every $u \in U$.

Proof We first reduce to the real Frobenius theorem, by noting that the conditions that E is holomorphic and that $[E, E] \subset E$ imply that the real distribution $\Re E \subset T_{X,\mathbb{R}}$ also satisfies the Frobenius integrability condition, and thus is integrable. X is thus covered by open sets U such that there exists a submersion

$$\phi_U : U \to V,$$

where V is open in $\mathbb{R}^{2(n-k)}$, satisfying

$$(\Re E)_u = \mathrm{Ker}\big(\phi_{*,u} : T_{U,u,\mathbb{R}} \to T_{\mathbb{R}^{2(n-k)},\phi(u)}\big), \quad \forall u \in U.$$

Next, we show that there exists a complex structure on the image of ϕ for which ϕ is holomorphic. For this, we first note that if $v = \phi(u)$, $T_{V,v} = T_{U,u}/\Re\, E$ and as $\Re\, E$ is stable under the endomorphism I corresponding to the almost complex structure on T_U, there is an induced complex structure on $T_{V,v}$. We then show that this complex structure does not depend on the point in the fibre of ϕ above v. Thus, there exists an almost complex structure on V for which the differential of ϕ is $\mathbb{C}$-linear at every point. Finally, to see that this almost complex structure is integrable, we take a complex submanifold of U transverse to the fibres of ϕ_U, which exists up to restricting U. Via ϕ_U, this submanifold becomes locally isomorphic to V, and this isomorphism is compatible with the almost complex structures. Thus, the almost complex structure on $\Im \phi_U$ is integrable, and it makes ϕ_U into a holomorphic map. □

Proof of theorem 2.24 in the real analytic case Theorem 2.26 implies the Newlander–Nirenberg theorem in the real analytic case as follows. Since everything is local, we may assume that X is an open set U of $\mathbb{R}^{2n}$ and that I is a real analytic map with values in End $\mathbb{R}^{2n}$, satisfying $I \circ I = -\mathrm{Id}$. Up to restricting U, we may assume that I is given by a convergent power series. This power series extends to the whole complex domain and gives a holomorphic map I from an open set $U_{\mathbb{C}}$ of $\mathbb{C}^{2n}$ (a neighbourhood of U) to End $\mathbb{C}^{2n}$. This map of course satisfies the condition $I \circ I = -1$. Now, this map I gives a holomorphic distribution $E_{\mathbb{C}}$ of rank n on $U_{\mathbb{C}}$, where we define

$$E_{\mathbb{C},u} \subset T^{1,0}_{U_{\mathbb{C}},u} \cong \mathbb{C}^{2n}$$

to be the eigenspace associated to the eigenvalue $-i$ of I. Note that by definition, along U, we have $E_{\mathbb{C},u} = T^{0,1}_{U,u} \subset T_{U,u} \otimes \mathbb{C} = \mathbb{C}^{2n}$.

By definition, the sections of $T^{0,1}_X$ on U are generated over $\mathbb{C}$ by the $\chi + iI\chi$, where χ is a real vector field over U. Similarly, the sections of $E_{\mathbb{C}}$ on $U_{\mathbb{C}}$ are generated by the $\chi + iI\chi$, where χ is a real or complex vector field on $U_{\mathbb{C}}$. It follows immediately that if I satisfies the integrability condition of theorem 2.24, then the holomorphic distribution $E_{\mathbb{C}}$ satisfies the integrability condition of theorem 2.26.

The distribution $E_{\mathbb{C}}$ is thus integrable, which gives (at least locally) a holomorphic submersion

$$\phi : U_{\mathbb{C}} \to \mathbb{C}^n$$

whose fibres are the integral holomorphic submanifolds of the distribution $E_{\mathbb{C}}$. We will show that the restriction of ϕ to U is a local diffeomorphism, and that the complex structure induced on U has an associated almost complex structure given precisely by I.

The first fact is clear: indeed, along U,

$$T_{U,u} \subset T_{U_{\mathbb{C}},u} \cong \mathbb{C}^{2n}$$

can be identified with $\mathbb{R}^{2n}$, while $\Re E_{\mathbb{C}}$ can be identified with $T^{0,1}_{U,u}$. These two spaces are thus transverse, and it follows that $\phi_{*|T_U}$ is an isomorphism, so that $\phi_{|U}$ is a diffeomorphism in the neighbourhood of u.

As for the second point, we need to check that the isomorphism

$$\phi_{*,\mathbb{R}} : T_{U,u} \to T_{\mathbb{C}^n,\phi(u)}$$

identifies I with the complex structure operator on $\mathbb{C}^n$. But this comes down to showing that the composition

$$T_{U,u} \hookrightarrow T_{U_\mathbb{C},u} \to T_{U_\mathbb{C},u}/(\Re E)_u$$

is $\mathbb{C}$-linear, where the left-hand term is equipped with the $\mathbb{C}$-linear structure given by I, while the complex structure on the right-hand term comes from that of $T_{U_\mathbb{C},u} \cong \mathbb{C}^{2n}$. Now, along U,

$$(\Re E)_u \subset T_{U_\mathbb{C},u} \cong T_{U,u} \otimes \mathbb{C}$$

is generated by the $\chi + iI\chi$, so that in the quotient $T_{U_\mathbb{C},u}/(\Re E)_u$, we have $\chi = -iI\chi$ for $\chi \in T_{U,u}$, i.e. $i\chi = I\chi$. □

2.3 The operators ∂ and ∂̄

2.3.1 Definition

Let (X, I) be an almost complex manifold; the decomposition (2.1) $T_{X,\mathbb{C}} = T_X^{1,0} \oplus T_X^{0,1}$ induces a dual decomposition

$$\Omega_{X,\mathbb{C}} = \Omega_X^{1,0} \oplus \Omega_X^{0,1}. \tag{2.2}$$

When X is a complex manifold, the bundle $\Omega_X^{1,0}$ of complex differential forms of type $(1, 0)$, i.e. $\mathbb{C}$-linear forms, is generated in holomorphic local coordinates $z_1, \ldots, z_n$ by the dz_i, i.e. a form α of type $(1, 0)$ can be written locally as $\alpha = \sum_i \alpha_i dz_i$, where the α_i are C^k functions if α is C^k. Since $d(dz_i) = 0$, it follows that

$$d\alpha = \sum_i d\alpha_i \wedge dz_i. \tag{2.3}$$

Furthermore, the decomposition (2.2) also induces the decomposition of the complex k-forms into forms of type (p, q), for $p + q = k$:

$$\bigwedge^k \Omega_{X,\mathbb{C}} = \sum_{p+q=k} \Omega_X^{p,q}, \tag{2.4}$$

where the bundle $\Omega_X^{p,q}$ is equal to

$$\bigwedge^p \Omega_X^{1,0} \otimes \bigwedge^q \Omega_X^{0,1}.$$

With this definition, formula (2.3) shows that if the almost complex structure is integrable and α is a differential form of type $(1, 0)$, then $d\alpha$ is a section of $\Omega_X^{2,0} \oplus \Omega_X^{1,1}$. In fact, using the formula

$$\chi(\alpha(\psi)) - \psi(\alpha(\chi)) = d\alpha(\chi, \psi) + \alpha([\chi, \psi])$$

where α is a 1-form and χ, ψ are vector fields, we easily see that this property is equivalent to the integrability condition of theorem 2.24, and thus to the integrability of the almost complex structure.

More generally, the bundle $\Omega_X^{p,q}$ admits as generators in holomorphic local coordinates $z_1, \dots, z_n$ the differential forms

$$dz_I \wedge d\overline{z}_J = dz_{i_1} \wedge \cdots \wedge dz_{i_p} \wedge d\overline{z}_{j_1} \wedge \cdots \wedge d\overline{z}_{j_q},$$

where I, J are sets of ordered indices $1 \leq i_1 < \cdots < i_p \leq n$ and $1 \leq j_1 < \cdots < j_q \leq n$. Note that these forms are closed, i.e. annihilated by the exterior differential operator d. A form α of type (p, q) can thus be written locally as $\alpha = \sum_{I,J} \alpha_{I,J} dz_I \wedge d\overline{z}_J$. It follows that

$$d\alpha = \sum_{I,J} d\alpha_{I,J} \wedge dz_I \wedge d\overline{z}_J$$

is the sum of a form of type $(p, q + 1)$ and a form of type $(p + 1, q)$.

Definition 2.27 *For a C^1 differential form α of type (p, q) on a complex manifold X, we define $\overline{\partial}\alpha$ to be the component of type $(p, q + 1)$ of $d\alpha$. Similarly, we define $\partial\alpha$ to be the component of type $(p + 1, q)$ of $d\alpha$.*

For $(p, q) = (0, 0)$, a form of type (p, q) is a function f. $\overline{\partial} f$ is then the $\mathbb{C}$-antilinear part of df, and thus it vanishes if and only if f is holomorphic. By definition, we have

$$df = \sum_i \frac{\partial f}{\partial z_i} dz_i + \sum_i \frac{\partial f}{\partial \overline{z}_i} d\overline{z}_i,$$

and thus

$$\overline{\partial} f = \sum_i \frac{\partial f}{\partial \overline{z}_i} d\overline{z}_i.$$

As mentioned above, a k-differential form α decomposes uniquely into components $\alpha^{p,q}$ of type (p, q), $p + q = k$. We then set

$$\overline{\partial}\alpha = \sum_{p,q} \overline{\partial}\alpha^{p,q}, \quad \partial\alpha = \sum_{p,q} \partial\alpha^{p,q}.$$

The following lemmas describe the essential properties of the operators ∂, $\overline{\partial}$.

Lemma 2.28 *The operator $\overline{\partial}$ satisfies Leibniz' rule*

$$\overline{\partial}(\alpha \wedge \beta) = \overline{\partial}\alpha \wedge \beta + (-1)^{d^0\alpha}\alpha \wedge \overline{\partial}\beta.$$

Similarly, the operator ∂ satisfies Leibniz' rule

$$\partial(\alpha \wedge \beta) = \partial\alpha \wedge \beta + (-1)^{d^0\alpha}\alpha \wedge \partial\beta.$$

Proof The second assertion follows from the first, since by definition of the operators ∂ and $\overline{\partial}$, we have the relation

$$\partial\alpha = \overline{\overline{\partial}\overline{\alpha}}.$$

As for the first relation, it suffices to prove it for α of type (p, q) and β of type (p', q'). We then obtain it immediately in this case, by taking the component of type $(p + p', q + q' + 1)$ of $d(\alpha \wedge \beta)$. □

Lemma 2.29 *We have the following relations between the operators ∂ and $\overline{\partial}$.*

$$\overline{\partial}^2 = 0, \quad \partial\overline{\partial} + \overline{\partial}\partial = 0, \quad \partial^2 = 0.$$

Proof This follows from the formulas

$$d \circ d = 0, \quad d = \partial + \overline{\partial}.$$

Indeed, these relations imply that

$$\partial^2 + \partial\overline{\partial} + \overline{\partial}\partial + \overline{\partial}^2 = d^2 = 0.$$

Now, if α is a form of type (p, q), then $\partial^2\alpha$ is of type $(p+2, q)$, $(\partial\overline{\partial} + \overline{\partial}\partial)\alpha$ is of type $(p+1, q+1)$ and $\overline{\partial}^2\alpha$ is of type $(p, q+2)$. Thus, $d^2\alpha = 0$ implies that

$$\partial^2\alpha = (\partial\overline{\partial} + \overline{\partial}\partial)\alpha = \overline{\partial}^2\alpha = 0.$$

□

2.3.2 Local exactness

The Poincaré lemma shows the local exactness of the operator d:

Lemma 2.30 (See Bott & Tu 1982) *Let α be a closed differential form of strictly positive degree on a differentiable manifold. Then, locally, there exists a differential form β such that $\alpha = d\beta$.*

We say that α is locally exact.

Now consider a complex manifold X. Let $\alpha = \overline{\partial}\beta$ be a form of type (p, q) which is $\overline{\partial}$-exact. Then we have $\overline{\partial}\alpha = 0$ by lemma 2.29. The following proposition is a partial converse which is the analogue of the Poincaré lemma for the operator $\overline{\partial}$.

Proposition 2.31 *Let α be a C^1 form of type (p, q) with $q > 0$. If $\overline{\partial}\alpha = 0$, then there locally exists on X a C^1 form β of type $(p, q-1)$ such that $\alpha = \overline{\partial}\beta$.*

Proof We first reduce to the case where $p=0$ by the following argument. Locally, we can write in holomorphic coordinates $z_1, \ldots, z_n$:

$$\alpha = \sum_{I,J} \alpha_{I,J} dz_I \wedge d\overline{z}_J,$$

where the sets of indices I are of cardinal p and the sets of indices J are of cardinal q. Then

$$\overline{\partial}\alpha = \sum_{I,J} \overline{\partial}\alpha_{I,J} \wedge dz_I \wedge d\overline{z}_J$$

by lemma 2.28. It follows that if $\overline{\partial}\alpha = 0$, for every I of cardinal p the form α_I of type $(0, q)$ defined by

$$\alpha_I = \sum_J \alpha_{I,J} d\overline{z}_J$$

is $\overline{\partial}$-closed. If the proposition is proved for forms of type $(0, q)$, then locally we have $\alpha_I = \overline{\partial}\beta_I$, and

$$\alpha = (-1)^p \overline{\partial}\Big(\sum_I dz_I \wedge \beta_I\Big).$$

It remains to show the proposition for forms of type $(0, q)$. Such a form can be written $\alpha = \sum_J \alpha_J d\overline{z}_J$. We do the proof by induction on the largest integer k such that there exists J with $k \in J$ and $\alpha_J \neq 0$. Necessarily $k \geq q$. If $k = q$, we have

$$\alpha = f d\overline{z}_1 \wedge \cdots \wedge d\overline{z}_q.$$

The condition $\overline{\partial}(\alpha) = 0$ is then equivalent to the fact that the function f is holomorphic in the variables z_l, $l > q$. We then apply theorem 1.28: note that its proof also gives the following.

Proposition 2.32 *Let $f(z_1, \dots, z_n)$ be a C^1 function which is holomorphic in the variables z_l, $l > q$. Then there locally exists a C^1 function g, holomorphic in the variables z_l, $l > q$, such that $\frac{\partial g}{\partial \bar z_q} = f$.*

If g is as in the proposition, then since $\frac{\partial g}{\partial \bar z_l} = 0$ for $l > q$, we have

$$\bar\partial(g d\bar z_1 \wedge \cdots \wedge d\bar z_{q-1}) = (-1)^{q-1} f d\bar z_1 \wedge \cdots \wedge d\bar z_q,$$

and thus the case $k = q$ is proved. Suppose the proposition is proved for $k - 1 \geq q$. Consider a form

$$\alpha = \alpha_1 + \alpha_2 \wedge d\bar z_k,$$

where only the coordinates of index strictly less than k appear in α_1 and α_2. Let $\alpha_2 = \sum_J \alpha_{2,J} d\bar z_J$, where the index sets J are of cardinal $q - 1$ and contained in $\{1, \dots, k - 1\}$.

The condition $\bar\partial\alpha = 0$ then implies that the functions $\alpha_{2,J}$ are holomorphic in the variables z_l, $l > k$. Proposition 2.32 then shows that we can write

$$\alpha_{2,J} = \frac{\partial \beta_{2,J}}{\partial \bar z_k}$$

with $\beta_{2,J}$ holomorphic in the variables z_l, $l > k$. Then

$$\bar\partial\Big(\sum_J \beta_{2,J} d\bar z_J\Big) = (-1)^{q-1}\alpha_2 \wedge d\bar z_k + \alpha_1',$$

where the form α_1' involves only the coordinates z_l, $l < k$. We have thus written

$$\alpha = \alpha_1'' + \bar\partial\beta$$

where only the coordinates of index strictly less than k appear in α_1''. So we have $\bar\partial\alpha_1'' = 0$, since the forms α and $\bar\partial\beta$ are $\bar\partial$-closed, and we can apply the induction hypothesis to α_1''. Since the form α_1'' is locally exact, the same holds for α. □

2.3.3 *Dolbeault complex of a holomorphic bundle*

Let E be a holomorphic vector bundle of rank k over a complex manifold X. Let $A^{0,q}(E)$ denote the space of C^∞ sections of the bundle $\Omega_X^{0,q} \otimes_{\mathbb{C}} E$. In a holomorphic trivialisation of E, $\tau_U : E_{|U} \cong U \times \mathbb{C}^k$, such a section can be written $(\alpha_1, \dots, \alpha_k)$, where the α_i are C^∞ forms of type $(0, q)$ on U. We then set

$$\bar\partial_U \alpha = (\bar\partial\alpha_1, \dots, \bar\partial\alpha_k);$$

it is a section of $\Omega_U^{0,q+1} \otimes_{\mathbb{C}} E$. We will show that this local definition in fact gives a form $\overline{\partial}\alpha \in A^{0,q+1}(E)$.

Lemma 2.33 *Let V be an open subset of X and $\tau_V : E_{|V} \cong V \times \mathbb{C}^k$ a holomorphic trivialisation of E over V. Then for $\alpha \in A^{0,q}(E)$, we have*

$$\overline{\partial}_U \alpha\big|_{U\cap V} = \overline{\partial}_V \alpha\big|_{U\cap V}.$$

Proof Let M_{UV} be the transition matrix, with holomorphic coefficients, which enables us to pass from the trivialisation τ_U to the trivialisation τ_V. Then, by definition, if α_U is a section of E over U, $\alpha_U = (\alpha_{1,U}, \dots, \alpha_{k,U})$ in the trivialisation τ_U, and α_V is a section of E over V, $\alpha_V = (\alpha_{1,V}, \dots, \alpha_{k,V})$ in the trivialisation τ_V, the sections α_U and α_V coincide on $U \cap V$ if and only if

$${}^t(\alpha_{1,V}, \dots, \alpha_{k,V}) = M_{UV}\, {}^t(\alpha_{1,U}, \dots, \alpha_{k,U}).$$

We can of course replace the functions α_i by differential forms. The form α can be written $(\alpha_{1,U}, \dots, \alpha_{k,U})$ in the trivialisation τ_U and $(\alpha_{1,V}, \dots, \alpha_{k,V})$ in the trivialisation τ_V, and we have, as above,

$${}^t(\alpha_{1,V}, \dots, \alpha_{k,V}) = M_{UV}\, {}^t(\alpha_{1,U}, \dots, \alpha_{k,U}).$$

To see that $\overline{\partial}_U \alpha|_{U\cap V} = \overline{\partial}_V \alpha|_{U\cap V}$, by the above and the definition of $\overline{\partial}_U$, $\overline{\partial}_V$, it suffices to show that

$${}^t(\overline{\partial}\alpha_{1,V}, \dots, \overline{\partial}\alpha_{k,V}) = M_{UV}\, {}^t(\overline{\partial}\alpha_{1,U}, \dots, \overline{\partial}\alpha_{k,U}).$$

But this follows immediately from the Leibniz formula lemma 2.28 and the fact that the matrix M_{UV} has holomorphic coefficients. □

Lemma 2.33 enables us to define an operator

$$\overline{\partial}_E : A^{0,q}(E) \to A^{0,q+1}(E) \tag{2.5}$$

by the condition $\overline{\partial}_E \alpha|_U = \overline{\partial}_U \alpha|_U$. Note that the meaning of this operator on the space $A^{0,0}(E)$ of $\mathcal{C}^\infty$ sections of E is the following.

Lemma 2.34 *The kernel*

$$\operatorname{Ker}\left(\overline{\partial}_E : A^{0,0}(E) \to A^{0,1}(E)\right)$$

contains exactly the holomorphic sections of E.

Proof This is clear, since the holomorphic sections are those which are given by n-tuples of holomorphic functions in local holomorphic trivialisations. But by definition, $\overline{\partial}_E$ acts like the operator $\overline{\partial}$ on these n-tuples, and we know that the functions annihilated by $\overline{\partial}$ are exactly the holomorphic functions. □

Naturally, this operator satisfies the same local properties as the operator $\overline{\partial}$ on the forms.

Lemma 2.35 *The operator $\overline{\partial}_E$ satisfies Leibniz' rule*

$$\overline{\partial}_E(\alpha \wedge \beta) = \overline{\partial}_E\alpha \wedge \beta + (-1)^q \alpha \wedge \overline{\partial}_E\beta.$$

Here, α is a differential form of type $(0, q)$, and β is a differential form of type $(0, q')$ with coefficients in E, so that $\alpha \wedge \beta$ is naturally a differential form of type $(0, q + q')$ with coefficients in E.

Clearly, the operator $\overline{\partial}_E$ also satisfies the property $\overline{\partial}_E^2 = 0$. Finally, the local exactness of $\overline{\partial}_E$ follows from that of the operator $\overline{\partial}$.

Proposition 2.36 *Let α be a form of type $(0, q)$ with coefficients in E, and $q > 0$. If $\overline{\partial}_E\alpha = 0$, then locally on X there exists a form β of type $(0, q - 1)$ with coefficients in E such that $\alpha = \overline{\partial}_E\beta$.*

Recall that a complex (of vector spaces for example) is a family of vector spaces V_i together with morphisms $d_i : V_i \to V_{i+1}$ satisfying $d_{i+1} \circ d_i = 0$. The standard example is the de Rham complex of a differentiable manifold, where $V_i = A^i(X)$ is the space of differential forms of degree i, and $d_i = d$.

Definition 2.37 *The complex*

$$\cdots A^{0,q-1}(E) \overset{\overline{\partial}_E}{\to} A^{0,q}(E) \overset{\overline{\partial}_E}{\to} A^{0,q+1}(E) \cdots$$

is called the Dolbeault complex of E.

2.4 Examples of complex manifolds

Riemann surfaces

Let us consider 2-dimensional differentiable manifolds. If we restrict ourselves to the compact oriented case, these manifolds are classified by their genus g: such a surface is diffeomorphic to the g-holed torus (Gramain 1971). Furthermore, we can always put complex structures on such surfaces X. Indeed, we first note that the Newlander–Nirenberg integrability condition is automatically satisfied by an almost complex structure on X, by the fact that the rank of the complex vector bundle $T_X^{1,0}$ is equal to 1, and the bracket of vector fields is alternating. Thus, every almost complex structure is induced by a complex structure. Moreover, the existence of almost complex structures follows from the existence of Riemannian metrics on X (i.e. of a Euclidean structure on each tangent space $T_{X,x}$, varying differentiably with x). Indeed, we have the following.

Lemma 2.38 *An almost complex structure on an oriented surface X is equivalent to a conformal structure on X, i.e. to a Riemannian metric on X defined up to multiplication by a positive function.*

Proof If g is a metric on $T_{X,x}$, we define the corresponding complex structure operator I_x on $T_{X,x}$ to be the unique element of the group $SO(T_{X,x}, g_x)$ satisfying the conditions: $I_x^2 = -1$ and (u, Iu) forms a positively oriented basis of $T_{X,x}$ for every $u \neq 0$. Clearly I_x depends only on g_x, up to a coefficient. Conversely, if I_x is a complex structure operator on $T_{X,x}$, there exists a single metric g_x on $T_{X,x}$ (up to a coefficient) which satisfies the condition $g(Iu, Iv) = g(u, v)$, $\forall u,\ v \in T_{X,x}$. Such a metric is necessarily the real part of a Hermitian metric on $(T_{X,x}, I_x)$. □

Compact Riemann surfaces are called curves in algebraic geometry. Indeed, they are 1-dimensional complex manifolds (varieties).

Complex projective space

The complex projective space $\mathbb{P}^n(\mathbb{C})$ is the set of complex lines of $\mathbb{C}^{n+1}$, or equivalently, the quotient of $\mathbb{C}^{n+1} - \{0\}$ by the equivalence relation identifying collinear vectors on $\mathbb{C}$. The topology is the quotient topology. The complex structure is obtained as follows. For each i, consider the open subset $\tilde{U}_i$ of $\mathbb{C}^{n+1} - \{0\}$ consisting of the points z such that $z_i \neq 0$. Let U_i be the image of $\tilde{U}_i$ in $\mathbb{P}^n(\mathbb{C})$. Each point $Z \in U_i$ admits a unique lifting z to $\tilde{U}_i$ which satisfies the condition $z_i = 1$. Thus, U_i is naturally homeomorphic to $\mathbb{C}^n$, which provides the holomorphic charts for $\mathbb{P}^n(\mathbb{C})$, which is covered by the U_i. It remains simply to check that the change of chart morphisms are holomorphic. But $U_i \cap U_j$ can obviously be identified with the classes of non-zero vectors $z \in \mathbb{C}^{n+1}$ such that $z_i \neq 0$ and $z_j \neq 0$. Given such a vector, the image of the representative of its class in the chart $U_i \cong \mathbb{C}^n$ is given by $(\frac{z_1}{z_i}, \ldots, 1, \ldots, \frac{z_{n+1}}{z_i})$, where the 1 is in the ith place, while the image of the representative of its class in the chart $U_j \cong \mathbb{C}^n$ is given by $(\frac{z_1}{z_j}, \ldots, 1, \ldots, \frac{z_{n+1}}{z_j})$, where the 1 is in the jth place. The transition morphism is thus given, up to the order of the coordinates, by

$$(\zeta_1, \dots, \zeta_n) \mapsto \left(\frac{1}{\zeta_j}, \frac{\zeta_1}{\zeta_j}, \dots, \frac{\zeta_n}{\zeta_j}\right) \tag{2.6}$$

on $\mathbb{C}^n - \{\zeta_j = 0\}$. As (2.6) is clearly holomorphic, we have equipped $\mathbb{P}^n(\mathbb{C})$ with a complex structure.

Complex tori

Let Γ be a lattice in $\mathbb{C}^n$, i.e. a free additive subgroup generated by a basis of $\mathbb{C}^n$ over $\mathbb{R}$. The group Γ acts by translation on $\mathbb{C}^n$, and the action is proper and fixed-point-free. The quotient $T = \mathbb{C}^n / \Gamma$ is compact. In fact, there exists a $\mathbb{R}$-linear automorphism of $\mathbb{C}^n = \mathbb{R}^{2n}$ sending Γ to $\mathbb{Z}^{2n}$, so that this quotient is naturally homeomorphic to $(\mathbb{R}/\mathbb{Z})^{2n} = (S^1)^{2n}$. Clearly, T admits a natural differentiable structure for which the quotient map is a local diffeomorphism. We then put an almost complex structure onto T by taking the holomorphic charts to be the local inverses of the quotient map. As these local inverses are defined up to translation by an element of Γ, the change of chart morphisms are given by these translations, which are obviously holomorphic. Thus, T is equipped with a holomorphic structure.

Exercises

1. *Double covers.* Let X be a complex manifold and L be a holomorphic line bundle on X. We assume that there exist a holomorphic line bundle K on X and an isomorphism

$$K^{\otimes 2} \cong L.$$

Assume we are given a non-zero holomorphic section σ of L. We denote by $\Sigma \subset L$ the image of σ.

(a) Show that the map

$$\phi : K \to L, \quad (x, \tau) \mapsto (x, \tau^2)$$

is a proper holomorphic map.

(Here x is a point of X and (x, τ) is a point of K over x.)

Let $R \subset X$ be the vanishing locus of σ. With the help of local trivialisations of L, R is locally defined by one holomorphic equation.

(b) Show that $Y := \phi^{-1}(\Sigma)$ is smooth if and only if R is smooth.

(c) Show that when R is smooth,

$$\phi : Y \to \Sigma \cong X$$

is ramified exactly along $\phi^{-1}(R) \cong R$. Show that the fibre $\phi^{-1}(x)$ consists of two distinct points when $x \notin R$.

2. *Degree of a map to* $\mathbb{P}^1$. Let X be a compact complex curve and let f be a non-constant meromorphic function on X.
(a) Show that we can view f as a holomorphic map from X to $\mathbb{P}^1$. Here we see $\mathbb{P}^1$ as the compactification of $\mathbb{C} = \mathbb{C} \times 1 \subset \mathbb{C}^2$ obtained by adding the point $(1, 0)$ (the point at infinity, which is denoted by ∞).
(b) Let t be a point of $\mathbb{P}^1$. Let D be a disk in $\mathbb{P}^1$ centred at t. Using exercise 3(c) of chapter 1, show that

$$n_t := \int_{f^{-1}(\partial D)} \frac{1}{2i\pi} \frac{df}{f - t}$$

is equal to the number of points of the fibre $f^{-1}(t)$, counted with the multiplicities given by the order of vanishing $k_x(f - t)$ of $f - t$ at x defined in that exercise.
(c) Show that n_t is independent of t. We shall denote it by n.
(d) Show that f is ramified at a point x if and only if the order of vanishing $k_x(f - t)$ of $f - t$ at x is at least equal to 2. Deduce from Sard's theorem that for t in a dense set of points, the fibre $f^{-1}(t)$ is a set of finite cardinality n. We call it the degree of the map n.
(e) Deduce from (c) the following result :
The divisor $f^{-1}(0) - f^{-1}(\infty)$ *of* f *(cf. exercise 3(b) of chapter 1) is of degree* 0,
where the degree of a formal finite sum $\sum_x n_x x$ is defined as the integer $\sum_x n_x$.

3. Show using the maximum principle that a connected compact complex manifold possesses no holomorphic functions other than the constant ones.

3
Kähler Metrics

In this chapter, we consider an additional structure on complex manifolds: a Kähler metric. A complex manifold X can always be equipped with a Hermitian metric, i.e. with a collection of Hermitian metrics, one on each tangent space $T_{X,x}$ varying differentiably with x; the tangent space for each point x is equipped here with the complex structure J_x induced by the complex structure of X. We show this by using partitions of unity and local trivialisations of the tangent bundle as a complex vector bundle.

A Hermitian structure h is a sesquilinear form written

$$h = g - i\omega,$$

where g is a Riemannian metric and ω is a 2-form called a Kähler form, which is of type $(1, 1)$ for the almost complex structure J, as follows from the invariance of h under J: $h(u, v) = h(Ju, Jv)$.

A Kähler manifold is a complex manifold equipped with a Hermitian metric whose Kähler form satisfies the condition $d\omega = 0$. It is very difficult to construct Kähler manifolds, or to decide whether or not a given complex manifold is Kähler. However, it is easy to see that the existence of a Kähler metric is very restrictive, and to give examples of non-Kähler complex manifolds (without recourse to Hodge theory).

In the second section of this chapter, we give a Riemannian characterisation of Kähler metrics. We begin by defining the Chern connection of a holomorphic vector bundle E equipped with a Hermitian structure. This is the unique connection which is compatible with the Hermitian metric, and whose part of type (0, 1) is equal to the operator $\overline{\partial}$ of E.

Theorem 3.1 *A Hermitian metric $h = g - i\omega$ on a complex manifold X is Kähler if and only if the Chern connection of the holomorphic tangent bundle T_X equipped with h coincides with the Levi-Civita connection for the metric g.*

Another, more intuitive characterisation is the fact that the almost complex structure operator J is flat for the Levi-Civita connection.

We will also prove a "normal form" theorem, which will considerably simplify the proof of the Kähler identities given later (cf. section 6.1.1). This theorem states that a Hermitian metric h is Kähler if and only if there exist holomorphic coordinates in the neighbourhood of each point x, centred at x, in which h can be identified with the standard Hermitian metric on $\mathbb{C}^n$ up to a term of order 2. This statement, among others, allows us to prove the theorem above, since the two connections are computed in local coordinates using the coefficients of the metric and their first derivatives.

We conclude this part with the construction of Kähler manifolds. First, we introduce the Chern form of a holomorphic bundle of rank 1 equipped with a Hermitian metric. Up to a coefficient, this is the curvature of its Chern connection. It so happens that these forms are closed real 2-forms of type (1, 1), and thus are candidates for Kähler forms. In the case of the projective space $\mathbb{P}^n$, for example, the Chern form of the tautological bundle $\mathcal{O}(1)$ (the dual of the tautological subbundle) equipped with the metric induced by a fixed Hermitian metric on $\mathbb{C}^{n+1}$, is a Kähler form, which gives the Fubini–Study metric and shows that $\mathbb{P}^n$ is a Kähler manifold, as are all of its complex submanifolds.

Kodaira's embedding theorem 7.11, which we will prove below (section 7.1.3), states that conversely, every compact Kähler manifold whose Kähler form is the Chern form of a holomorphic line bundle can be realised as a complex submanifold of projective space. Thus, this construction produces only projective manifolds.

There exists no general technique for constructing Kähler manifolds. However, we will show the following result.

Theorem 3.2 *The blowup of a compact Kähler manifold along a smooth complex submanifold is a Kähler manifold.*

3.1 Definition and basic properties

3.1.1 Hermitian geometry

Let V be a complex vector space, which we can also consider as a real vector space equipped with an endomorphism I of complex structure. Let $W = \mathrm{Hom}(V, \mathbb{R})$. Then $W_{\mathbb{C}} := \mathrm{Hom}_{\mathbb{R}}(V, \mathbb{C})$ admits the decomposition (already introduced in the preceding chapters)

$$W_{\mathbb{C}} = W^{1,0} \oplus W^{0,1}$$

into $\mathbb{C}$-linear and $\mathbb{C}$-antilinear forms. Let

$$W^{1,1} = W^{1,0} \otimes W^{0,1} \subset \bigwedge^2 W_{\mathbb{C}}$$

and $W_{\mathbb{R}}^{1,1} := W^{1,1} \cap \bigwedge^2 W_{\mathbb{R}}$. We then have

Lemma 3.3 *There is a natural identification between the Hermitian forms on* $V \times V$ *and the elements of* $W_{\mathbb{R}}^{1,1}$ *given by*

$$h \mapsto \omega = -\Im h.$$

Here, h is a complex-valued bilinear form which is $\mathbb{C}$-linear on the left and $\mathbb{C}$-antilinear on the right, and satisfies $h(u, v) = \overline{h(v, u)}$.

Proof Firstly, if h is Hermitian, then h satisfies $h(u, v) = \overline{h(v, u)}$, and thus the bilinear form ω on V defined by $\omega(u, v) = -\Im h(u, v)$ is alternating. Thus it is an element of $\bigwedge^2 W_{\mathbb{R}}$, and we need to check that it is also an element of $W^{1,1}$. But by definition, ω is in $W^{1,1}$ if and only if the natural extension of ω (by $\mathbb{C}$-bilinearity) to a 2-form on $V_{\mathbb{C}}$ vanishes on the bivectors $(u, v), u,\ v \in V^{1,0}$ and on the bivectors $(u, v), u,\ v \in V^{0,1}$, the second property following from the first by using complex conjugation. Now, $V^{1,0}$ is generated by the $\tilde{v} = v - iIv,\ v \in V$. Let $v,\ u \in V$: we have

$$\omega(\tilde{u}, \tilde{v}) = \omega(u, v) - \omega(Iu, Iv) - i(\omega(u, Iv) + \omega(Iu, v)).$$

As h is $\mathbb{C}$-linear on the left and $\mathbb{C}$-antilinear on the right, we have $h(Iu, Iv) = h(u, v)$ and thus $\omega(u, v) = \omega(Iu, Iv)$. Similarly, the condition

$$h(u, Iv) = -h(Iu, v)$$

implies that $\omega(u, Iv) = -\omega(Iu, v)$. Thus $\omega(\tilde{u}, \tilde{v}) = 0$.

Conversely, let us start from $\omega \in W_{\mathbb{R}}^{1,1}$, and set

$$g(u, v) = \omega(u, Iv), \quad h(u, v) = g(u, v) - i\omega(u, v).$$

We have

$$h(u, Iv) = g(u, Iv) - i\omega(u, Iv) = -\omega(u, v) - ig(u, v) = -ih(u, v).$$

As ω is alternating, we have $\Im h(u, v) = -\Im h(v, u)$. Moreover, as $\omega(u, Iv) = -\omega(Iu, v)$, we have $g(u, v) = g(v, u)$ and thus $h(u, v) = \overline{h(v, u)}$. Thus h is Hermitian.

We check immediately that these two constructions are inverses of each other. □

Definition 3.4 *Let us say that a real alternating form ω of type $(1, 1)$ on V is positive if the corresponding Hermitian form h is positive definite.*

Take $\mathbb{C}$-linear coordinates $z_1, \ldots, z_n$ on V. Then for $z = (t_1, \ldots, t_n)$, $z' = (t'_1, \ldots, t'_n)$, we have $h(z, z') = \sum_{i,j} \alpha_{ij} t_i \overline{t'}_j$, with $\alpha_{ij} = \overline{\alpha}_{ji} = h(e_i, e_j)$, where the e_i form the basis of V dual to (z_i). Thus, we have

$$\omega(z, z') = \frac{i}{2} \sum_{i,j} \alpha_{ij} \left(t_i \overline{t'}_j - t'_i \overline{t}_j\right).$$

In other words, we have the equality of bilinear forms on V:

$$\omega = \frac{i}{2} \sum_{i,j} \alpha_{ij} z_i \wedge \overline{z}_j \in W^{1,1}. \tag{3.1}$$

The proof of lemma 3.3 shows that we can also identify such Hermitian forms h with the symmetric bilinear forms associated to them by the relation $g(u, v) = \Re h(u, v)$. The forms g obtained in this way are exactly those satisfying the condition $g(Iu, Iv) = g(u, v)$. In what follows, we consider the Hermitian case, where h and thus also g are positive definite.

Remark 3.5 *It is obvious by the relation $g(u, v) = \omega(u, Iv)$ that if g is non-degenerate, then ω is also non-degenerate. In the Hermitian case, the real vector space V is then equipped with both a Euclidean structure and a symplectic structure.*

3.1.2 Hermitian and Kähler metrics

Let (M, I), $I : T_M \to T_M$ be an almost complex manifold. A Hermitian metric on M is a collection of Hermitian metrics h_x on each tangent space $T_{M,x}$, seen as a complex vector space via I_x. We say that h is continuous (resp. differentiable) if in local coordinates $x_1, \ldots, x_n$ for M, the functions

$$x \mapsto h_x\left(\frac{\partial}{\partial x_i}, \frac{\partial}{\partial x_j}\right)$$

are continuous (resp. differentiable). Lemma 3.3 can be applied with parameters to associate to such a metric h a real 2-form of type $(1, 1)$:

$$\omega = -\Im h \in \Omega^{1,1}_M \cap \Omega^2_{M,\mathbb{R}}.$$

This form is called the Kähler form of the metric h.

Definition 3.6 *We say that the Hermitian metric h is Kähler if I is integrable and the 2-form ω is closed.*

Remark 3.7 *In particular, by remark 3.5, the manifold M equipped with the form ω is a symplectic manifold, i.e. is equipped with a closed 2-form which is everywhere non-degenerate.*

3.1.3 Basic properties

Volume form

An almost complex manifold of dimension $2n$ equipped with a Hermitian structure is, in particular, a Riemannian manifold. Moreover, it is canonically oriented, the positive orientation on each tangent space $T_{M,m}$ being given by the following rule: if $u_1, \ldots, u_n$ is a basis of $T_{M,m}$ over $\mathbb{C}$, then $u_1, Iu_1, \ldots, u_n, Iu_n$ is an oriented basis of $T_{M,m}$ over $\mathbb{R}$.

Such an oriented Riemannian manifold has a canonical volume form, i.e. an everywhere non-zero section of the bundle $\Omega^{2n}_{M,\mathbb{R}}$. Its value at the point m is the unique form which is positive on every oriented basis of $T_{M,m}$ and has norm 1 for the induced metric on $\Omega^{2n}_{M,m,\mathbb{R}}$. In the Hermitian case, we have the following lemma.

Lemma 3.8 *The volume form associated to a Hermitian metric h on M is equal to $\frac{\omega^n}{n!}$.*

Proof Let $e_1, \ldots, e_n$ be a basis of $T_{M,m}$ over $\mathbb{C}$ such that $h(e_i, e_j) = \delta_{ij}$. Then $e_1, Ie_1, \ldots, e_n, Ie_n$ is a real basis of $T_{M,m}$, orthonormal for g_m and with positive orientation. The volume form of (M, g) at the point m is then the unique form which has value 1 on $e_1 \wedge Ie_1 \wedge \cdots \wedge e_n \wedge Ie_n$. Thus, it suffices to check that we have

$$\frac{\omega^n}{n!}(e_1 \wedge Ie_1 \wedge \cdots \wedge e_n \wedge Ie_n) = 1. \tag{3.2}$$

Now, if $dx_1, dy_1, \ldots, dx_n, dy_n$ is the dual basis of $\Omega_{M,m,\mathbb{R}}$, and $dz_j = dx_j + i dy_j$, then by formula (3.1), we have

$$\omega_m = \frac{i}{2}\sum_i dz_i \wedge d\bar{z}_i.$$

Thus, $\frac{\omega^n}{n!} = (\frac{i}{2})^n \prod_1^n dz_i \wedge d\bar{z}_i$. Now, $\frac{i}{2} dz_i \wedge d\bar{z}_i = dx_i \wedge dy_i$. Thus

$$\frac{\omega^n}{n!} = dx_1 \wedge dy_1 \wedge \cdots \wedge dx_n \wedge dy_n,$$

which proves (3.2). □

It follows from this lemma that the volume of M, which by definition is the integral over M of the volume form (which is defined only if M is compact, and is then strictly positive) is also equal to the integral over M of the $2n$-form $\frac{\omega^n}{n!}$. In particular, we have the following corollary in the Kähler case.

Corollary 3.9 *If M is a compact Kähler manifold, then for every integer k between 1 and n, the closed form ω^k is not exact.*

Proof If $\omega^k = d\gamma$, then we also have $\omega^n = d(\omega^{n-k} \wedge \gamma)$. But then Stokes' formula (theorem 1.10) implies that $\int_M \omega^n = 0$, which is not the case since it is the volume of M. □

This enables us to show the existence of non-Kähler complex manifolds. Examples of this are given by compact complex manifolds whose de Rham cohomology group $H^{2k}(M) = \{\text{closed forms of degree } 2k\}/\{\text{exact forms}\}$ (cf. following chapter) is zero. Indeed, we may rephrase corollary 3.9 by saying that the de Rham class of ω^k is non-zero in $H^{2k}(M)$.

The de Rham class of ω is called the Kähler class of the Kähler metric.

Submanifolds

A complex submanifold N of a complex manifold M is a differentiable submanifold whose tangent space at each point is stable under the almost complex structure operator I of M. We first note that the induced almost complex structure on N is integrable, since (N, I) also satisfies the Newlander–Nirenberg integrability criterion. Thus, we can also see N as the image of a holomorphic immersion j, and applying the local holomorphic inversion theorem, we see that N is locally defined by $n - k$ holomorphic equations with independent differentials on $\mathbb{C}$, where $n = \dim_{\mathbb{C}} M$ and $k = \dim_{\mathbb{C}} N$.

Suppose now that M is Kähler, with Kähler form ω_M. Then the Hermitian metric on M induces a Hermitian metric on N, and by definition, the corresponding Kähler form ω_N is equal to $j^*\omega_M$. As ω_N is closed, we see that N is also a Kähler manifold. If, moreover, N is compact, then by the argument given above applied to N, we have the formula

$$0 < \operatorname{Vol} N = \int_N j^*(\omega_M^k).$$

In particular, we have

Lemma 3.10 *If N is a complex compact submanifold of a Kähler manifold M, N is not a boundary in M.*

Proof If $N = \phi(\partial\Gamma)$ for a differentiable map $\phi : \Gamma \to M$ of a manifold with boundary Γ, then by Stokes' formula we have

$$\int_N j^*(\omega_M^k) = \int_\Gamma d(\phi^*(\omega_M^k)) = 0.$$

□

This observation allowed Hironaka to construct other examples of non-Kähler complex manifolds (see Hartshorne 1977).

3.2 Characterisations of Kähler metrics

3.2.1 Background on connections

Recall (Bott & Tu 1982) that a connection ∇ on a real differentiable vector bundle E of rank k and class $\mathcal{C}^\infty$ is an operator which makes it possible to "differentiate the sections of E", i.e. an $\mathbb{R}$-linear map

$$\nabla : \mathcal{C}^\infty(E) \to A^1(E),$$

where $A^1(E)$ is the space of the $\mathcal{C}^\infty$ sections of the bundle

$$\Omega_{X,\mathbb{R}} \otimes E = \operatorname{Hom}(T_X, E)$$

satisfying Leibniz' rule

$$\nabla(f\sigma) = df \otimes \sigma + f\nabla\sigma.$$

If ψ is a vector field on X, and σ a section of E, we use the notation

$$\nabla_\psi(\sigma) = \nabla\sigma(\psi).$$

In a local trivialisation of E, $E_U \cong U \times \mathbb{R}^k$, the sections of E can be identified with k-tuples of functions $(f_1, \ldots, f_k)$, and Leibniz' rule shows that the connection can be written in such an open set in the form

$$\nabla(f_1, \ldots, f_k) = (df_1, \ldots, df_k) + (f_1, \ldots, f_k) \cdot M,$$

where the matrix M is a matrix of type (k, k) whose coefficients are 1-forms. This matrix is called the matrix of the connection in the trivialisation under consideration. We can give analogous definitions for $\mathcal{C}^k$ bundles and complex bundles (in which case we require ∇ to be $\mathbb{C}$-linear).

Let (M, g) be a Riemannian manifold. Consider the real tangent bundle. There exists on T_M a connection uniquely determined by g, called the Levi-Civita connection (Milnor 1963). A connection ∇ on a vector bundle E equipped with a metric g is said to be compatible with the metric if for two differentiable sections χ, ψ of E, we have

$$d(g(\chi, \psi)) = g(\chi, \nabla\psi) + g(\nabla\chi, \psi).$$

(The right-hand term is a 1-form, if we define $g(u, \alpha \otimes v) = \alpha g(u, v)$ for α a 1-form and u, v sections of E.)

Proposition 3.11 *If (M, g) is a Riemannian manifold, there exists a unique connection*

$$\nabla : \mathcal{C}^\infty(T_M) \to A^1(T_M)$$

on the tangent bundle T_M satisfying the properties:

- *∇ is compatible with g.*
- *∇ is without torsion, i.e. satisfies*

$$\nabla_\chi \psi - \nabla_\psi \chi = [\chi, \psi],$$

for all vector fields χ, ψ over M.

This connection is called the Levi-Civita connection of (M, g). Let us now consider a holomorphic vector bundle E over a complex manifold X. Suppose that E is equipped with a Hermitian metric ($\mathcal{C}^\infty$ to simplify). In other words, we assume that each fibre E_x is equipped with a Hermitian metric h_x whose matrix in each holomorphic trivialisation is $\mathcal{C}^\infty$ in x. In the preceding chapter, we defined the operator

$$\overline{\partial}_E : \mathcal{C}^\infty(E) \to A^{0,1}(E),$$

which is almost a connection, since it satisfies Leibniz' rule with respect to the operator $\overline{\partial}$ on the functions. Now let ∇ be a complex connection on E. Then the operator

$$\nabla^{0,1} : \mathcal{C}^{\infty}(E) \to A^{0,1}(E)$$

obtained by composing ∇ with the projection $A^1(E) \to A^{0,1}(E)$ also satisfies Leibniz' rule with respect to the operator $\overline{\partial}$ on the functions.

We will say that a connection ∇ on E is compatible with h if for two sections σ, τ of E, we have

$$d(h(\sigma, \tau)) = h(\nabla\sigma, \tau) + h(\sigma, \nabla\tau), \tag{3.3}$$

where as above the right-hand term is a 1-form. (One must pay attention, however, to the fact that h is sesquilinear, which leads us to define $h(e, \alpha \otimes f) = \overline{\alpha} h(e, f)$ for $e, f \in E_x$ and $\alpha \in \Omega_{X,x}$.)

Proposition 3.12 *There exists a unique connection ∇ on E satisfying the following properties:*
- *∇ is compatible with h.*
- *We have the equality*

$$\nabla^{0,1} = \overline{\partial}_E.$$

This connection is called the Chern connection of (E, h).

Proof If we take the part of type $(1, 0)$ of the equality (3.3), then, writing $\nabla^{1,0}$ for the part of type $(1, 0)$ of E, we obtain

$$\partial(h(\sigma, \tau)) = h(\nabla^{1,0}\sigma, \tau) + h(\sigma, \overline{\partial}\tau), \tag{3.4}$$

since $\nabla^{0,1} = \overline{\partial}$. As h is non-degenerate, the matrix of $\nabla^{1,0}$ in a local holomorphic basis $\sigma_1, \ldots, \sigma_k$ of E is determined by (3.4), which becomes

$$h(\nabla^{1,0}\sigma_i, \sigma_j) = \partial(h(\sigma_i, \sigma_j)).$$

□

3.2.2 Kähler metrics and connections

Let us now consider the case of the holomorphic tangent bundle T_X of X. Let h be a Hermitian metric on X, i.e. on T_X. Then as mentioned above, h determines a Riemannian metric $g = \Re\, h$ on X. We thus have two connections on T_X: indeed, recall that as a real differentiable bundle, the holomorphic tangent bundle T_X is canonically isomorphic to the real tangent bundle $T_{X,\mathbb{R}}$. We thus have the

Levi-Civita connection of $(T_{X,\mathbb{R}}, g)$ and the Chern connection of (T_X, h). The next theorem gives a characterisation of Kähler metrics.

Theorem 3.13 *The following properties are equivalent:*

(i) The metric h is Kähler.

(ii) The complex structure endomorphism I is flat for the Levi-Civita connection. This means that it satisfies

$$\nabla(I\chi) = I\nabla\chi, \quad \forall\chi \in A^0(T_{X,\mathbb{R}}).$$

(iii) The Chern connection and the Levi-Civita connection coincide on T_X, identified with $T_{X,\mathbb{R}}$ via the map $\Re$.

Proof The implication (iii)⇒(ii) is obvious since the Chern connection is $\mathbb{C}$-linear by definition, and the map $\Re$ identifies multiplication by i with the operator I.

(ii)⇒(i) is proved as follows. First, we have the relation $g(u, v) = \omega(u, Iv)$ between the metric and the Kähler form ω. By definition of the Levi-Civita connection, we have

$$d(g(\chi, \psi)) = g(\nabla\chi, \psi) + g(\chi, \nabla\psi),$$

so when ∇ commutes with I, we also have

$$d(\omega(\chi, \psi)) = \omega(\nabla\chi, \psi) + \omega(\chi, \nabla\psi).$$

This means that for three vector fields $\phi,\ \chi,\ \psi$, we have

$$\phi(\omega(\chi, \psi)) = \omega(\nabla_\phi\chi, \psi) + \omega(\chi, \nabla_\phi\psi). \tag{3.5}$$

We now have the formula

$$\begin{aligned} d\omega(\phi, \chi, \psi) &= \phi(\omega(\chi, \psi)) - \chi(\omega(\phi, \psi)) + \psi(\omega(\phi, \chi)) \\ &\quad - \omega([\phi, \chi], \psi) + \omega(\phi, [\chi, \psi]) + \omega([\phi, \psi], \chi), \end{aligned}$$

which follows easily from lemma 2.17. Replacing the brackets $[\chi, \phi]$ by $\nabla_\chi\phi - \nabla_\phi\chi$ etc. in this formula, we immediately see that (3.5) implies $d\omega = 0$ and thus that h is Kähler.

To prove (i)⇒(iii), we first note that the Levi-Civita and Chern connections coincide for the Kähler metric associated to the Hermitian metric $h = \sum_i dz_i d\overline{z}_i$ with constant coefficients over $\mathbb{C}^n$. Indeed, clearly in both cases the connections are the unique connections which annihilate the vector fields with constant coefficients, i.e. whose connection matrix is zero in the natural trivialisation of the tangent bundle of $\mathbb{C}^n$. Moreover, we note that the matrices of the two connections at a point, in local coordinates and in the induced trivialisation of the tangent bundle, depend only on the matrices of the metrics to the first order in the neighbourhood of this point. This is clear for the Chern connection by the proof of proposition 3.12, and it is a standard fact for the Levi-Civita connection.

To deduce the identity of the two connections in the Kähler case, we then show the following proposition, which will be very useful later on.

Proposition 3.14 *Let X be an n-dimensional complex manifold, and let h be a Kähler metric on X. Then, in the neighbourhood of each point x of X, there exist holomorphic coordinates $z_1, \ldots, z_n$, centred at x, such that the matrix $h_{ij} = h\left(\frac{\partial}{\partial z_i}, \frac{\partial}{\partial \overline{z}_j}\right)$ of h in these coordinates is equal to $I_n + O\left(\sum_i |z_i|^2\right)$.*

This proposition shows that in the neighbourhood of each point, a Kähler metric is isomorphic to a constant metric to the first order. Thus, the matrices of its Levi-Civita and Chern connections coincide at each point, as they do for a metric with constant coefficients, which concludes the proof of the implication (i)⇒(iii). □

Proof of proposition 3.14 Take holomorphic coordinates $z_1, \ldots, z_n$ centred at x. Up to a linear change of coordinates, we can of course assume that the matrix of h in the basis $\frac{\partial}{\partial z_i}$ is equal to I_n at the point x where the coordinates vanish. We thus have

$$h = \sum_i dz_i d\overline{z}_i + \sum_{i,j} \epsilon_{ij} dz_i d\overline{z}_j + O(|z|^2),$$

where the matrix ϵ_{ij} is a Hermitian matrix whose coefficients are linear forms in the $z_i, \overline{z}_i$. Let us write

$$\epsilon_{ij} = \epsilon_{ij}^{\mathrm{hol}} + \epsilon_{ij}^{\mathrm{antihol}}$$

(decomposition into $\mathbb{C}$-linear and antilinear parts). Note that we obviously have

$$\epsilon_{ij}^{\mathrm{antihol}} = \overline{\epsilon_{ji}^{\mathrm{hol}}}, \tag{3.6}$$

since ϵ_{ij} is Hermitian. The form ω can be written to the first order in the neighbourhood of x as

$$\omega = \frac{i}{2}\left(\sum_i dz_i \wedge d\overline{z}_i + \sum_{i,j} \epsilon_{ij}^{\mathrm{hol}} dz_i \wedge d\overline{z}_j + \sum_{i,j} \epsilon_{ij}^{\mathrm{antihol}} dz_i \wedge d\overline{z}_j\right) + O(|z|^2).$$

The fact that ω is closed at the point x implies that the form

$$\sum_{i,j} \epsilon_{ij}^{\text{hol}} dz_i \wedge d\overline{z}_j$$

is ∂-closed at the point x, and thus in fact everywhere, since it is a form with linear coefficients. We thus have

$$\frac{\partial \epsilon_{ij}^{\text{hol}}}{\partial z_k} = \frac{\partial \epsilon_{kj}^{\text{hol}}}{\partial z_i}.$$

This implies that there exist holomorphic functions $\phi_j(z_1, \ldots, z_n)$, which we may assume vanish at 0, such that

$$\epsilon_{ij}^{\text{hol}} = \frac{\partial \phi_j}{\partial z_i}.$$

Set $z_i' = z_i + \phi_i(z)$. As ϕ_i vanishes to order 1 at 0, up to restricting the neighbourhood, the z_i' provide coordinates centred at x. It remains to see that the metric h is constant to the first order in these coordinates. But we have

$$\begin{aligned} dz_i' &= dz_i + \sum_k \frac{\partial \phi_i}{\partial z_k} dz_k \\ &= dz_i + \sum_k \epsilon_{ki}^{\text{hol}} dz_k. \end{aligned}$$

Thus we obtain

$$\begin{aligned} &\sum_i dz_i' \wedge d\overline{z'}_i \\ &= \sum_i dz_i \wedge d\overline{z}_i + \sum_{i,k} \left(\epsilon_{ki}^{\text{hol}} dz_k \wedge d\overline{z}_i + \overline{\epsilon_{ki}^{\text{hol}}} dz_i \wedge d\overline{z}_k\right) + O(|z|^2) \\ &= \sum_i dz_i \wedge d\overline{z}_i + \sum_{i,k} \left(\epsilon_{ki}^{\text{hol}} dz_k \wedge d\overline{z}_i + \epsilon_{ik}^{\text{antihol}} dz_i \wedge d\overline{z}_k\right) + O(|z|^2) \\ &= \sum_i dz_i \wedge d\overline{z}_i + \sum_{i,k} \epsilon_{ki} dz_k \wedge d\overline{z}_i + O(|z|^2). \end{aligned}$$

Now, this is equal to $\frac{2}{i}\omega + O(|z|^2)$. Thus, we have

$$\omega = \frac{i}{2} \sum_i dz_i' \wedge d\overline{z'}_i + O(|z'|^2),$$

and the analogous formula for h. Proposition 3.14 is thus proved. □

3.3 Examples of Kähler manifolds

In this section, we propose to give some examples of Kähler metrics. Among the examples of complex manifolds given in the preceding chapter, Riemann surfaces and complex tori provide Kähler manifolds in a very easy way. Indeed, every Hermitian metric on a Riemann surface is Kähler, since the Kähler form, which must be of degree 2 on a 2-dimensional manifold, is necessarily closed. As for complex tori, they have flat Kähler metrics obtained by considering metrics with constant coefficients on $\mathbb{C}^n$. Such a metric is invariant under translation, and thus gives a metric on each quotient $T = \mathbb{C}^n/\Gamma$, where Γ is a discrete subgroup.

3.3.1 Chern form of line bundles

Let X be a complex manifold and L a holomorphic line bundle over X, i.e. a holomorphic vector bundle of rank 1. Let U_i, $i \in I$, be an open cover of X such that $L_{|U_i}$ admits a holomorphic trivialisation $L_{|U_i} \cong U_i \times \mathbb{C}$. Such a trivialisation is equivalent to giving an everywhere non-zero holomorphic section σ_i of L on U_i (the one which can be identified with the constant section equal to 1 in the trivialisation). The transition matrices g_{ij} corresponding to these trivialisations are given by invertible holomorphic functions on $U_i \cap U_j$. Obviously, we have

$$\sigma_i = g_{ij}\sigma_j$$

on $U_i \cap U_j$. Now, let h be a Hermitian metric on L. For $x \in X$, h_x is clearly determined by its value on any non-zero element of L_x, since $h_x(\lambda u) = |\lambda|^2 h_x(u)$. Set $h_i = h(\sigma_i)$. It is a strictly positive function on U_i, and we have $h_i = |g_{ij}|^2 h_j$ on $U_i \cap U_j$. The 2-forms

$$\omega_i = \frac{1}{2i\pi}\partial\overline{\partial}\log h_i$$

on U_i thus coincide on $U_i \cap U_j$, since $\partial\overline{\partial}\log |g_{ij}|^2 = 0$, and provide a 2-form ω on X. Up to the coefficient $\frac{i}{2\pi}$, this 2-form, which is called the Chern form, is the curvature of the Chern connection (proposition 3.12) on the Hermitian bundle (L, h) (see Milnor (1963, section 9.2.1) for the notion of curvature of a connection).

The forms constructed in this way are clearly closed, since they are locally exact, and real of type $(1, 1)$. When the bundle L satisfies certain positivity conditions, we can construct Hermitian metrics h on L, whose associated form ω is positive, i.e. corresponds to a Hermitian metric on X.

3.3.2 Fubini–Study metric

Consider $\mathbb{P}^n(\mathbb{C})$: there is a natural holomorphic line bundle S over $\mathbb{P}^n(\mathbb{C})$, whose fibre at $\Delta \in \mathbb{P}^n(\mathbb{C})$ is the rank 1 vector subspace $\Delta \subset \mathbb{C}^{n+1}$. To see that this bundle has a natural holomorphic structure, we note that in the open sets U_i introduced in section 2.4, S is trivialised by the section σ_i which to Δ associates the unique generating vector z of Δ satisfying $z_i = 1$. On $U_i \cap U_j$, we then obviously have $\sigma_j = \frac{Z_i}{Z_j}\sigma_i$, where the function $\frac{Z_i}{Z_j}$ is meromorphic (i.e. can be written locally as the quotient of two holomorphic functions) on $\mathbb{P}^n(\mathbb{C})$, and holomorphic invertible on $U_i \cap U_j$. Thus, we do have a holomorphic structure on S (here the Z_i are the coordinates on $\mathbb{C}^{n+1}$.)

Definition 3.15 *Let $\mathcal{O}_{\mathbb{P}^n}(1)$ denote the dual of S.*

Now let h be the standard Hermitian metric on $\mathbb{C}^{n+1}$. By restriction, the inclusion of vector bundles

$$S \subset \mathbb{P}^n(\mathbb{C}) \times \mathbb{C}^{n+1}$$

gives a Hermitian metric h on S, as well as on its dual $\mathcal{O}_{\mathbb{P}^n}(1)$. The real closed 2-form ω_i of type (1, 1) associated to this metric h^* is, by definition, equal on U_i to

$$\frac{1}{2i\pi}\partial\bar{\partial}\log h^*(\sigma_i^*),$$

where σ_i^* is the section dual to σ_i on U_i. Now, we have $h^*(\sigma_i^*) = \frac{1}{h(\sigma_i)}$. Finally, for the natural identification $U_i \cong \mathbb{C}^n$, the section σ_i of S, which we can consider as a holomorphic, $\mathbb{C}^{n+1}$-valued map, is given by $\sigma_i(z_1, \ldots, z_n) = (z_1, \ldots, 1, z_i, \ldots, z_n)$, where the 1 is in the ith position. We thus obtain

$$h(\sigma_i) = 1 + \sum_i |z_i|^2,$$

and

$$\omega_i = \frac{1}{2i\pi}\partial\bar{\partial}\log\left(\frac{1}{1+\sum_i |z_i|^2}\right).$$

Lemma 3.16 *The form ω defined this way on $\mathbb{P}^n(\mathbb{C})$ is positive.*

Proof We have

$$\bar{\partial}\log\left(\frac{1}{1+\sum_i |z_i|^2}\right) = -\frac{\bar{\partial}(1+\sum_i |z_i|^2)}{1+\sum_i |z_i|^2} = -\frac{\sum_i z_i d\bar{z}_i}{1+\sum_i |z_i|^2},$$

so that

$$\omega_i = \frac{i}{2\pi}\frac{(1+\sum_i |z_i|^2)\sum_i dz_i \wedge d\overline{z}_i - (\sum \overline{z}_i dz_i) \wedge (\sum_i z_i d\overline{z}_i)}{(1+\sum_i |z_i|^2)^2}.$$

At the point 0, we thus have

$$\omega_i = \frac{i}{2\pi}\sum_i dz_i \wedge d\overline{z}_i,$$

which is positive, and as it is clear by construction that ω is invariant under the transitive (and holomorphic) action of $SU(n+1)$ on $\mathbb{P}^n$, ω is positive everywhere. □

The Kähler metric defined in this way on $\mathbb{P}^n$ is called the Fubini–Study metric. We also obtain, as a corollary, that every complex projective manifold (i.e. complex submanifold of projective space) is Kähler.

This construction generalises to projective bundles, and makes it possible to show that a projective bundle (coming from a vector bundle) over a Kähler manifold X is also a Kähler manifold.

Definition 3.17 *Let E be a holomorphic vector bundle of rank $r+1$ over a complex manifold X. The manifold $\mathbb{P}(E)$, which is the quotient of E minus the zero section by the natural action of $\mathbb{C}^*$, is called the projective bundle associated to E.*

The complex structure on $\mathbb{P}(E)$ is obvious: $\mathbb{P}(E)$ admits a natural morphism π to X, which can be deduced from that of E by passing to the quotient. On open sets U_i of a trivialisation of E, we have

$$\pi^{-1}(U_i) \cong_i U_i \times \mathbb{P}^r,$$

and the identifications between

$$\pi^{-1}(U_i \cap U_j) \cong_i U_i \cap U_j \times \mathbb{P}^r$$

and

$$\pi^{-1}(U_i \cap U_j) \cong_j U_i \cap U_j \times \mathbb{P}^r$$

are given by the projective morphisms induced by the transition matrices of E, and are thus holomorphic.

There is a natural relative version of the line bundle $\mathcal{O}_{\mathbb{P}^n}(1)$ defined above. Let S be the line subbundle of π^*E over $\mathbb{P}(E)$ whose fibre at a point $(x, \Delta \subset E_x)$

is the rank 1 vector subspace $\Delta \subset E_x$. We then define $\mathcal{O}_{\mathbb{P}(E)}(1)$ as the dual of S. On each fibre of π, naturally isomorphic to $\mathbb{P}^r$, the restriction of $\mathcal{O}_{\mathbb{P}(E)}(1)$ is naturally isomorphic to $\mathcal{O}_{\mathbb{P}^r}(1)$.

Let h be a Hermitian metric on the bundle E. Then h induces a Hermitian metric on π^*E, and thus, by restriction, a metric on S and its dual $\mathcal{O}_{\mathbb{P}(E)}(1)$. The Chern form ω_E associated to this metric is not necessarily positive on $\mathbb{P}(E)$, but its restriction to each fibre $\pi^{-1}(x)$ is positive, since it is equal to the Fubini–Study metric on $\mathbb{P}(E_x)$ associated to the metric h_x on E_x. Suppose now that X is Kähler, and let ω_X be a Kähler form on X. If X is compact, it is easy to see that for $\lambda \gg 0$, the real closed form of type (1, 1)

$$\omega = \omega_E + \lambda \pi^* \omega_X$$

is positive on $\mathbb{P}(E)$. Thus we have shown the following.

Proposition 3.18 *If X is compact Kähler and E is a holomorphic bundle over X, then the manifold $\mathbb{P}(E)$ is Kähler.*

Remark 3.19 *$\mathbb{P}(E)$ is also obviously compact, as a quotient of the bundle of unit spheres of E for any Hermitian metric on E.*

3.3.3 Blowups

Let X be a complex manifold, and $Y \subset X$ a complex submanifold of codimension k. Locally along Y, there exist holomorphic functions $f_1, \dots, f_k$ with independent differentials, such that $Y = \{z \mid f_i(z) = 0\}$. These equations are not unique, but we have the following.

Lemma 3.20 *If $g_1, \dots, g_k$ form another system of local equations for Y, then locally in the neighbourhood of Y, there exists a matrix M_{ij} of holomorphic functions such that*

$$g_i = \sum_j M_{ji} f_j. \tag{3.7}$$

Moreover, the matrix M_{ij} is invertible along Y, and its restriction to Y is uniquely determined by the f_i, g_j.

Proof It suffices to prove the lemma in the case where the $f_i(z)$ are the first k coordinates. The functions g_i then have the property of vanishing on $\{z \mid z_1 = 0, \dots, z_k = 0\}$. Taking the power series expansion of g_i, we see immediately that we must have $g_i = \sum_{j \le k} M_{ji} z_j$.

The fact that $M_{|Y}$ is uniquely determined can be shown by taking the differentials of (3.7) along Y, which gives the relations

$$dg_i = \sum_{j \leq k} M_{ji} df_j, \tag{3.8}$$

and using the fact that the df_i are independent along Y. The invertibility of M along Y, and thus in a neighbourhood of Y, also follows from the equations (3.8). □

Remark 3.21 *If we cover a neighbourhood of Y by open sets U in which Y is defined by k holomorphic equations, we obtain transition matrices M^{UV} which are invertible matrices with holomorphic coefficients defined in the neighbourhood of Y. These matrices satisfy the condition*

$$f_i^U = \sum_{j,i} M_{ji}^{UV} f_j^V, \tag{3.9}$$

where the f_i^U are equations for $Y \cap U$. The restrictions of these matrices to Y are uniquely determined by these equations. These matrices are the transition matrices for the conormal bundle $N^*_{Y/X}$ of Y in X, whose fibre at y consists of the complex linear forms on $T_{X,y}$ which vanish on $T_{Y,y}$. Indeed, this follows from differentiating the equations (3.9), which yields $df_i^U = \sum_{j,i} M_{ji}^{UV} df_j^V$.

Let U be an open set of X, on which there exist functions $f_1, \ldots, f_k$ with independent differentials such that $Y \cap U = \{z \in U \mid f_i(z) = 0,\ i = 1, \ldots, k\}$. Now set

$$\tilde{U}_Y = \{(Z, z) \in \mathbb{P}^{k-1} \times U \mid Z_i f_j(z) = Z_j f_i(z), \forall i, j \leq k\}. \tag{3.10}$$

Here, $Z = (Z_1, \ldots, Z_k)$ is a representative vector of the corresponding point of $\mathbb{P}^{k-1}$. We easily check that $\tilde{U}_Y$ is a smooth complex submanifold of $\mathbb{P}^{k-1} \times U$. We have a map $\tau = \mathrm{pr}_2 : \tilde{U}_Y \to U$, which is an isomorphism over $U - Y \cap U$, the inverse being given by $z = (z_1, \ldots, z_n) \mapsto ((f_1(z), \ldots, f_k(z)), z)$. Above $Y \cap U$, the fibre of τ is equal to $\mathbb{P}^{k-1}$. We will now use lemma 3.20 to glue the blown up open sets $\tilde{U}_Y$ together to construct the blowup of X along Y.

Lemma 3.22 *Let U, V be two open sets of X, in which Y is defined by equations $f_1^U, \ldots, f_k^U, f_1^V, \ldots, f_k^V$ respectively, with independent differentials. Then if $\tau_U : \tilde{U}_Y \to U$ and $\tau_V : \tilde{V}_Y \to V$ are the blowups of U and V along $U \cap Y$ and $V \cap Y$ respectively, there exists a natural isomorphism*

$$\phi_{UV} : {\tau_U}^{-1}(U \cap V) \cong {\tau_V}^{-1}(U \cap V)$$

such that $\tau_U = \tau_V \circ \phi_{UV}$.

Proof It suffices to construct the isomorphism locally in the neighbourhood of $\tau_U^{-1}(Y \cap U)$, since such an isomorphism is certainly unique by continuity, and is already defined outside Y. Thus, up to restricting U, we can assume that we have a holomorphic invertible matrix M_{UV} which sends the equations f_i^U to the equations f_i^V, i.e.

$$f_i^U = \sum_{j,i} M_{ji}^{UV} f_j^V.$$

Let $P_{UV} = {}^t M_{UV}^{-1}$. Then the holomorphic diffeomorphism

$$\psi_{UV} : \mathbb{P}^{k-1} \times U \cap V \to \mathbb{P}^{k-1} \times U \cap V$$

defined by $\psi_{UV}(Z, z) = (P_{UV}(z) \cdot Z, z)$ (where Z is considered as a column vector) clearly sends $\tau_U{}^{-1}(U \cap V)$ to $\tau_V{}^{-1}(U \cap V)$, and the inverse map is given by the inverse diffeomorphism. □

Definition 3.23 *The manifold $\tilde{X}_Y$ obtained by gluing the manifolds $\tilde{U}_Y$ above the intersections $U \cap V$, using the identifications given by lemma 3.22, is called the blowup of X along Y.*

We have a blowup map $\tau : \tilde{X}_Y \to X$, equal to τ_U over $\tilde{U}_Y$. It is an isomorphism above $X - Y$. We also have $\tau^{-1}(Y) \cong \mathbb{P}(N_{Y/X})$, since the matrices P_{UV} which give the transition morphisms for the projective bundle $\tau^{-1}(Y)$ are the transition matrices for the normal bundle $N_{X/Y} = T_{X|Y}/T_Y$, which is the dual of the conormal bundle encountered above. We easily see that $\tau^{-1}(Y) \subset \tilde{X}_Y$ is a smooth hypersurface, i.e. a smooth complex submanifold of codimension 1. In fact, consider the local definition (3.10) of the blowup. If $(y, (Z_1, \ldots, Z_k)) \in \tilde{U}_Y$ with $y \in Y$, then there exists i such that $Z_i \neq 0$. The function $f_i \circ \tau$ gives a local holomorphic equation for $\tau^{-1}(Y)$ in $\tilde{U}_Y$ in the neighbourhood of $(y, (Z_1, \ldots, Z_k))$. Indeed, in the neighbourhood of $(y, (Z_1, \ldots, Z_k))$, the relations $Z_j f_i \circ \tau = Z_i f_j \circ \tau$ on $\tilde{U}_Y$ give $\frac{Z_j}{Z_i} f_i \circ \tau = f_j \circ \tau$, and thus $f_i \circ \tau(u) = 0 \Rightarrow f_j \circ \tau(u) = 0$, $\forall j$ in the neighbourhood of $(y, (Z_1, \ldots, Z_k))$.

The following proposition gives other examples of Kähler manifolds.

Proposition 3.24 *If X is Kähler and $Y \subset X$ is a compact complex submanifold of X, the blown up manifold $\tilde{X}_Y$ is Kähler, and it is compact if X is.*

Proof It is clear by the local description of τ that τ is proper, so that $\tilde{X}_Y$ is compact if X is.

Let ω_X be a Kähler form on X; then $\tau^*(\omega_X)$ is a real closed form of type $(1, 1)$ which is positive outside $\tau^{-1}(Y)$, but only semi-positive along $\tau^{-1}(Y)$.

Clearly the kernel of this form along $\tau^{-1}(Y)$ consists of the tangent space to the fibres of τ.

Suppose we have a real closed form λ of type $(1, 1)$ on $\tilde{X}_Y$, which is zero outside a compact neighbourhood of Y and strictly positive on the fibres of τ_*. Then the compactness of Y easily implies that for $C \gg 0$ the form $C\tau^*\omega_X + \lambda$ is positive, and equips $\tilde{X}_Y$ with a Kähler structure.

We know that $\tau^{-1}(Y)$ is isomorphic to the projective bundle $\mathbb{P}(N_{Y/X}) \xrightarrow{\tau} Y$, and that this bundle is equipped with a real closed form λ_1 of type $(1, 1)$ which is strictly positive on the fibres of τ_*. This form is obtained as the Chern form of the line bundle $\mathcal{O}_{\mathbb{P}(N_{Y/X})}(1)$ for a Hermitian metric h induced by a Hermitian metric on $N_{Y/X}$.

Lemma 3.25 *There exists a holomorphic line bundle $\mathcal{L}$ over $\tilde{X}_Y$, trivial outside $\tau^{-1}(Y)$ and whose restriction to $\tau^{-1}(Y)$ is isomorphic to $\mathcal{O}_{\mathbb{P}(N_{Y/X})}(1)$.*

Temporarily admitting this lemma, we finish the proof of proposition 3.24 by noting that by a partition of unity argument, the metric h on $\mathcal{O}_{\mathbb{P}(N_{Y/X})}(1)$ extends to a metric h_L on $\mathcal{L}$ which, outside a compact neighbourhood of Y, is the flat metric for the given trivialisation of $\mathcal{L}$ over $\tilde{X}_Y - \tau^{-1}(Y)$. Then the Chern form ω_L is zero outside a compact neighbourhood of Y, and its restriction to $\mathbb{P}(N_{Y/X})$ is equal to λ_1. Thus, we can choose $\lambda = \omega_L$, and apply the preceding argument. □

Proof of lemma 3.25 Let D be a hypersurface of a complex manifold, i.e. D is locally defined by a holomorphic equation which is unique up to multiplication by an invertible function. Let us take a covering of X by open sets U such that $U \cap D$ is defined by an equation $f_U = 0$ in U, for a holomorphic function f_U on U. We can of course take $U = X - D$ with $f_U = 1$ for one of our open sets. On the intersections $U \cap V$, the function $g_{UV} = \frac{f_U}{f_V}$ is invertible. We can then construct a holomorphic line bundle whose transition functions are the invertible functions g_{UV}, since they satisfy the cocycle condition (cf. the next chapter). This holomorphic bundle, which we denote by $\mathcal{O}_X(-D)$, is clearly trivial outside D.

Let us now apply this construction to the hypersurface $D_Y = \tau^{-1}(Y) \cong \mathbb{P}(N_{Y/X})$ of $\tilde{X}_Y$. The proof of lemma 3.25 can be finished using the following result.

Lemma 3.26 *The restriction of $\mathcal{O}_{\tilde{X}_Y}(-D_Y)$ to $D_Y = \mathbb{P}(N_{Y/X})$ is isomorphic to $\mathcal{O}_{\mathbb{P}(N_{Y/X})}(1)$.*

Proof If $D \subset X$ is a smooth hypersurface, then obviously $\mathcal{O}_X(-D)_{|D}$ is isomorphic to the conormal bundle of D in X. Indeed, by differentiation, the local equations for D in X give local trivialisations of $N^*_{D/X}$. Moreover, if we have the relation $f_U = g_{UV} f_V$ in $U \cap V$ between two equations for D, then we also have the relation $df_U = g_{UV} df_V$ along D.

Applying this to the hypersurface $D_Y = \tau^{-1}(Y) \cong \mathbb{P}(N_{Y/X})$ of $\tilde{X}_Y$, we are reduced to showing that $N_{D_Y/\tilde{X}_Y}$ is isomorphic to the tautological subbundle of $\tau^*(N_{Y/X})$. First, we have a natural map $N_{D_Y/\tilde{X}_Y} \to \tau^* N_{Y/X}$ induced by the differential $\tau_* : T_{\tilde{X}_Y} \to \tau^* T_X$ along D_Y. Using the explicit local description of the blowup, we check that this map gives an isomorphism of $N_{D_Y/\tilde{X}_Y}$ with the tautological subbundle $S \subset \tau^*(N_{Y/X})$. □

Remark 3.27 *The converse of proposition 3.24 is false. It is possible that a non-Kähler compact complex manifold can become Kähler or even projective after blowing up a submanifold. Examples of this can be obtained by considering the small resolutions of a 3-dimensional projective manifold having an ordinary double point (Clemens 1983b).*

Exercises

1. Let X be a connected complex manifold, and h be a Kähler metric. Let ω be the associated Kähler form. Show that if $\dim X \geq 2$, the wedge product with ω is injective on 1-forms.
   Show that if $\dim X \geq 2$ and ϕ is a differentiable function with values in $\mathbb{R}^+$ such that ϕh is also a Kähler metric, then ϕ is constant.
2. Let E be a holomorphic vector bundle of rank r over a complex manifold X. Show that if L is a holomorphic line bundle on X, then $\mathbb{P}(E^* \otimes L^*) = \mathbb{P}(E^*)$ but the line bundle $\mathcal{O}_{\mathbb{P}(E^* \otimes L^*)}(1)$ is isomorphic to $\mathcal{O}_{\mathbb{P}(E^*)}(1) \otimes \pi^* L$, where $\pi : \mathbb{P}(E^*) \to X$ is the structural morphism.

4

Sheaves and Cohomology

In this chapter, we introduce several very general objects, which will be used to interpret the results of Hodge theory concerning the de Rham cohomology of a Kähler manifold, and to apply them from a more theoretical and conceptual point of view. First, we need to introduce the notion of a sheaf (of abelian groups, rings, modules, etc.) over a topological space X. A sheaf $\mathcal{F}$ is the following collection of data: a group (ring, module, etc.) $\mathcal{F}(U)$ of sections of $\mathcal{F}$ on U, for each open set U of X, together with restriction maps $\mathcal{F}(U) \to \mathcal{F}(V)$ for $V \subset U$. We require that a section of $\mathcal{F}$ on U is determined by its restrictions to the open sets V of a covering of U, and conversely, that a section can be constructed by gluing together sections of the open sets of a covering, under the condition that these coincide on the intersection of two arbitrary open sets of the covering. The sheaves which will interest us the most in this book are the constant sheaves, whose sections on U are locally constant maps with values in a fixed group G, and the sheaves of (continuous, differentiable, holomorphic) sections of a (topological, differentiable, holomorphic) vector bundle over a topological space, a differentiable manifold or a complex manifold. Let $\mathcal{A}$ denote the sheaf of continuous, differentiable or holomorphic functions over X, and let the term "of class $\mathcal{A}$" mean continuous, differentiable or holomorphic according to $\mathcal{A}$. For every sheaf of functions of class $\mathcal{A}$, we will show that the correspondence which to a vector bundle of class $\mathcal{A}$ associates its sheaf of sections of class $\mathcal{A}$, which is a sheaf of free $\mathcal{A}$-modules, is an equivalence of categories. In fact, in algebraic geometry, it is typical to use the same notation for an algebraic vector bundle and its sheaf of algebraic sections, but this can be somewhat dangerous when one vector bundle is considered simultaneously as a differentiable bundle and a holomorphic bundle, for then there are two distinct sheaves corresponding to the bundle, namely the sheaf of differentiable sections and the sheaf of holomorphic sections.

The category of sheaves of abelian groups over a topological space X is an abelian category (in the first section of this chapter, we construct the quotients and kernels of morphisms of sheaves). There is a natural functor Γ of global sections, which to $\mathcal{F}$ associates $\Gamma(\mathcal{F}) = \mathcal{F}(X)$. This functor has values in the category of abelian groups. It is left-exact but not right-exact, i.e. a surjective morphism $\phi : \mathcal{F} \to \mathcal{G}$ of sheaves does not necessarily induce a surjective morphism on the level of the global sections. Sheaf cohomology is a theory which is used to compute and understand this defect in exactness via the use of invariants, namely the images under derived functors $R^i\Gamma$, which are written $H^i(X, \cdot)$, of the sheaves $\operatorname{Ker}\phi$, $\mathcal{F}$ and $\mathcal{G}$.

More generally, we explain how to compute the derived objects $R^i F(M)$, where F is a left-exact functor from an abelian category $\mathcal{A}$ having sufficiently many injective objects (see section 4.2.2 below) to an abelian category $\mathcal{B}$. These objects can be computed using an injective resolution, and the choice of another resolution gives a canonically isomorphic object. Another important result is the fact that we need only consider F-acyclic resolutions to compute these derived functors.

In the final section, we return to sheaves over a differentiable or complex manifold, and to the functor Γ. We give examples of Γ-acyclic sheaves (flasque and fine sheaves). We deduce from this the following theorem.

Theorem 4.1 *If X is a differentiable manifold, then the cohomology $H^i(X, \mathbb{C})$ of X with values in the constant sheaf of stalk $\mathbb{C}$ is equal to the de Rham cohomology*

$$H^i_{DR}(X, \mathbb{C}) = \frac{\operatorname{Ker}\left(d : A^i_{\mathbb{C}}(X) \to A^{i+1}_{\mathbb{C}}(X)\right)}{\operatorname{Im}\left(d : A^{i-1}_{\mathbb{C}}(X) \to A^i_{\mathbb{C}}(X)\right)}.$$

We give other examples of acyclic resolutions (namely singular or Čech resolutions) of constant sheaves, which are used to prove other results on these cohomology groups (finiteness in the compact case for example), and to compare them with the singular cohomology groups. All of these versions will be useful in the rest of the book. We also prove the following theorem.

Theorem 4.2 *(Dolbeault) If X is a complex manifold, and E is a holomorphic vector bundle over X, then the cohomology $H^i(X, E)$ of X with values in the sheaf of holomorphic sections of E is equal to the cohomology of the Dolbeault complex*

$$H^i_{\mathrm{Dolb}}(X, E) = \frac{\operatorname{Ker}\left(\overline{\partial} : A^{0,i}_X(E) \to A^{0,i+1}_X(E)\right)}{\operatorname{Im}\left(\overline{\partial} : A^{0,i-1}_X(E) \to A^{0,i}_X(E)\right)}.$$

4.1 Sheaves

4.1.1 Definitions, examples

All the sheaves considered here are sheaves of abelian groups. It goes without saying that the general definitions are in fact valid in a much more general context (sheaves of sets, see Godement (1958)). Let X be a topological space. A presheaf $\mathcal{F}$ of abelian groups over X is given by an abelian group $\mathcal{F}(U)$ for each open set U of X, together with a restriction morphism

$$\rho_{UV} : \mathcal{F}(U) \to \mathcal{F}(V)$$

for each pair of open sets $V \subset U$, which is a morphism of abelian groups. We require that $\mathcal{F}(\emptyset) = \{0\}$. Furthermore, for any three open sets $W \subset V \subset U$ of X, we also require the following compatibility condition:

$$\rho_{UW} = \rho_{VW} \circ \rho_{UV} : \mathcal{F}(U) \to \mathcal{F}(W).$$

In general, we will denote $\rho_{UV}(\sigma)$ by $\sigma|_V$.

Definition 4.3 *A sheaf of abelian groups is a presheaf satisfying the following condition: for every open set U of X, and for every covering of U by open sets $V \in \mathcal{V}$, the natural map*

$$\Pi_V \rho_{UV} : \mathcal{F}(U) \to \Pi_V \mathcal{F}(V)$$

induces an isomorphism of $\mathcal{F}(U)$ onto

$$\{(\sigma_V)_{V\in\mathcal{V}} \mid \sigma_{V|W\cap V} = \sigma_W|_{W\cap V}, \forall V,\ W \in \mathcal{V}\}.$$

The elements of $\mathcal{F}(U)$ are called the sections of $\mathcal{F}$ on U. The sheaf condition means that giving a section of $\mathcal{F}$ on U is equivalent to giving a collection of sections of $\mathcal{F}$ on each open set of the covering, whose restrictions coincide on the intersections.

A morphism of presheaves $\phi : \mathcal{F} \to \mathcal{G}$ is a collection, for each open set U, of morphisms

$$\phi_U : \mathcal{F}(U) \to \mathcal{G}(U)$$

such that for $\sigma \in \mathcal{F}(U)$ and $V \subset U$, we have

$$\phi_U(\sigma)|_V = \phi_V(\sigma|_V).$$

Lemma 4.4 *For every presheaf $\mathcal{F}$ over X, there exists a unique sheaf $\mathcal{F}_f$ over X satisfying the following conditions:*

- *There exists a morphism of presheaves*

$$\phi : \mathcal{F} \to \mathcal{F}_f.$$

- *For every morphism of presheaves*

$$\psi : \mathcal{F} \to \mathcal{G}$$

where $\mathcal{G}$ *is a sheaf, there exists a unique morphism of sheaves* $\chi : \mathcal{F}_f \to \mathcal{G}$ *such that* $\psi = \chi \circ \phi$.

Proof Let us define the sheaf $\mathcal{F}_f$ as follows. First, let $\mathcal{F}_1$ be the presheaf defined by

$$\mathcal{F}_1(U) = \mathcal{F}(U)/\mathcal{F}_0(U)$$

where

$$\mathcal{F}_0(U) = \{\sigma \in \mathcal{F}(U) \mid \exists \mathcal{V}, \ \sigma_{|V} = 0 \ \text{ for all } \ V \in \mathcal{V}\}.$$

Here $\mathcal{V}$ belongs to the set of coverings by open sets of U. For such a covering, set

$$A_{\mathcal{V}}(U) = \{(\sigma_V)_{V \in \mathcal{V}}, \ \sigma_V \in \mathcal{F}_1(V) \mid \sigma_{V|W \cap V} = \sigma_{W|W \cap V}, \ \forall V, \ W \in \mathcal{V}\}.$$

We say that a covering $\mathcal{V}$ of U is finer than a covering $\mathcal{V}'$, if for every open set $V \in \mathcal{V}$, there exists $V' \in \mathcal{V}'$ such that $V \subset V'$. This order relation satisfies the condition that two coverings always admit a common refinement. If $\mathcal{V}'$ is finer than $\mathcal{V}$, and if we have a refining map $\sigma : \mathcal{V}' \to \mathcal{V}$ such that $V \subset \sigma(V)$, $\forall V \in \mathcal{V}'$, then we have the obvious restriction map

$$\rho_{\mathcal{V},\mathcal{V}',\sigma} : A_{\mathcal{V}} \to A_{\mathcal{V}'}.$$

Note that in fact, by the definition of $A_{\mathcal{V}}$, this restriction map does not depend on the choice of σ.

We then set

$$\mathcal{F}_f(U) = \lim_{\substack{\to \\ R}} A_{\mathcal{V}}$$

This direct limit of the groups $A_{\mathcal{V}}$ equipped with the restriction maps $\rho_{\mathcal{V},\mathcal{V}'}$, on the directed set of coverings R, is the group consisting of the $(\sigma_{\mathcal{V}})_{\mathcal{V} \in R}$ satisfying the property that there exists a covering $\mathcal{V}$ such that

$$\sigma_{\mathcal{V}'} = \rho_{\mathcal{V},\mathcal{V}'}(\sigma_{\mathcal{V}})$$

for $\mathcal{V}'$ finer than $\mathcal{V}$, quotiented by the subgroup consisting of the $(\sigma_{\mathcal{V}})_{\mathcal{V}\in R}$ such that for some $\mathcal{V}$, we have $\sigma_{\mathcal{V}'} = 0$, for every covering $\mathcal{V}'$ which is finer than $\mathcal{V}$.

We have a natural map $\phi_{\mathcal{F}}$ from $\mathcal{F}$ to $\mathcal{F}_f$ given by restriction and passage to the quotient.

Given a morphism of presheaves

$$\psi : \mathcal{F} \to \mathcal{G},$$

there obviously exists an associated morphism

$$\psi_f : \mathcal{F}_f \to \mathcal{G}_f,$$

where $\mathcal{G}_f$ is defined similarly. This morphism satisfies

$$\psi_f \circ \phi_{\mathcal{F}} = \phi_{\mathcal{G}} \circ \psi.$$

Now, if $\mathcal{G}$ is a sheaf, then clearly we have $\mathcal{G} \overset{\phi_G}{\cong} \mathcal{G}_f$, which ensures the existence of χ such that $\psi = \chi \circ \phi$. The uniqueness can be shown just as easily. To conclude, it remains to show that $\mathcal{F}_f$ is a sheaf, which is also easy. □

Example 4.5 *On a locally connected space X, consider the constant presheaf equal to G, where G is a fixed abelian group. To every non-empty open subset U of X, this presheaf associates the group G, and the restriction morphisms are only the identity. This presheaf is not a sheaf, since if we are given an element of G on each connected component of X, where the set of the connected components is taken as a covering, these elements of course coincide on the intersections of the open sets of the covering (the intersections are empty), and thus we have a section of the sheaf associated to this constant presheaf equal to G. The group of sections obtained in this way is equal to G only if X is connected. More generally, we can easily show that the sheaf associated to this presheaf is the sheaf of locally constant functions with values in G which to an open set U associates $G^{C(U)}$, where $C(U)$ is the set of connected components of U. The associated sheaf is still called the constant sheaf of stalk G.*

Apart from the constant sheaves introduced above, the sheaves we will consider here are the sheaves of sections of vector bundles. If X is a topological space, we first have the structure sheaf, which to U associates the continuous (real or complex) functions on U. We denote it by $\underline{\mathcal{C}}^0$. It is a sheaf of rings, i.e. the restriction morphisms are ring morphisms.

Similarly, we can introduce the sheaves $\underline{\mathcal{C}}^k$ of differentiable $\mathcal{C}^k$ functions on a given $\mathcal{C}^k$ manifold. If X is a complex manifold, we usually write $\mathcal{O}_X$ for the sheaf of holomorphic functions which to U associates the ring of holomorphic functions on U.

Definition 4.6 *Let $\mathcal{A}$ be a sheaf of rings over X. A sheaf of $\mathcal{A}$-modules over X is a sheaf $\mathcal{F}$ such that each $\mathcal{F}(U)$ is equipped with the structure of an $\mathcal{A}(U)$-module compatible with its group structure. The restriction morphisms*

$$\mathcal{F}(U) \to \mathcal{F}(V)$$

are morphisms of $\mathcal{A}(U)$-modules, where $\mathcal{F}(V)$ is equipped with the structure of an $\mathcal{A}(U)$-module via the restriction morphism

$$\mathcal{A}(U) \to \mathcal{A}(V).$$

The typical example of such a sheaf of modules is given by the sheaf of sections of a vector bundle $E \xrightarrow{\pi} X$. If E is a topological vector bundle, the presheaf $\mathcal{E}$

$$U \mapsto \{\text{continuous sections } \sigma : U \to E_{|U}\}$$

is a sheaf of modules over the sheaf of real continuous functions. Addition and multiplication by functions come from the vector space structure on each fibre. Similarly, to a complex vector bundle, we can associate a sheaf of modules over the sheaf of complex continuous functions. In the differentiable case, the differentiable sections of given class provide a sheaf of modules over the sheaf of differentiable functions of the same class. Finally, if E is a holomorphic vector bundle over a complex manifold X, the holomorphic sections of E form a sheaf of $\mathcal{O}_X$-modules.

Definition 4.7 *If $\mathcal{A}$ is a sheaf of rings, a sheaf $\mathcal{F}$ of $\mathcal{A}$-modules is said to be a sheaf of free $\mathcal{A}$-modules if there exists an integer n such that $\mathcal{F}$ is locally isomorphic to $\mathcal{A}^n$ as a sheaf of $\mathcal{A}$-modules. The integer n is then called the rank of $\mathcal{F}$.*

This definition allows us to characterise the sheaves associated to vector bundles as above.

Lemma 4.8 *Let $\mathcal{A}$ be one of the sheaves of functions mentioned above. Then the correspondence*

$$E \mapsto \mathcal{E}$$

establishes a bijection (in fact an equivalence of categories) between vector bundles and sheaves of free $\mathcal{A}$-modules.

Proof It suffices to construct the inverse correspondence. Let $\mathcal{E}$ be a sheaf of free $\mathcal{A}$-modules. There exists a covering of X by open sets U on which there exists an isomorphism of sheaves of $\mathcal{A}$-modules

$$\tau_U : \mathcal{E}_{|U} \cong \mathcal{A}^n_U.$$

Thus, on $U \cap V$, we have an isomorphism of sheaves of free $\mathcal{A}$-modules

$$\tau_V \circ \tau_U^{-1} : \mathcal{A}^n_{U\cap V} \cong \mathcal{A}^n_{U\cap V}.$$

Clearly such an isomorphism is given by an $n \times n$ matrix M_{UV} of elements of $\mathcal{A}_{U\cap V}$, invertible at every point. The bundle E we associate to $\mathcal{E}$ is the vector bundle of rank n, trivial on the open sets U of the covering, and whose transition matrices are the M_{UV}. In other words, E is obtained by gluing the $U \times \mathbb{R}^n$ (or $U \times \mathbb{C}^n$ in the complex case) via the identification of

$$U \cap V \times \mathbb{R}^n \subset U \times \mathbb{R}^n$$

with

$$U \cap V \times \mathbb{R}^n \subset V \times \mathbb{R}^n$$

given by $\mathrm{Id} \times M_{UV}$.

An important point in this construction of E by gluings is the fact that the transition matrices M_{UV} are not arbitrary. They satisfy the cocycle condition

$$M_{UV} M_{VW} M_{WU} = \mathbb{I}_n,$$

which implies that the identification described above is compatible on the triple intersections $U \cap V \cap W$. Another way to see these gluings consists in noting that thanks to the cocycle condition on the matrices M_{UV}, the above identifications establish an equivalence relation on the disjoint union of the $U \times \mathbb{R}^n$, and that the quotient is exactly E.

The sheaf of $\mathcal{A}$-modules corresponding to E is clearly isomorphic to $\mathcal{E}$. □

4.1.2 Stalks, kernels, images

Definition 4.9 *The stalk $\mathcal{F}_x$ of a sheaf or of a presheaf $\mathcal{F}$ over X at a point $x \in X$ is equal to*

$$\lim_{\substack{\to \\ x\in U}} \mathcal{F}(U).$$

An element of $\mathcal{F}_x$ is called a germ of sections of $\mathcal{F}$ at x.

Here the direct limit is taken over the set of open sets containing x, ordered by the inclusion relation, where the morphisms $\mathcal{F}(U) \to \mathcal{F}(V)$ for $V \subset U$ are the restriction morphisms.

If $\phi : \mathcal{F} \to \mathcal{G}$ is a morphism of sheaves, then ϕ induces a morphism of abelian groups

$$\phi_x : \mathcal{F}_x \to \mathcal{G}_x$$

at each point.

Definition 4.10 *The morphism ϕ is injective (resp. surjective) if for every $x \in X$ the morphism ϕ_x is injective (resp. surjective).*

We have the following facts.

Lemma 4.11 *Let $\phi : \mathcal{F} \to \mathcal{G}$ be a morphism of sheaves. Then the presheaf*

$$U \mapsto \mathrm{Ker}\,(\phi_U : \mathcal{F}(U) \to \mathcal{G}(U))$$

is a sheaf, written $\mathrm{Ker}\,\phi$, *which is zero if and only ϕ is injective.*

Proof Let U be an open set of X, and let the U_i be open sets covering U. For each i, let $\sigma_i \in \mathrm{Ker}\,\phi : \mathcal{F}(U_i) \to \mathcal{G}(U_i)$ be such that $\sigma_{i\,|U_i \cap U_j} = \sigma_{j\,|U_i \cap U_j}$. Then by the sheaf property for $\mathcal{F}$, there exists a unique section $\sigma \in \mathcal{F}(U)$ such that $\sigma_{|U_i} = \sigma_i$. The section $\phi(\sigma) \in \mathcal{G}(U)$ then vanishes on the U_i, and thus it is zero by the sheaf property for $\mathcal{G}$. Thus, $\sigma \in \mathrm{Ker}\,\phi : \mathcal{F}(U) \to \mathcal{G}(U)$, and we have shown that this presheaf is a sheaf.

The second assertion follows from the fact that if a sheaf has stalks equal to zero, then it is itself zero. Indeed, if σ is a section of such a sheaf, then σ vanishes on each stalk, which implies that σ vanishes in the neighbourhood of each point, and thus on the open sets of a covering. Thus σ is zero. □

The analogous statement does not hold for images. However, we have the following.

Lemma 4.12 *Let $\phi : \mathcal{F} \to \mathcal{G}$ be a morphism of sheaves. Then the sheaf associated to the presheaf*

$$U \mapsto \mathrm{Im}\,(\phi_U : \mathcal{F}(U) \to \mathcal{G}(U)), \tag{4.1}$$

written $\mathrm{Im}\,\phi$, *is equal to $\mathcal{G}$ if and only ϕ is surjective.*

Proof It is clear by the universal property of the sheaf associated to a presheaf that there exists a natural map

$$j : \operatorname{Im}\phi \to \mathcal{G}.$$

This map is injective, since it is already injective on the presheaf (4.1). Suppose that ϕ_x is surjective; then j_x is surjective. If σ is a section of $\mathcal{G}$ on U, there thus exists a covering of U by open sets V, and sections τ_V of $\operatorname{Im}\phi$, such that $j(\tau_V) = \sigma_{|V}$. As j is injective, the τ_V coincide on the intersections, so there exists a section τ of $\operatorname{Im}\phi$ such that $\tau_{|V} = \tau_V$. Then $\sigma = j(\tau)$ and

$$j : \operatorname{Im}\phi(U) \to \mathcal{G}(U)$$

is surjective. Therefore j is an isomorphism. The converse is immediate. □

Example 4.13 *Let $\mathcal{F}$ be the sheaf of continuous complex-valued maps. Let $\mathcal{G}$ be the multiplicative sheaf of continuous invertible complex-valued maps, and finally, let ϕ be the exponential map. Then ϕ is surjective, since every continuous invertible function is locally the exponential of a continuous function. However, the map*

$$\phi_U : \mathcal{F}(U) \to \mathcal{G}(U)$$

is not in general surjective. For example, if $X = \mathbb{C}^*$ and f is the invertible function $z \mapsto z$, then f is not the exponential of a continuous function. In this example, the presheaf (4.1) is not a sheaf.

We can also define the cokernel of a morphism of sheaves of abelian groups $\phi : \mathcal{F} \to \mathcal{G}$. It is the sheaf associated to the presheaf

$$U \mapsto \operatorname{Coker}(\phi_U : \mathcal{F}(U) \to \mathcal{G}(U)).$$

4.1.3 Resolutions

Let $\mathcal{F}$, $\mathcal{G}$, $\mathcal{H}$ be three sheaves, and let $\phi : \mathcal{F} \to \mathcal{G}$, $\psi : \mathcal{G} \to \mathcal{H}$ be morphisms of sheaves such that $\psi \circ \phi = 0$.

Definition 4.14 *The sequence*

$$\mathcal{F} \xrightarrow{\phi} \mathcal{G} \xrightarrow{\psi} \mathcal{H}$$

is said to be exact in the middle if we have the equality of sheaves $\operatorname{Im}\phi = \operatorname{Ker}\psi$.

Definition 4.15 *A complex of sheaves is a collection of sheaves $\mathcal{F}^i, i \in \mathbb{Z}$, together with morphisms of sheaves $d_i : \mathcal{F}^i \to \mathcal{F}^{i+1}$ such that $d_{i+1} \circ d_i = 0$.*

Now let $\mathcal{F}$ be a sheaf, and $\mathcal{F}^i$, $i \in \mathbb{N}$ a complex of sheaves. Let $j : \mathcal{F} \to \mathcal{F}^0$ be a morphism.

Definition 4.16 *The complex $\mathcal{F}^\cdot$ is called a resolution of $\mathcal{F}$ if for every $i \geq 0$, the sequence*

$$\mathcal{F}^i \xrightarrow{\phi_i} \mathcal{F}^{i+1} \xrightarrow{\phi_{i+1}} \mathcal{F}^{i+2}$$

is exact in the middle, and j is injective with $j(\mathcal{F}) = \operatorname{Ker} \phi_0$.

Note that since j is injective, $\operatorname{Im} j$ is isomorphic to $\mathcal{F}$. The remainder of this section will devoted to the description of some important resolutions which will be used later.

The Čech resolution

Let $\mathcal{F}$ be a sheaf over X, and let U_i, $i \in \mathbb{N}$ be a countable covering by open sets of X. For each finite set $I \subset \mathbb{N}$, set

$$U_I = \bigcap\nolimits_{i \in I} U_i.$$

If $V \xrightarrow{j} X$ is the inclusion of an open set, then whenever $\mathcal{G}$ is a sheaf over V, we define the sheaf $j_*\mathcal{G}$ by the formula

$$j_*\mathcal{G}(U) = \mathcal{G}(V \cap U).$$

We also introduce the sheaf $j^*\mathcal{F}$, sometimes written $\mathcal{F}_V$; it is called the restriction of $\mathcal{F}$ to V. To an open set $U \subset V$, this sheaf associates $\mathcal{F}(U)$. For every open set U_I of X, let j_I be the inclusion of U_I in X, and let

$$\mathcal{F}_I := j_{I*}\mathcal{F}_{|U_I}.$$

We then define

$$\mathcal{F}^k = \bigoplus\nolimits_{|I|=k+1} \mathcal{F}_I$$

and $d : \mathcal{F}^k \to \mathcal{F}^{k+1}$ by the formula

$$(d\sigma)_{j_0,\dots,j_{k+1}} = \sum_i (-1)^i \sigma_{j_0,\dots,\hat{j}_i,\dots,j_{k+1}\,|U \cap U_{j_0,\dots,j_{k+1}}}, \quad j_0 < \dots < j_{k+1},$$

which is valid for $\sigma = (\sigma_I)$, $I \subset \mathbb{N}$, $\sigma_I \in \mathcal{F}_I(U) = \mathcal{F}(U \cap U_I)$. We easily check that $d \circ d = 0$. Let us also define $j : \mathcal{F} \to \mathcal{F}^0$ by $j(\sigma)_i = \sigma_{|U \cap U_i}$ for $\sigma \in \mathcal{F}(U)$.

Proposition 4.17 *The complex*

$$0 \to \mathcal{F}^0 \xrightarrow{d} \mathcal{F}^1 \xrightarrow{d} \cdots \mathcal{F}^n \xrightarrow{d} \mathcal{F}^{n+1} \cdots \tag{4.2}$$

is a resolution of $\mathcal{F}$.

We call this resolution the Čech resolution of $\mathcal{F}$ associated to the covering U_i of X. The functorial nature of this resolution renders it very useful.

Proof The injectivity of j is due to the property of uniqueness of the sections of $\mathcal{F}$ having given restriction to the U_i. The fact that Im j can be identified with the kernel of d on $\mathcal{F}^0$ is exactly equivalent to the fact that sections of $\mathcal{F}$ on $U \cap U_i$ which coincide on the intersections glue together to form a section of $\mathcal{F}$ on U. The exactness in general can be checked stalk by stalk, as follows. Let $x \in X$, and let i be such that $x \in U_i$. We then define

$$\delta : \mathcal{F}_x^k \to \mathcal{F}_x^{k-1}$$

for $k \geq 1$ by the following formula. An element $\sigma \in \mathcal{F}_x^k$ is represented by a series of germs $\sigma_I \in \mathcal{F}(V_I \cap U_I)$ for $|I| = k+1$, where V_I is an open set containing x which we can assume is contained in U_i. We then define $\delta\sigma$ by

$$(\delta\sigma)_{i_0,\ldots,i_{k-1}} = \epsilon\sigma_{i,i_0,\ldots,i_{k-1}}, \quad i_0 < \cdots < i_{k-1}, \tag{4.3}$$

where ϵ is the signature of the permutation reordering the set $\{i, i_0, \ldots, i_{k-1}\}$. We use the convention that $\sigma_{i,i_0,\ldots,i_{k-1}} = 0$ if $i \in \{i_0, \ldots, i_{k-1}\}$. To see that (4.3) makes sense, we need to see that the right-hand term defines a germ of a section of $j_{i_0,\ldots,i_{k-1}*}\mathcal{F}$ on the neighbourhood of x. But as each V_I is contained in U_i, we have $V_{i,i_0,\ldots,i_{k-1}} \cap U_{i_0,\ldots,i_{k-1}} = V_{i,i_0,\ldots,i_{k-1}} \cap U_{i,i_0,\ldots,i_{k-1}}$, so that $\sigma_{i,i_0,\ldots,i_{k-1}}$ can be seen as a section of $j_{i_0,\ldots,i_{k-1}*}\mathcal{F}$ on $V_{i,i_0,\ldots,i_{k-1}}$. We immediately check that $d \circ \delta + \delta \circ d = \mathrm{Id}$ on $\mathcal{F}_x^k$ for $k \geq 1$. This implies the exactness of the complex (4.2) at the point x. □

The de Rham resolution

Let X be a $\mathcal{C}^\infty$ differentiable manifold. The constant sheaf of stalk $\mathbb{R}$ is naturally included in the sheaf of $\mathcal{C}^\infty$ functions. Let $\mathcal{A}^k$ be the sheaf of $\mathcal{C}^\infty$ differential forms, i.e. the sheaf of sections of the bundle $\Omega^k_{X,\mathbb{R}}$. The exterior differential is a morphism of sheaves

$$d : \mathcal{A}^k \to \mathcal{A}^{k+1}.$$

Poincaré's lemma says that a closed form of degree $k > 0$ is locally exact, which means that the sequence

$$\mathcal{A}^{k-1} \xrightarrow{d} \mathcal{A}^k \xrightarrow{d} \mathcal{A}^{k+1}$$

is exact in the middle for $k \geq 1$. Finally, the kernel of $d : \mathcal{A}^0 \to \mathcal{A}^1$ consists precisely of the locally constant functions, so that we have the following result.

Proposition 4.18 *The complex*

$$0 \to \mathcal{A}^0 \xrightarrow{d} \mathcal{A}^1 \cdots \to \mathcal{A}^n \to 0, \tag{4.4}$$

where $n = \dim X$, *is a resolution of the constant sheaf* $\mathbb{R}$.

Naturally, we obtain a similar resolution of the constant sheaf $\mathbb{C}$ by using complex differential forms.

The Dolbeault resolution

Let X be a complex manifold and $E \to X$ a holomorphic vector bundle. Let $\mathcal{E}$ be the associated sheaf of free $\mathcal{O}_X$-modules. Let $\mathcal{A}^{0,q}(E)$ be the sheaf of C^∞ sections of $\Omega^{0,q} \otimes E$. In (2.5), we defined the operator

$$\overline{\partial} : \mathcal{A}^{0,q}(E) \to \mathcal{A}^{0,q+1}(E).$$

We know (cf. lemma 2.34 and proposition 2.36) that this operator satisfies:

- The kernel of $\overline{\partial} : \mathcal{A}^{0,0}(E) \to \mathcal{A}^{0,1}(E)$ is equal to the sheaf of holomorphic sections of E, i.e. to $\mathcal{E}$ (here $\mathcal{A}^{0,0}(E)$ is the sheaf of C^∞ sections of E).
- For $q > 0$, a section of $\mathcal{A}^{0,q}(E)$ is $\overline{\partial}$-closed if and only if it is locally $\overline{\partial}$-exact. In other words, we have the following.

Proposition 4.19 *The complex*

$$0 \to \mathcal{A}^{0,0}(E) \xrightarrow{\overline{\partial}} \mathcal{A}^{0,1}(E) \cdots \xrightarrow{\overline{\partial}} \mathcal{A}^{0,n}(E) \to 0, \tag{4.5}$$

where $n = \dim_{\mathbb{C}} X$, *is a resolution of the sheaf* $\mathcal{E}$.

4.2 Functors and derived functors

4.2.1 Abelian categories

A category $\mathcal{C}$ is given by a set of objects, called Ob $\mathcal{C}$, together with sets written Hom$(\cdot, \cdot)$ of maps, called morphisms, between these objects, which can be composed in such a way that the composition satisfies the usual associativity properties. For every object X, there is an element $\mathbb{I}_X \in \mathrm{Hom}(X, X)$, which is the identity with respect to right or left composition of the maps. An abelian category $\mathcal{C}$ is a category satisfying the following conditions:

- For every pair of objects $A,\ B$ of $\mathcal{C}$, Hom$\,(A, B)$ is an abelian group, and the composition of morphisms

$$\mathrm{Hom}\,(A, B) \times \mathrm{Hom}\,(B, C) \to \mathrm{Hom}\,(A, C)$$

is bilinear for these abelian group structures.

Every morphism $\phi : A \to B$ admits a kernel and a cokernel; the kernel of ϕ is an object C written Ker ϕ, equipped with a morphism $\chi : C \to A$, such that for every object M of $\mathcal{C}$, left composition with χ induces an isomorphism

$$\mathrm{Hom}\,(M, C) \cong \{\psi \in \mathrm{Hom}\,(M, A) \mid \phi \circ \psi = 0\}.$$

Similarly, the cokernel of ϕ is an object D, written Coker ϕ, equipped with a morphism $\chi : B \to D$, such that for every object M of $\mathcal{C}$, right composition with χ induces an isomorphism

$$\mathrm{Hom}\,(D, M) \cong \{\psi \in \mathrm{Hom}\,(B, M) \mid \psi \circ \phi = 0\}.$$

A morphism $\phi : A \to B$ is said to be injective if Hom (Ker ϕ, A) $= \{0\}$.

The image of a morphism ϕ can be defined as the cokernel of its kernel or as the kernel of its cokernel.

- Direct sums exist; the direct sum $A \oplus B$ is such that for every object M of $\mathcal{C}$, we have

$$\mathrm{Hom}\,(M, A \oplus B) = \mathrm{Hom}\,(M, A) \oplus \mathrm{Hom}\,(M, B),$$
$$\mathrm{Hom}\,(A \oplus B, M) = \mathrm{Hom}\,(A, M) \oplus \mathrm{Hom}\,(B, M).$$

The standard examples of abelian categories are the category of abelian groups and their morphisms, and the category of modules over a given ring. If X is a topological space, the category of sheaves of abelian groups or sheaves of $\mathcal{A}$-modules, where $\mathcal{A}$ is a sheaf of rings over X, is also abelian.

A functor F from a category $\mathcal{C}$ to a category $\mathcal{C}'$ is a map $A \mapsto F(A)$ from the objects of $\mathcal{C}$ to the objects of $\mathcal{C}'$, together with a map $\phi \mapsto F(\phi)$ from

Hom (A, B) to Hom $(F(A), F(B))$ compatible with composition for every pair of objects A, B of $\mathcal{C}$. A functor F between abelian categories is such that the maps F from Hom (A, B) to Hom $(F(A), F(B))$ are morphisms of abelian groups. We also require F to respect direct sums. Such a functor is called left-exact if for every morphism $\phi : A \to B$, the kernel Ker $F(\phi)$ of the corresponding morphism $F(\phi) : F(A) \to F(B)$ is equal to $F(\text{Ker}\,\phi)$.

Example 4.20 *If M is an abelian group, the functor $A \mapsto \text{Hom}\,(M, A)$ of the category of abelian groups to itself is left-exact.*

The following example is the main one used throughout the remainder of this text.

Example 4.21 *Let X be a topological space. The functor Γ from the category of sheaves of abelian groups over X to the category of abelian groups, which to a sheaf $\mathcal{F}$ of abelian groups over X associates the group of its global sections $\mathcal{F}(X)$, is left-exact.*

Let A, B, C be three objects of $\mathcal{C}$, and let $\phi : A \to B, \psi : B \to C$ be morphisms.

Definition 4.22 *We say that the sequence*

$$0 \to A \stackrel{\phi}{\to} B \stackrel{\psi}{\to} C \to 0$$

is a short exact sequence if $A \stackrel{\phi}{\to} B$ is isomorphic to the kernel of ψ and $B \stackrel{\psi}{\to} C$ is isomorphic to the cokernel of ϕ. (The kernel Ker ψ of ψ is an object of $\mathcal{C}$ equipped with a morphism $\chi : \text{Ker}\,\psi \to B$. The isomorphism above is an isomorphism $i : A \cong \text{Ker}\,\psi$ such that $\chi \circ i = \phi$. The analogous notion holds for the cokernel.)

4.2.2 Injective resolutions

Definition 4.23 *An object I of an abelian category is called injective if for every injective morphism $A \stackrel{j}{\to} B$ and for every morphism $\phi : A \to I$, there exists a morphism $\psi : B \to I$ such that $\psi \circ j = \phi$.*

The injective objects in the category of abelian groups are the divisible groups G, i.e. those such that for every $g \in G$ and every $n \in \mathbb{N}^*$, there exists $g' \in G$ such that $ng' = g$.

A complex in an abelian category is a sequence of objects M_i, $i \in \mathbb{Z}$ and maps $d^i : M^i \to M^{i+1}$ such that $d^{i+1} \circ d^i = 0$.

A morphism of complexes $\phi^\cdot : (M^\cdot, d_M) \to (N^\cdot, d_N)$ is a collection of morphisms $\phi^i : M^i \to N^i$ such that $d_N \circ \phi^i = \phi^{i+1} \circ d_M$.

Definition 4.24 *The degree i cohomology of a complex (M, d_M) is the object*

$$H^i(M^\cdot) := \operatorname{Coker}\left(d_M^{i-1} : M^{i-1} \to \operatorname{Ker} d_M^i\right).$$

Clearly, for every i, a morphism of complexes $\phi^\cdot : (M^\cdot, d_M) \to (N^\cdot, d_N)$ induces morphisms $H^i(\phi^\cdot) : H^i(M^\cdot) \to H^i(N^\cdot)$. A morphism $\phi^\cdot$ of complexes is called a quasi-isomorphism if the induced morphisms $H^i(\phi^\cdot)$ are isomorphisms for every i.

A homotopy H between two morphisms of complexes $\phi^\cdot : (M^\cdot, d_M) \to (N^\cdot, d_N)$ and $\psi^\cdot : (M^\cdot, d_M) \to (N^\cdot, d_N)$ is a collection of morphisms

$$H^\cdot : M^\cdot \to N^{\cdot-1},$$

satisfying

$$H^{i+1} \circ d_M^i + d_N^{i-1} \circ H^i = \phi^i - \psi^i, \quad \forall i \geq 0. \tag{4.6}$$

If there exists a homotopy between two morphisms of complexes $\phi^\cdot : (M^\cdot, d_M) \to (N^\cdot, d_N)$ and $\psi^\cdot : (M^\cdot, d_M) \to (N^\cdot, d_N)$, then the induced morphisms $H^i(\phi^\cdot)$ and $H^i(\psi^\cdot)$ are equal. Indeed, relation (4.6) shows that $\phi^i - \psi^i : \operatorname{Ker} d_M^i \to N^i$ factors through d_N^{i-1}, and thus induces 0 in $\operatorname{Hom}(\operatorname{Ker} d_M^i, \operatorname{Coker} d_N^{i-1})$.

Definition 4.25 *A complex M^i, $i \geq 0$ is called a resolution of an object A of $\mathcal{C}$ if $\operatorname{Im} d^i = \operatorname{Ker} d^{i+1}$ for $i \geq 0$ and there exists an injective morphism $j : A \to M^0$ such that $j : A \to M^0$ is isomorphic to $\operatorname{Ker} d^0$.*

We say that the abelian category $\mathcal{C}$ has *sufficiently many injective objects* if every object A of $\mathcal{C}$ admits an injective morphism $j : A \to I$, where I is injective.

Lemma 4.26 *If $\mathcal{C}$ has sufficiently many injective objects, then every object of $\mathcal{C}$ admits an injective resolution, i.e. a resolution $I^\cdot$ by a complex all of whose objects are injective objects.*

Proof We construct such a resolution by induction, by choosing an injective morphism

$$j : A \to I^0,$$

then an injective morphism $j^1 : \mathrm{Coker}\, j \to I^1$. We can then define d^0 as the composition of j^1 with the natural morphism $I^0 \to \mathrm{Coker}\, j$. Having constructed the resolution to the kth level, we choose an injective morphism

$$j_{k+1} : \mathrm{Coker}\, d^{k-1} \to I^{k+1}$$

and define

$$d^k : I^k \to I^{k+1}$$

to be the composition of the morphism $I^k \to \mathrm{Coker}\, d^{k-1}$ with j_{k+1}. □

An essential point is the uniqueness up to homotopy of such a resolution.

Proposition 4.27 *Let $I^\cdot, A \xrightarrow{i} I^0$, and $J^\cdot, B \xrightarrow{j} J^0$ be resolutions of A, B respectively, and let $\phi : A \to B$ be a morphism. Then if the second resolution is injective, there exists a morphism of complexes $\phi^\cdot : I^\cdot \to J^\cdot$ satisfying $\phi^0 \circ i = j \circ \phi$. Moreover, if we have two such morphisms $\phi^\cdot$ and $\psi^\cdot$, there exists a homotopy $H^\cdot$ between $\phi^\cdot$ and $\psi^\cdot$.*

Proof The morphism $\phi^0 : I^0 \to J^0$ can be obtained as the extension of $j \circ \phi : A \to J^0$ to I^0, which exists since J^0 is injective. The morphism $\phi^1 : I^1 \to J^1$ can be obtained by noting that $d_J^0 \circ \phi^0 \circ i = 0$. We then construct ϕ^1 as the extension to I^1 of the morphism induced by ϕ^0:

$$d_J^0 \circ \phi^0 : \mathrm{Coker}\, i \to J^1,$$

and this extension exists since J^1 is injective. In general, we note that once ϕ^{k-1} is constructed, we have $d_J^{k-1} \circ \phi^{k-1} \circ d_I^{k-2} = 0$ and we construct ϕ^k as the extension to I^k of the morphism induced by ϕ^{k-1}:

$$d_J^{k-1} \circ \phi^{k-1} : \mathrm{Im}\, d_I^{k-1} \cong \mathrm{Coker}\, d_I^{k-2} \to J^k;$$

this extension exists since J^k is injective.

If we have two morphisms $\psi^\cdot$ and $\phi^\cdot$ satisfying the conditions above, we construct the homotopy H in the same way: $H^1 : I^1 \to J^0$ is constructed as the extension to I^1 of the morphism $\phi^0 - \psi^0 : \mathrm{Coker}\, i \to J^0$ defined on $\mathrm{Coker}\, i \overset{d^0}{\hookrightarrow} I^1$. The construction of H^k is analogous. □

In particular, applying this proposition to the case where $I^\cdot$ and $J^\cdot$ are two injective resolutions of A, we obtain morphisms $\phi^\cdot : I^\cdot \to J^\cdot$ and $\psi^\cdot : J^\cdot \to I^\cdot$ such that $\psi^\cdot \circ \phi^\cdot$ and $\phi^\cdot \circ \psi^\cdot$ are morphisms of complexes (from $I^\cdot$ to itself and from $J^{\cdot}$ to itself respectively), which are both homotopic to the identity. We then say that $\phi^{\cdot}$ is a homotopy equivalence. Thus, we see that an injective resolution is unique up to homotopy equivalence.

4.2.3 Derived functors

Let $\mathcal{C}$ and $\mathcal{C}'$ be two abelian categories, and let F be a left-exact functor from $\mathcal{C}$ to $\mathcal{C}'$. Assume that $\mathcal{C}$ has sufficiently many injective objects.

Theorem 4.28 *For every object M of $\mathcal{C}$, there exist objects $R^iF(M), i \geq 0$ in $\mathcal{C}'$, determined up to isomorphism, satisfying the following conditions.*

- *We have $R^0F(M) = F(M)$.*
- *For every short exact sequence*

$$0 \to A \xrightarrow{\phi} B \xrightarrow{\psi} C \to 0$$

in $\mathcal{C}$, we can construct a long exact sequence (i.e. an exact complex) in $\mathcal{C}'$:

$$0 \to F(A) \xrightarrow{\phi} F(B) \xrightarrow{\psi} F(C) \to R^1F(A) \to R^1F(B) \to R^1F(C) \to \cdots. \tag{4.7}$$

- *For every injective object I of $\mathcal{C}$, we have $R^iF(I) = 0,\ i > 0$.*

Proof For every object A of $\mathcal{C}$, choose an injective resolution $I^{\cdot}$ of A. We then have a complex $F(I^{\cdot})$ in $\mathcal{C}'$. Define $R^iF(A)$ as the ith cohomology of this complex

$$R^iF(A) = H^i(F(I^{\cdot})). \tag{4.8}$$

Note that if the objects $R^iF(M)$ exist and satisfy the three properties above, we must have the equality (4.8), as we see by splitting the complex $0 \to A \to I^{\cdot}$ into short exact sequences. This implies the uniqueness of R^iF up to isomorphism.

As F is left-exact, we have $R^0F(A) = \mathrm{Ker}\,(d^0 : F(I^0) \to F(I^1)) = F(A)$. Furthermore, the $R^iF(A)$ do not depend, up to isomorphism, on the choice of the injective resolution by proposition 4.27. Indeed, if $I^{\cdot}$ and $J^{\cdot}$ are two such resolutions, there exists a homotopy equivalence between $I^{\cdot}$ and $J^{\cdot}$, i.e. morphisms of complexes

$$\phi^{\cdot} : I^{\cdot} \to J^{\cdot}, \quad \psi^{\cdot} : J^{\cdot} \to I^{\cdot},$$

and homotopies

$$H^{\cdot} : I^{\cdot} \to I^{\cdot -1}, \quad K^{\cdot} : J^{\cdot} \to J^{\cdot -1}$$

between $\psi^{\cdot} \circ \phi^{\cdot}$ and Id and between $\phi^{\cdot} \circ \psi^{\cdot}$ and Id. The functor F applied to the morphisms of complexes $\phi^{\cdot}$ and $\psi^{\cdot}$ gives morphisms of complexes $F(\phi^{\cdot})$ and $F(\psi^{\cdot})$ between the complexes $F(I^{\cdot})$ and $F(J^{\cdot})$.

The functor F applied to $H^{\cdot}$ and to $K^{\cdot}$ gives a homotopy equivalence between $F(\phi^{\cdot})$ and $F(\psi^{\cdot})$. Thus, the morphisms $H^i(F(\phi))$ and $H^i(F(\psi))$ are inverse morphisms of each other, and the two resolutions give isomorphic objects $R^i F(A)$. Moreover, the isomorphisms constructed above are canonical, since $\phi^{\cdot}$ is well-defined up to homotopy.

It remains to see the last two points. Firstly, it is obvious that if I is injective, then $R^i F(I) = 0$ for $i > 0$, since the resolution $I^0 = I$, $I^i = 0$, $i > 0$ is an injective resolution of I. To show that we have a long exact sequence of derived functors associated to a short exact sequence in $\mathcal{C}$, we note the two following facts.

Lemma 4.29 *If*

$$0 \to A \overset{\phi}{\to} B \overset{\psi}{\to} C \to 0$$

is a short exact sequence in $\mathcal{C}$, then there exist injective resolutions $I^{\cdot}$, $J^{\cdot}$, $K^{\cdot}$ of A, B, C respectively, and an exact sequence of complexes

$$0 \to I^{\cdot} \overset{\phi^{\cdot}}{\to} J^{\cdot} \overset{\psi^{\cdot}}{\to} K^{\cdot} \to 0$$

with $\phi^0 \circ i = j \circ \phi$, $\psi^0 \circ j = k \circ \psi$.

Proof We first show, by refining proposition 4.27, that there exist injective resolutions $I^{\cdot}$ and $J^{\cdot}$ of A and B respectively, and an *injective* morphism $\phi^{\cdot}$: $I^{\cdot} \to J^{\cdot}$ satisfying the condition $\phi^0 \circ i = j \circ \phi$. As the cokernel of an injective morphism between two injective objects is injective, it suffices to take for $K^{\cdot}$ the cokernel Coker $\phi^{\cdot}$, which is a resolution of C. □

Next, we note that every short exact sequence of injective objects $0 \to I^i \overset{\phi^i}{\to} J^i \overset{\psi^i}{\to} K^i \to 0$ is split. This means that the morphisms $0 \to I^i \overset{\phi^i}{\to} J^i$ admit left inverses, called retractions. Such inverses are obtained by the universal property of injective objects, using the fact that the objects $I^{\cdot}$ are injective. Such a retraction σ_i provides an isomorphism $J^i \cong I^i \oplus \operatorname{Ker} \sigma_i$ with $\operatorname{Ker} \sigma_i \cong K_i$. It follows that by applying the functor F, we again obtain an exact sequence of complexes

$$0 \to F(I^\cdot) \overset{\phi^\cdot}{\to} F(J^\cdot) \overset{\psi^\cdot}{\to} F(K^\cdot) \to 0$$

such that each short exact sequence

$$0 \to F(I^k) \overset{\phi^k}{\to} F(J^k) \overset{\psi^k}{\to} F(K^k) \to 0$$

is split. It is then a standard fact that such an exact sequence of complexes gives a long exact sequence of cohomology (4.7). This concludes the proof of theorem 4.28. □

The $R^i F$ are not functors from $\mathcal{C}$ to $\mathcal{C}'$, since the objects $R^i F(M)$ are only defined up to isomorphism in $\mathcal{C}'$. However, by their construction, the objects $R^i F(M)$ are canonically defined by the choice of an injective resolution of M, and if we have two such resolutions, there exists a *canonical* isomorphism between the objects $R^i F(M)$ computed via each of the two resolutions.

The $R^i F$ also have the following functorial property.

Proposition 4.30 *If $\phi : A \to B$ is a morphism in $\mathcal{C}$, and $I^\cdot$, $J^\cdot$ are injective resolutions of A and B respectively, then there exists a canonical morphism induced by ϕ,*

$$R^i F(\phi) : R^i F(A) \to R^i F(B),$$

where the derived objects are computed using the chosen resolutions.

Proof Proposition 4.27 enables us to associate to a morphism $\phi : A \to B$ a morphism of complexes

$$F(\phi^\cdot) : F(I^\cdot) \to F(J^\cdot)$$

which is well-defined up to homotopy, where $I^\cdot$, $J^\cdot$ are injective resolutions of A and B respectively. We thus have an induced morphism

$$R^i F(\phi) : R^i F(A) \to R^i F(B)$$

which does not depend on the choice of $\phi^\cdot$. □

In practice, injective resolutions are difficult to manipulate. The following result shows how to replace injective resolutions by resolutions satisfying a weaker condition.

Definition 4.31 *We say that an object M of $\mathcal{C}$ is acyclic for the functor F (or F-acyclic) if we have $R^i F(M) = 0$ for all $i > 0$.*

Proposition 4.32 *Let $M^\cdot, i : A \to M^0$ be a resolution of A, where the M^i are acyclic for the functor F. Then $R^i F(A)$ is equal to the cohomology $H^i(F(M^\cdot))$ of the complex $F(M^\cdot)$.*

Proof The proof is by induction on i. We have a short exact sequence

$$0 \to A \to M^0 \to B \to 0, \tag{4.9}$$

where B is the cokernel of d^0. By d^0, B admits the shifted resolution

$$0 \to B \xrightarrow{d^0} M^1 \to M^2 \cdots.$$

The exact sequence (4.9) induces a long exact sequence (4.7) of derived objects, and as M^0 is acyclic, this long exact sequence can be summarised by

$$R^i F(B) = R^{i+1} F(A) \text{ for } i \geq 1,$$

$$R^1 F(A) = \operatorname{Coker}(F(M^0) \to F(B)).$$

As F is left-exact, we have $F(B) = \operatorname{Ker}(d^0 : F(M^1) \to F(M^2))$, and the second equality means exactly that $R^1 F(A) = H^1(F(M^\cdot))$. As for the first equality, it enables us to apply the induction hypothesis to B, to conclude. □

4.3 Sheaf cohomology

From now on, we consider the category of sheaves of abelian groups over a topological space X, and the functor Γ of "global sections" which to $\mathcal{F}$ associates $\Gamma(X, \mathcal{F}) = \mathcal{F}(X)$, with values in the category of abelian groups. It is not difficult to see that the category of sheaves of abelian groups has sufficiently many injective objects: indeed, the category of abelian groups has sufficiently many injective objects, and if $\mathcal{F}$ is a sheaf of abelian groups, we can embed $\mathcal{F}$ into the sheaf $U \mapsto \bigoplus_{x \in U} I_x$, where I_x is an injective group containing the stalk $\mathcal{F}_x$. This implies the existence of the derived functors $R^i\Gamma$. They are generally written

$$R^i\Gamma(\mathcal{F}) =: H^i(X, \mathcal{F}).$$

In practice, other resolutions are used, which is possible by proposition 4.32; they make it possible to prove finiteness and comparison theorems.

4.3.1 Acyclic resolutions

An important category of acyclic sheaves is that of flasque sheaves.

Definition 4.33 *A sheaf $\mathcal{F}$ is said to be flasque if for every pair of open sets $V \subset U$, the restriction map $\rho_{UV} : \mathcal{F}(U) \to \mathcal{F}(V)$ is surjective.*

Proposition 4.34 *Flasque sheaves are acyclic for the functor Γ.*

Proof Let $\mathcal{F}$ be a flasque sheaf; as noted above, there exists an inclusion $\mathcal{F} \subset I$ with I injective and flasque, and thus a short exact sequence $\mathcal{G}$.

$$0 \to \mathcal{F} \to I \to \mathcal{G} \to 0. \tag{4.10}$$

We first show that for every open set U of X, the map

$$I(U) \to \mathcal{G}(U) \tag{4.11}$$

is surjective. For this, let σ be a section of $\mathcal{G}$ on U. First let $V,\ W$ be two open sets of U on which there exist sections $\tau_V \in I(V),\ \tau_W \in I(W)$ respectively lifting σ to I. Consider the difference

$$\tau_{V|V\cap W} - \tau_{W|V\cap W} \in \mathcal{F}(V \cap W).$$

As $\mathcal{F}$ is flasque, there exists a section $\chi_V \in \mathcal{F}(V)$ such that $\chi_{V|V\cap W} = \tau_V - \tau_W$. Then if $\tau'_V = \tau_V - \chi_V$, we have $\tau'_{V|V\cap W} = \tau_{W|V\cap W}$. Thus, there exists a section $\tau \in I(V \cup W)$ such that $\tau_{|V} = \tau'_V,\ \tau_{|W} = \tau_W$. Clearly τ is sent to $\sigma_{|V\cup W}$.

Let us now introduce a pair (W, τ) which is maximal (for the obvious order relation), where W is an open set of U, and τ is a section of I lifting σ to W. Noting that σ lifts locally in I, and using the preceding result, we immediately see that $W = U$.

But then the long exact sequence associated to (4.10), and the fact that $H^k(X, I) = 0,\ k > 0$, imply that $H^1(X, \mathcal{F}) = 0$ and

$$H^k(X, \mathcal{F}) = H^{k-1}(X, \mathcal{G}), \quad k - 1 \geq 1.$$

Furthermore, the surjectivity of the maps (4.11) implies that $\mathcal{G}$ is also flasque. It follows by induction that $H^k(X, \mathcal{F}) = 0,\ k > 0$. □

Proposition 4.34 allows us to use the Godement resolutions, which have the advantage of being canonical and functorial, to compute the cohomology of a sheaf. The Godement resolution of $\mathcal{F}$ is constructed by considering the inclusion of $\mathcal{F}$ into the sheaf $\mathcal{F}_{\text{God}}$

$$U \mapsto \mathcal{F}_{\text{God}}(U) = \bigoplus_{x \in U} \mathcal{F}_x,$$

where the sum is actually the infinite direct product. This sheaf is obviously flasque. We then inject the quotient $\mathcal{F}_{\text{God}}/\mathcal{F}$ into the flasque sheaf $(\mathcal{F}_{\text{God}}/\mathcal{F})_{\text{God}}$, and so on.

We have the following definition.

Definition 4.35 *A fine sheaf $\mathcal{F}$ over X is a sheaf of $\mathcal{A}$-modules, where $\mathcal{A}$ is a sheaf of rings over X satisfying the property:*
For every open cover U_i, $i \in I$ of X, there exists a partition of unity f_i, $i \in I$, $\sum f_i = 1$ (where the sum is locally finite), subordinate to this covering.

The following result enables us to construct reasonable acyclic resolutions for the functor Γ.

Proposition 4.36 *If $\mathcal{F}$ is a fine sheaf, we have $H^i(X, \mathcal{F}) = 0$, $\forall i > 0$.*

Proof First, using Godement resolutions, we show that $\mathcal{F}$ admits a flasque resolution $I^{\cdot}$, $\mathcal{F} \subset I^0$, each of whose terms I^k is a sheaf of $\mathcal{A}$-modules, where the differentials are morphisms of sheaves of $\mathcal{A}$-modules. Then, by propositions 4.32 and 4.34, we have

$$H^k(X, \mathcal{F}) = \text{Ker}\,(\Gamma(I^k) \to \Gamma(I^{k+1}))/\text{Im}\,(\Gamma(I^{k-1}) \to \Gamma(I^k)).$$

But let $\alpha \in \text{Ker}\,(\Gamma(I^k) \to \Gamma(I^{k+1}))$. The local exactness of the complex $I^{\cdot}$ in degree > 0 shows that locally α comes from I^{k-1}, i.e. there exists an open cover U_i of X such that $\alpha_{|U_i} = d\beta_i$, where $\beta_i \in I^{k-1}(U_i)$. Let f_i be a partition of unity subordinate to the covering U_i, and let

$$\beta = \sum_i f_i \beta_i,$$

where the sum is locally finite. Here $f_i\beta_i$ is a section of I^{k-1} on X; its value in I_x^{k-1} is equal to 0 for x outside of U_i, and to $f_i\beta_i$ for $x \in U_i$. As f_i has support in U_i, this defines a section of I^{k-1}. The right-hand sum is then well-defined since it is a locally finite sum. We have $d\beta = \alpha$ since $\alpha = \sum_i f_i \alpha_{|U_i}$. Thus $\text{Ker}\,(\Gamma(I^k) \to \Gamma(I^{k+1})) = \text{Im}\,(\Gamma(I^{k-1}) \to \Gamma(I^k))$ and $H^k(X, \mathcal{F}) = 0$ for $k > 0$. □

This applies particularly to the case of differentiable manifolds.

Corollary 4.37 *Let X be a C^∞ manifold. Then*

$$H^k(X,\mathbb{R}) = \mathrm{Ker}\,(d : A^k(X) \to A^{k+1}(X))/\mathrm{Im}\,(d : A^{k-1}(X) \to A^k(X)),$$

where $A^i(X)$ is the real vector space of differential forms of degree i. A similar statement holds for the complex cohomology.

Proof We use the de Rham resolution (4.4) of $\mathbb{R}$ (respectively $\mathbb{C}$). Proposition 4.36 shows that it is an acyclic resolution since the $\mathcal{A}^k$ are sheaves of $\underline{C}^\infty$-modules. Proposition 4.32 then implies that $H^k(X,\mathbb{R})$ (resp. $H^k(X,\mathbb{C})$) is equal to the cohomology of the complex of the global sections of the real (resp. complex) de Rham complex. □

Corollary 4.38 *Let E be a holomorphic vector bundle over a complex manifold X, and let $\mathcal{E}$ be the sheaf of holomorphic sections of E. Then*

$$H^q(X,\mathcal{E}) = \mathrm{Ker}\,(\overline{\partial} : A^{0,q}(E) \to A^{0,q+1}(E))/\mathrm{Im}\,(\overline{\partial} : A^{0,q-1}(E) \to A^{0,q}(E)).$$

Proof We reason as in the preceding corollary, using the Dolbeault resolution (4.5) of $\mathcal{E}$. □

A useful consequence of this statement is the following.

Corollary 4.39 *If E is as above, we have $H^q(X,E) = 0$ for $q > n = \dim X$.*

We will now introduce the Čech cohomology, which is extremely useful in practice, since it gives a uniform way of computing cohomology groups, unlike the de Rham type resolutions, which specifically concern constant sheaves over manifolds. Let $\mathcal{F}$ be a sheaf of abelian groups over a topological space X. Let $\mathcal{U} = (U_i)_{i\in\mathbb{N}}$ be a countable ordered open covering of X.

Definition 4.40 *Define $\check{H}^q(\mathcal{U},\mathcal{F})$ to be the qth cohomology group of the complex of global sections*

$$C^q(\mathcal{U},\mathcal{F}) = \bigoplus\nolimits_{|I|=q+1} \mathcal{F}(U_I)$$

of the Čech complex (4.2) associated to the covering $\mathcal{U}$.

We then have the following.

Theorem 4.41 *If the open sets $U_I = \bigcap_{i \in I} U_i$ satisfy $H^q(U_I, \mathcal{F}) = 0$ for all $q > 0$, then*

$$H^q(X, \mathcal{F}) = \check{H}^q(\mathcal{U}, \mathcal{F}), \quad \forall q \geq 0.$$

For the proof, we need to have recourse to a construction which will play an important role later on: that of the simple complex associated to a double complex.

Definition 4.42 *A double complex in an abelian category $\mathcal{A}$ is given by a collection of objects $K^{p,q}$, $(p, q) \in \mathbb{Z}^2$, together with morphisms (called differentials)*

$$D_1 : K^{p,q} \to K^{p+1,q}, \quad D_2 : K^{p,q} \to K^{p,q+1},$$

satisfying the relations

$$D_1 \circ D_1 = 0, \quad D_2 \circ D_2 = 0, \quad D_1 \circ D_2 = D_2 \circ D_1.$$

Suppose now that the double complex $K^{\cdot,\cdot}$ satisfies the following finiteness condition: There exist p_0, $q_0 \in \mathbb{Z}$ such that $K^{p,q} = 0$ for $p \geq p_0$ or $q \geq q_0$. We then construct the following complex:

$$K^n = \bigoplus_{p+q=n} K^{p,q}, \quad D = D_1 + (-1)^p D_2 \text{ over } K^{p,q} \subset K^n.$$

The finiteness condition ensures that the direct sum is finite. There is an element of arbitrariness in the choice of the signs of the differential D. The sign $(-1)^p$ is placed here in order to ensure that $D \circ D = 0$. There is a variation on this construction, in the case where the differentials D_i anticommute instead of commuting: we then set $D = D_1 + D_2$.

Definition 4.43 *The complex $(K^\cdot, D)$ is the simple complex associated to the double complex $(K^{\cdot,\cdot}, D_1, D_2)$.*

Proof of theorem 4.41 Take a flasque resolution $I^\cdot$ of $\mathcal{F}$. For each I^l, $l \geq 0$ we can consider the Čech resolution $I^{l,\cdot}$ of I^l associated to the covering $\mathcal{U}$. By the functoriality of the Čech resolution, the $I^{\cdot,\cdot}$ form a double complex. Moreover, the $I^{l,\cdot}$ are acyclic, since we have

$$I^{l,k} = \bigoplus_{|J|=k+1} j_{J*}(I^l_{|U_J}),$$

so that $I^{l,k}$ is flasque, and thus acyclic by proposition 4.34. The simple complex $K^\cdot = \bigoplus_{p+q=k} I^{p,q}$ associated to the double complex $I^{\cdot,\cdot}$ is thus an acyclic resolution of $\mathcal{F}$. Thus, by proposition 4.32, we have

$$H^q(X, \mathcal{F}) = H^q(\Gamma(K^\cdot)).$$

But the complex $\Gamma(K^\cdot)$ is the simple complex associated to the double complex $\Gamma(I^{\cdot,\cdot})$. The lines $\Gamma(I^{\cdot,l})$ of this complex are the complexes which compute the cohomology groups of $\mathcal{F}$ over the disjoint union of the U_I for $|I| = l + 1$. By hypothesis, these cohomology groups are zero in positive degree. We then deduce by lemma 8.5, to be proved later, that the cohomology of the complex $\Gamma(X, K^\cdot)$ is equal to the cohomology of the complex

$$\mathrm{Ker}\,(\Gamma(X, I^{0,l}) \to \Gamma(X, I^{1,l})) = \Gamma(X, \mathcal{F}^l)$$

equipped with the Čech differential. Thus, we have

$$\check{H}^q(\mathcal{U}, \mathcal{F}) = H^q(\Gamma(X, \mathcal{F}^\cdot)) = H^q(\Gamma(X, K^\cdot)) = H^q(X, \mathcal{F}).$$

□

Furthermore, note that by proposition 4.27 giving the universal property of injective resolutions, if $\mathcal{F}$ is a sheaf over X, $\mathcal{U}$ a countable open cover of X, and $I^\cdot$ an injective resolution of $\mathcal{F}$, then we have a morphism of complexes

$$\mathcal{F}^\cdot \to I^\cdot,$$

where $\mathcal{F}^\cdot = \underline{C}^\cdot(\mathcal{U}, \mathcal{F})$ is the Čech resolution (4.2) of $\mathcal{F}$. Thus, we have a canonical morphism

$$\check{H}^q(\mathcal{U}, \mathcal{F}) \to H^q(X, \mathcal{F}). \tag{4.12}$$

We will need the following result (Godement 1958), whose proof we omit.

Theorem 4.44 *If X is separable, then by passage to the direct limit, the morphisms (4.12) induce an isomorphism*

$$\lim_{\vec{\mathcal{U}}} \check{H}^q(\mathcal{U}, \mathcal{F}) \cong H^q(X, \mathcal{F}).$$

Remark 4.45 *To give meaning to this direct limit, we need the following remark. We have seen the notion of the refinement of an open cover, which puts the structure of a directed set onto the set of open covers. When an open cover $\mathcal{V}$ is finer than an open cover $\mathcal{U}$, we have a natural restriction map, from the Čech complex relative to $\mathcal{U}$ to the Čech complex relative to $\mathcal{V}$, on condition that we specify a refining map. It happens that the map induced on the cohomology*

$$\check{H}^q(\mathcal{U}, \mathcal{F}) \to \check{H}^q(\mathcal{V}, \mathcal{F})$$

does not depend on the choice of the refining map. This is why the groups $\check{H}^q(\mathcal{U}, \mathcal{F})$ *form a directed system associated with the directed set of countable coverings.*

4.3.2 The de Rham theorems

The preceding section allowed us to identify the de Rham cohomology group

$$H^q_{\mathrm{DR}}(X, \mathbb{R}) = \frac{\mathrm{Ker}\,(d : A^q(X) \to A^{q+1}(X))}{\mathrm{Im}\,(d : A^{q-1}(X) \to A^q(X))}$$

of a differentiable manifold X with the qth cohomology group of X with values in the constant sheaf $\mathbb{R}$, called the Betti cohomology group $H^q(X, \mathbb{R})$. Theorem 4.41 enables us to compute these groups in a combinatorial way. Indeed, for a ball U of $\mathbb{R}^n$, we have $H^q(U, \mathbb{R}) = 0$ by Poincaré's lemma and corollary 4.37. Now, a manifold admits a covering by open sets U_i which are homeomorphic to balls, as are all their intersections U_I. Thus, we can compute $H^q(X, \mathbb{R})$ as the qth cohomology group of the Čech complex associated to this covering. In fact, this result also holds for cohomology with integral coefficients.

Remark 4.46 *This shows that for a compact manifold X, the cohomology groups* $H^q(X, \mathbb{Z})$ *are of finite type, since we can compute them as the Čech cohomology groups relative to a finite covering by contractible open sets whose multi-intersections are contractible, so that each* $C^q(\mathcal{U}, \mathbb{Z})$ *is of finite type.*

There also exists another notion of the cohomology of a topological space, defined over $\mathbb{Z}$ as the Betti cohomology, called the singular cohomology, and written $H^q_{\mathrm{sing}}(X, \mathbb{Z})$. (The Betti cohomology is defined over $\mathbb{Z}$ in the sense that, as we will see later, we have

$$H^q(X, \mathbb{R}) = H^q(X, \mathbb{Z}) \otimes_{\mathbb{Z}} \mathbb{R}.)$$

The singular cohomology (Spanier 1966) is the cohomology of the complex of the singular cochains $C^q_{\mathrm{sing}}(X, \mathbb{Z})$, i.e. the dual of the complex of the singular chains

$$(C_q(X, \mathbb{Z}), \partial).$$

The group of singular q-chains $C_q(X, \mathbb{Z})$ is the free abelian group generated by the continuous maps of the simplex

$$\Delta_q = \left\{(t_1, \ldots, t_{q+1}) \in [0, 1]^{q+1} \mid \sum_i t_i = 1\right\}$$

of dimension q in X, and the boundary ∂ is given by

$$\partial\phi = \sum_i (-1)^i \phi|_{\Delta_q^i},$$

where $\Delta_{q-1} \cong \Delta_q^i$ is the ith face of Δ:

$$\Delta_q^i = \{(t_1, \ldots, t_{q+1}) \in \Delta_q \mid t_i = 0\}.$$

There is another acyclic resolution of $\mathbb{Z}$, which enables us to identify the singular cohomology and the Betti cohomology.

Theorem 4.47 *Let X be a locally contractible topological space. Then we have a canonical isomorphism*

$$H^q_{\mathrm{sing}}(X, \mathbb{Z}) \cong H^q(X, \mathbb{Z}).$$

The same result holds with $\mathbb{Z}$ replaced by any commutative ring G. (We then consider the cohomology of X with coefficients in the constant sheaf of stalk G on the right, and the singular cohomology with coefficients in G on the left.)

Proof Let us consider the sheaf $\mathcal{C}^q_{\mathrm{sing}}$ of singular cochains, which is associated to the presheaf

$$U \mapsto C^q_{\mathrm{sing}}(U, \mathbb{Z}). \tag{4.13}$$

The differential ∂ on each $C^q_{\mathrm{sing}}(U, \mathbb{Z})$ gives a differential

$$\partial : \mathcal{C}^q_{\mathrm{sing}} \to \mathcal{C}^{q+1}_{\mathrm{sing}}.$$

The complex constructed in this way is a resolution of the constant sheaf $\mathbb{Z}$. Indeed, the complex $C^q_{\mathrm{sing}}(U, \mathbb{Z})$ is exact in positive degree on each contractible open set U, since the cohomology of $(C^q(U, \mathbb{Z}), \partial)$ is the singular cohomology of U, which is zero for a contractible space. This shows that the complex of presheaves (4.13) is exact in positive degree at the level of the stalks, and thus the complex $\mathcal{C}^q_{\mathrm{sing}}$ is exact in positive degree. Also, since X is locally pathwise connected, we have

$$\mathrm{Ker}\,(\partial : \mathcal{C}^0_{\mathrm{sing}} \to \mathcal{C}^1_{\mathrm{sing}}) = \mathbb{Z}$$

(i.e. the constant sheaf equal to $\mathbb{Z}$). Thus, this complex is a resolution of $\mathbb{Z}$. To conclude, we note that this resolution is acyclic by proposition 4.34, since it is flasque. To see this, note that the presheaf (4.13) is obviously flasque, and check the equality

$$\mathcal{C}^q_{\mathrm{sing}}(U, \mathbb{Z}) = C^q_{\mathrm{sing}}(U, \mathbb{Z}) / C^q_{\mathrm{sing}}(U, \mathbb{Z})_0, \tag{4.14}$$

where $C^q_{\text{sing}}(U, \mathbb{Z})_0$ is the set of $\alpha \in C^q_{\text{sing}}(U, \mathbb{Z})$ such that there exists a covering $\mathcal{V}$ of U with $\alpha_{|C_q(V,\mathbb{Z})} = 0$ for all $V \in \mathcal{V}$.

Thus, we can apply proposition 4.32 to conclude that

$$H^q(X, \mathbb{Z}) = H^q(\Gamma(\mathcal{C}^{\cdot}_{\text{sing}})).$$

It remains only to show that the complex

$$\Gamma(\mathcal{C}^{\cdot}_{\text{sing}}) = C^{\cdot}_{\text{sing}}(X)/C^{\cdot}_{\text{sing}}(X)_0$$

is quasi-isomorphic to the complex $C^{\cdot}_{\text{sing}}(X)$, which is essentially the theorem of small chains (Spanier 1966). □

Remark 4.48 *The singular cohomology of X with real coefficients can be identified with* $\text{Hom}(H_q^{\text{sing}}(X), \mathbb{R})$*, where the singular homology is the homology of the complex of singular chains* $(C_q(X, \mathbb{Z}), \partial)$*. (We speak of homology rather than cohomology here, because the differential* ∂ *decreases the degree.) Indeed, this follows from the isomorphism of complexes*

$$C^q_{\text{sing}}(X, \mathbb{R}) \to \text{Hom}(C_q(X, \mathbb{Z}), \mathbb{R}).$$

The composition

$$H^q_{DR}(X) \cong H^q_{\text{sing}}(X, \mathbb{R}) \to \text{Hom}\big(H_q^{\text{sing}}(X, \mathbb{Z}), \mathbb{R}\big),$$

where the first isomorphism is given by theorem 4.47 and corollary 4.37, is described as follows. Let ω *be a q-form on X. Then the linear form*

$$\int \omega : \phi \mapsto \int_{\Delta_q} \phi^* \omega$$

corresponds to ω*; it is defined at least on the subgroup of* $C_q(U, \mathbb{Z})$ *generated by the differentiable maps from* Δ_q *to X. When* ω *is closed, Stokes' formula (theorem 1.10) proves that this linear form induces a form on* $H_q(X, \mathbb{Z})$*, i.e. vanishes on the boundaries. This is the linear form associated to the class of the closed from* ω*. Indeed, by Stokes' formula, the map* $\omega \mapsto \int \omega$ *actually gives a morphism (of sheaves) from the de Rham complex to the complex of singular (differentiable) cochains, and this morphism, which is a morphism of acyclic resolutions, necessarily induces the above isomorphism* $H^q_{DR}(X) \cong H^q_{\text{sing}}(X, \mathbb{R})$.

4.3.3 Interpretations of the group H^1

Let $\mathcal{F}$ be a sheaf of abelian groups over a separable topological space X, and let $\alpha \in H^1(X, \mathcal{F})$. Then α can be represented by a Čech cocycle for a suitable open cover of X. This actually holds for all cohomology groups, by theorem 4.44, but in the case of the H^1, one can see it immediately as follows.

Let $\mathcal{F} \subset I$ be an inclusion into an injective sheaf. We then have a short exact sequence

$$0 \to \mathcal{F} \to I \to \mathcal{G} \to 0,$$

and as I is injective, so acyclic, the exact sequence (4.7) gives an isomorphism

$$H^1(X, \mathcal{F}) \cong \operatorname{Coker}(H^0(X, I) \to H^0(X, \mathcal{G})).$$

As the map $I \to \mathcal{G}$ is surjective, a section $\beta \in H^0(\mathcal{G})$ lifts locally, on the open sets U_i of a cover $\mathcal{U}$ of X which we may assume countable, to sections $\beta_i \in \Gamma(I_{|U_i})$. Then, on $U_i \cap U_j$, $\beta_{ij} = \beta_i - \beta_j$ is a section of $\mathcal{F}$. Clearly β_{ij} satisfies the cocycle condition

$$\beta_{ij} + \beta_{jk} - \beta_{ik} = 0$$

in $\Gamma(U_i \cap U_j \cap U_k, \mathcal{F})$, and thus determines a class in the Čech cohomology group of $\mathcal{F}$ relative to the covering $\mathcal{U}$. We see immediately that this class does not depend on the liftings, nor on the representative β.

This representation of the H^1 by Čech cocycles enables us to give the following interpretation of the groups $H^1(X, \mathcal{A}^*)$, where $\mathcal{A}$ is one of the sheaves of rings $\mathbb{C}$, $\mathcal{C}^0_{X,\mathbb{C}}$, $\mathcal{O}_X$ (the last one in the case where X is a complex manifold), and $\mathcal{A}^*$ is the sheaf of corresponding multiplicative groups.

Theorem 4.49 *The group $H^1(X, \mathcal{A}^*)$ is in bijection with the set of isomorphism classes of sheaves of free rank 1 modules over $\mathcal{A}$, and also with the isomorphism classes of rank 1 complex vector bundles equipped with flat, continuous, or holomorphic structures according to $\mathcal{A}$.*

Here, a flat structure means that for suitable trivialisations, the transition matrices have constant coefficients, and a holomorphic structure means that for suitable trivialisations, the transition matrices have holomorphic coefficients.

Proof Let U_i be an open cover of X, and let $\beta_{ij} \in \Gamma(U_{ij}, \mathcal{A}^*)$ be a Čech cocycle. Set

$$\mathcal{L}_\beta(U) = \{(f_i)_{i\in I} \mid f_i \in \Gamma(\mathcal{A}_{|U\cap U_i}),\ f_{i\,|U\cap U_i\cap U_j} = \beta_{ij} f_{j\,|U\cap U_i \cap U_j}\}.$$

Clearly, $\mathcal{L}$ is a sheaf of $\mathcal{A}$-modules. The cocycle condition on β guarantees that this sheaf has sufficiently many local sections: indeed, on each U_i, we have the section $(f_j = \beta_{ji},\ j \neq i,\ f_i = 1)$. We easily check that $\mathcal{L}_\beta$ is in fact generated, as an $\mathcal{A}$-module, by this section on U_i, and thus $\mathcal{L}_\beta$ is a sheaf of free rank 1 $\mathcal{A}$-modules, trivial on the open sets U_i.

If we change the cocycle β_{ij} by a coboundary to a cocycle $\beta'_{ij} = \frac{\beta_i}{\beta_j}\beta_{ij}$, the sheaf $\mathcal{L}_\beta$ is isomorphic to the sheaf $\mathcal{L}_{\beta'}$ via multiplication by β_i on U_i (in the trivialisations given above). Conversely, an isomorphism

$$\gamma : \mathcal{L}_\beta \cong \mathcal{L}_{\beta'}$$

is given by invertible functions γ_i on U_i such that $\beta'_{ij} = \frac{\gamma_i}{\gamma_j}\beta_{ij}$. We thus have a bijection between the isomorphism classes of sheaves of free rank 1 $\mathcal{A}$-modules which are trivial on the open sets U_i, and the group $\check{H}^1(\mathcal{U}, \mathcal{A}^*)$. Passing to the limit on the coverings, we obtain theorem 4.49. □

The group $H^1(X, \mathcal{F})$ also has another geometric interpretation, which is particularly important in the study of complex vector bundles: it parametrises the "extensions of $\mathcal{F}$ by a trivial bundle". In the case where $\mathcal{F}$ is the sheaf of holomorphic sections of a holomorphic vector bundle over a complex manifold X, we have for example the following.

Theorem 4.50 *The group $H^1(X, \mathcal{F})$ parametrises the isomorphism classes of extensions of F by the trivial bundle, i.e. of holomorphic vector bundles G containing F as a holomorphic vector subbundle and such that the quotient bundle is the trivial bundle of rank* 1.

Here the notion of an isomorphism class is the following: we say that G is isomorphic to G' as extension of F by the trivial bundle if there exists an isomorphism of holomorphic vector bundles between G and G' which induces the identity on F and on the quotient bundle $X \times \mathbb{C}$.

Proof Given an extension G of F by the trivial bundle, we have an exact sequence of the corresponding sheaves of holomorphic sections:

$$0 \to \mathcal{F} \to \mathcal{G} \to \mathcal{O}_X \to 0.$$

Theorem 4.28 then gives a map

$$\delta : H^0(X, \mathcal{O}_X) \to H^1(X, \mathcal{F}),$$

and $\delta(1)$ gives the desired cohomology class.

Conversely, if $\beta \in H^1(X, \mathcal{F})$, let β_{ij} be a Čech representative of β relative to an open cover $(U_i)_{i\in I}$ of X. Then, for $U \subset X$, set

$$\mathcal{G}(U) = \{(\sigma_i, f_i)_{i\in I} \mid \sigma_i \in \mathcal{F}(U \cap U_i),\ f_i \in \mathcal{O}_X(U \cap U_i),$$
$$\sigma_i = \sigma_j + f_i\beta_{ij} \in \mathcal{F}(U \cap U_i \cap U_j),\ f_j = f_i \in \mathcal{O}_X(U \cap U_i \cap U_j)\}.$$

Clearly $\mathcal{G}$ is a sheaf of $\mathcal{O}_X$-modules. The cocycle condition for β guarantees that this sheaf is isomorphic to $\mathcal{F} \oplus \mathcal{O}_X$ over $U_i \cap U_j$. Thus, it is a sheaf of free $\mathcal{O}_X$-modules. It obviously contains $\mathcal{F}$ (set $f_i = 0$), and we easily see that the quotient is trivial. □

Exercises

1. Using theorem 1.28, and the arguments given in the proof of proposition 2.31, show that any form of type $(0, i)$ which is $\overline{\partial}$-closed on $\mathbb{C}^n$ is $\overline{\partial}$-exact for $i > 0$. Deduce from this that

$$H^i(\mathbb{C}^n, \mathcal{O}_{\mathbb{C}^n}) = 0, \quad \forall i > 0.$$

2. *Residue formula.* Let X be a compact complex curve.
Let μ be a volume form on X. We can consider μ as a closed form of type $(1, 1)$ on X.
(a) By considering the integral $\int_X \mu$, show that μ is not $\overline{\partial}$-exact. Deduce from this that $H^1(X, K_X)$ is different from $\{0\}$ and admits a surjective map

$$\mathrm{Tr} : H^1(X, K_X) \to \mathbb{C}, \quad \omega \mapsto \int_X \omega.$$

Let $D = \sum_i x_i$ be a divisor of X (all of whose multiplicities are equal to 1). We denote by

$$K_X(D)$$

the holomorphic line bundle (or sheaf of free $\mathcal{O}_X$-modules of rank 1), whose sections are the holomorphic forms of degree 1 on any open set not containing any of the x_i's, and in the neighbourhood of each x_i, the meromorphic forms which can be written as

$$\phi(z_i)\frac{dz_i}{z_i}$$

where z_i is a local coordinate centred at x_i and ϕ is holomorphic.

(b) Show that we have an exact sequence

$$0 \to K_X \to K_X(D) \stackrel{\text{Res}}{\to} \sum_i \mathbb{C}_{x_i} \to 0, \qquad (4.15)$$

where each sheaf $\mathbb{C}_{x_i}$ is a "skyscraper" sheaf supported at x_i, whose group of sections over any open set not containing x_i is $\{0\}$ and whose group of sections on an open set containing x_i is $\mathbb{C}$. Here the map $\text{Res}_i : K_X(D) \to \mathbb{C}_{x_i}$ maps a meromorphic form ω to its residue at x_i, defined as

$$\text{Res}_i(\omega) = \int_{\partial D_i} \frac{1}{2i\pi}\omega,$$

where D_i is a disk centred at x_i not containing any of the x_j's, $i \neq j$.
(c) Let $\delta : \bigoplus_i \mathbb{C} = H^0(X, \bigoplus_i \mathbb{C}_{x_i}) \to H^1(X, K_X)$ be the arrow appearing in the long exact sequence associated to the short exact sequence (4.15). Show that $\delta(1_{x_i})$ is the class in $H^1(X, K_X)$ of the form $\overline{\partial}\mu_i$, where μ_i is a differential form of type (1, 0), which is C^∞ away from x_i, and equal to $\frac{dz_i}{z_i}$ in a neighbourhood of x_i.
(d) Show that

$$\int_X \delta\mu_i = -2i\pi.$$

Deduce from the long exact sequence associated to the short exact sequence (4.15) the following result:
If ω is a meromorphic 1-form on X having poles of order at most 1 at each x_i, and holomorphic otherwise, then

$$\sum_i \text{Res}_i \omega = 0.$$

Part II

The Hodge Decomposition

5

Harmonic Forms and Cohomology

In the preceding chapter, we showed that the de Rham cohomology groups of a differentiable manifold X were topological invariants. We will now show that if we also have a Riemannian structure on the manifold X (which we assume compact), it is possible to exhibit representatives, which are particular closed differential forms, for the de Rham cohomology classes. These differential forms, which are called harmonic forms, are not only closed, but satisfy another first order differential equation: they are coclosed, i.e. annihilated by the formal adjoint d^* of the operator d.

Since the manifold X is compact, the metric on X provides a metric $(\cdot, \cdot)_{L^2}$, the L^2 metric, on the spaces $A^k(X)$ of $\mathcal{C}^\infty$ differential forms. Using Stokes' formula, one can easily prove the existence of a formal adjoint d^* of the differential operator $d : A^k(X) \to A^{k+1}(X)$. The differential operator $d^* : A^{k+1}(X) \to A^k(X)$ thus satisfies the formal adjunction property

$$(\alpha, d\beta)_{L^2} = (d^*\alpha, \beta)_{L^2}, \quad \alpha \in A^{k+1}(X), \quad \beta \in A^k(X).$$

More generally, we will see that every differential operator

$$P : \mathcal{C}^\infty(X, F) \to \mathcal{C}^\infty(X, E)$$

between two vector bundles equipped with metrics, over a compact manifold X equipped with a volume form, admits a formal adjoint

$$P^* : \Gamma(X, E) \to \Gamma(X, F)$$

satisfying the above adjunction property for the L^2 metric on the spaces of sections of E and F.

Since the Laplacian $\Delta_d : A^k(X) \to A^k(X)$ is defined by $\Delta_d = dd^* + d^*d$, we easily see that a form is both closed and coclosed if and only if it is annihilated by the Laplacian. The Laplacian is an elliptic differential operator of order 2.

For a differential operator P of order k between two bundles F and E over X, being elliptic means that the symbol of P, which is a section σ_P of the bundle $\mathrm{Sym}^k T_X \otimes \mathrm{Hom}\,(F, E)$, satisfies the property that

$$\sigma_P(u) : F_x \to E_x$$

is injective for $0 \neq u \in \Omega_{X,x}$.

The property which we use here (without proof) is the following.

Theorem 5.1 *If*

$$P : \mathcal{C}^\infty(X, F) \to \mathcal{C}^\infty(X, E)$$

is an elliptic differential operator between two vector bundles of the same rank equipped with metrics, with X compact, then we have a decomposition

$$\mathcal{C}^\infty(X, F) = \mathrm{Ker}\, P \oplus \mathrm{Im}\, P^*.$$

Moreover Ker P *is finite-dimensional.*

This theorem easily implies the following result.

Theorem 5.2 *If X is a compact Riemannian manifold, we have a decomposition as an orthogonal direct sum*

$$A^k(X) = \mathcal{H}^k \oplus \mathrm{Im}\, d \oplus \mathrm{Im}\, d^*,$$

where $\mathcal{H}^k$ is the space of harmonic forms of degree k.

This decomposition shows that the space $\mathcal{H}^k \subset \mathrm{Ker}\, d \subset A^k(X)$ is in bijection with $H^k(X, K)$ ($K = \mathbb{R}$ or $\mathbb{C}$), which is the main result we will be using below.

We also apply these considerations to the operators

$$\overline{\partial}_E : A_X^{0,k}(E) \to A_X^{0,k+1}(E)$$

of a holomorphic vector bundle E over a complex manifold, which enables us to construct a Laplacian $\Delta_{\overline{\partial}}$ as well as $\Delta_{\overline{\partial}}$-harmonic representatives for the cohomology $H^k(X, E)$.

As a first application of the theorem of representation of cohomology classes by harmonic forms, we prove the Poincaré duality theorem (with real or complex coefficients) and the Serre duality theorem.

5.1 Laplacians

5.1.1 The L^2 metric

Let X be a compact differentiable manifold equipped with a metric g. We then have an induced metric $(,)$ on each vector bundle $\Omega^k_{X,\mathbb{R}}$; if $e_1, \ldots, e_n$ is an orthonormal basis for $(T_{X,x}, g_x)$ and e_i^* is the dual basis, the $e_{i_1}^* \wedge \cdots \wedge e_{i_k}^*$ form an orthonormal basis for the metric $(,)_x$ on $\Omega^k_{X,x}$.

Assume now that X is compact and oriented, and let Vol be the volume form of X relative to g. The L^2 metric on the space $A^k(X)$ of differential forms on X is defined by

$$(\alpha, \beta)_{L^2} = \int_X (\alpha, \beta)\mathrm{Vol}, \tag{5.1}$$

where (α, β) is the function $x \mapsto (\alpha_x, \beta_x)_x$ on X, which is continuous whenever α, β and g are continuous.

Let $n = \dim X$. For each $x \in X$, we have a natural isomorphism, given by the right exterior product

$$p : \bigwedge^{n-k} \Omega_{X,x} \cong \mathrm{Hom}\left(\bigwedge^k \Omega_{X,x}, \bigwedge^n \Omega_{X,x}\right),$$

where $\bigwedge^n \Omega_{X,x}$ is a 1-dimensional vector space. When $\Omega_{X,x}$ is equipped with a metric and is oriented, $\bigwedge^n \Omega_{X,x}$ is canonically isomorphic to $\mathbb{R}$, thanks to the volume form. Moreover, the metric $(,)_x$ also gives an isomorphism

$$m : \bigwedge^k \Omega_{X,x} \cong \mathrm{Hom}\left(\bigwedge^k \Omega_{X,x}, \mathbb{R}\right).$$

We can thus define the operator

$$*_x = p^{-1} \circ m : \bigwedge^k \Omega_{X,x} \cong \bigwedge^{n-k} \Omega_{X,x},$$

which varies differentiably with x when g is differentiable, and which is of the same class as g.

Definition 5.3 *Let $*$ denote the isomorphism of vector bundles*

$$* : \Omega^k_{X,\mathbb{R}} \cong \Omega^{n-k}_{X,\mathbb{R}}$$

constructed in this way. Let $*$ also denote the induced morphism on the level of sections, i.e. of differential forms:

$$* : A^k(X) \cong A^{n-k}(X).$$

The operator $*$ is called the Hodge operator. Its essential property is the following.

Lemma 5.4 *For* $\alpha, \beta \in A^k(X)$, *we have*

$$(\alpha, \beta)_{L^2} = \int_X \alpha \wedge *\beta.$$

Proof By definition, for every $x \in X$, we have

$$(\alpha_x, \beta_x)_x \mathrm{Vol}_x = \alpha_x \wedge *\beta_x.$$

□

Lemma 5.5 *The operator* $*$ *satisfies the identity*

$$*^2 = (-1)^{k(n-k)} \text{ on } A^k(X).$$

Proof For $\alpha_x, \beta_x \in \Omega^k_{X,x}$, we have the equality $(\alpha_x, \beta_x)_x \mathrm{Vol}_x = \alpha_x \wedge * \beta_x$ which characterises $*$. But clearly $*_x : \Omega^k_{X,x} \to \Omega^{n-k}_{X,x}$ preserves metrics, so we also have

$$\begin{aligned}(\alpha_x, \beta_x)_x \mathrm{Vol}_x &= (*\alpha_x, *\beta_x)_x \mathrm{Vol}_x = *\beta_x \wedge * * \alpha_x \\ &= (-1)^{k(n-k)} * *\alpha_x \wedge *\beta_x = \alpha_x \wedge *\beta_x.\end{aligned}$$

Since this holds for every β_x, it follows that $(-1)^{k(n-k)} * *\alpha_x = \alpha_x$. □

We extend $*$ by $\mathbb{C}$-linearity to complex-valued forms. If we also extend the metrics (,) to Hermitian metrics on the complexified bundles $\Omega^k_{X,\mathbb{C}}$, we have the property

$$(\alpha_x, \beta_x)\mathrm{Vol}_x = \alpha_x \wedge \overline{*\beta}_x.$$

On the subject of the Hermitian metrics induced on the complexified bundles, let us note the following fact. Let V be a complex vector space, and let h be a Hermitian metric on V. Let $W = \mathrm{Hom}\,(V, \mathbb{R})$. We have the decomposition

$$W_{\mathbb{C}} = W^{1,0} \oplus W^{0,1}, \quad \bigwedge^k W_{\mathbb{C}} = \bigoplus_{p+q=k} W^{p,q};$$

each component $W^{p,q} = \bigwedge^p W^{1,0} \otimes \bigwedge^q W^{0,1}$ has a Hermitian metric $h^{p,q}$ induced by h on $W^{1,0} \cong \mathrm{Hom}_{\mathbb{C}}(V, \mathbb{C})$, $W^{0,1} \cong \mathrm{Hom}_{\overline{\mathbb{C}}}(V, \mathbb{C})$ and their tensor products. Furthermore, we have the Hermitian metric h_k induced by g on $\bigwedge^k W_{\mathbb{C}} \cong \bigwedge^k W \otimes \mathbb{C}$.

Lemma 5.6 *We have the relation* $2^k h_k = \sum h^{p,q}$ *on* $\bigwedge^k W_{\mathbb{C}}$, *where the right-hand sum is the direct sum of the metrics* $h^{p,q}$.

Proof By homogeneity, it suffices to check this equality in degree 1, which is very easy. □

5.1.2 Formal adjoint operators

Let $A^k(X)$ be the space of $\mathcal{C}^\infty$ forms on a manifold X equipped with a $\mathcal{C}^\infty$ metric. When X is oriented, we can define an operator

$$d^* : A^k(X) \to A^{k-1}(X)$$

by the formula $d^* = (-1)^k *^{-1} d*$ on $A^k(X)$. This operator is the formal adjoint of d for the L^2 metric on forms, in the following sense.

Lemma 5.7 *Either assume that X is compact, or consider only integrals of forms with compact support. Then the operator d^* satisfies the formal adjunction relation*

$$(\alpha, d^*\beta)_{L^2} = (d\alpha, \beta)_{L^2}. \tag{5.2}$$

Proof We have $(d\alpha, \beta)_{L^2} = \int_X d\alpha \wedge *\beta$. But we also have

$$d(\alpha \wedge *\beta) = d\alpha \wedge *\beta + (-1)^{d^0\alpha}\alpha \wedge d * \beta.$$

Stokes' formula (1.5) thus gives

$$(d\alpha, \beta)_{L^2} = -\int_X (-1)^{d^0\alpha}\alpha \wedge d * \beta.$$

As we have

$$(\alpha, d^*\beta)_{L^2} = (-1)^{d^0\beta} \int_X \alpha \wedge d * \beta,$$

the equality (5.2) follows from $d^0\alpha + 1 = d^0\beta$. □

Note that in particular, if n is even, then by lemma 5.5, we have the equality $d^* = - * d *$.

5.1.3 Adjoints of the operators $\overline{\partial}$

If X is a complex manifold, we have the operators ∂ and $\overline{\partial}$ defined on complex differential forms, satisfying the relation $d = \partial + \overline{\partial}$.

Lemma 5.8 *The operators $\partial^* = - * \overline{\partial} *$ and $\overline{\partial}^* = - * \partial *$ are formal adjoints of ∂ and $\overline{\partial}$ respectively, for the Hermitian L^2 metric on complex forms.*

Proof For complex forms, we have the equality

$$(\alpha, \beta)_{L^2} = \int_X \alpha \wedge \overline{*\beta}.$$

In particular, $(\overline{\partial}\alpha, \beta)_{L^2} = \int_X \overline{\partial}\alpha \wedge \overline{*\beta}$. But we have $\int_X \overline{\partial}\phi = 0$ for every form ϕ of degree $2n-1$, $n = \dim X$, and thus

$$(\overline{\partial}\alpha, \beta)_{L^2} = -\int_X (-1)^{d^0\alpha}\alpha \wedge \overline{\partial}\,\overline{*\beta}.$$

As $\overline{\partial}\,\overline{*\beta} = \overline{\partial *\beta}$, this is equal to $-\int_X (-1)^{d^0\alpha}\alpha \wedge \overline{* *^{-1}\,\partial *\beta}$.

As we have $d^0\partial * \beta = 2n - d^0\alpha$, we have $*^{-1}\partial * \beta = (-1)^{d^0\alpha} * \partial * \beta$, and thus

$$-\int_X (-1)^{d^0\alpha}\alpha \wedge \overline{* *^{-1}\,\partial *\beta} = -\int_X \alpha \wedge \overline{* * \partial * \beta} = (\alpha, \overline{\partial}^*\beta)_{L^2}.$$

The statement concerning ∂ is proved similarly. □

We can do the same construction with the operator $\overline{\partial}_E$ of a holomorphic vector bundle E over a complex manifold X. Suppose that E and X are equipped with a Hermitian metric. Then each vector bundle $\Omega_X^{p,q} \otimes E$ is equipped with a Hermitian metric. Now, $\Omega_X^{n,n} = \Omega_{X,\mathbb{C}}^{2n}$ is trivialised by the volume form Vol. Thus, $\Omega_X^{0,q} \otimes E$ and $\Omega_X^{n,n-q} \otimes E^*$ are naturally dual as complex vector bundles. Furthermore, we have a $\mathbb{C}$-antilinear isomorphism given by the Hermitian metric

$$\Omega_X^{0,q} \otimes E \to \left(\Omega_X^{0,q} \otimes E\right)^*.$$

We deduce an antilinear isomorphism

$$*_E : \Omega_X^{0,q} \otimes E \to \Omega_X^{n,n-q} \otimes E^*.$$

Remark 5.9 *The bundle $\Omega_X^{n,n-q} \otimes E^*$ is of course isomorphic to $\Omega_X^{0,n-q} \otimes \Omega_X^{n,0} \otimes E^*$. The bundle $\Omega_X^{n,0} = \bigwedge^n \Omega_X^{1,0}$ is a holomorphic vector bundle of rank 1, called the canonical bundle of X, and written K_X.*

If X is compact, let $(,)_{L^2}$ be the Hermitian metric on the space $A^{0,q}(E)$ of differential forms of type $(0, q)$ defined by

$$(\alpha, \beta)_{L^2} = \int_X (\alpha, \beta)\mathrm{Vol}.$$

Clearly, for $x \in X$ and $\alpha_x, \beta_x \in \Omega_{X,x}^{0,q} \otimes E_x$, we have

$$(\alpha_x, \beta_x)_x \mathrm{Vol} = \alpha_x \wedge *_E \beta_x,$$

where in the right-hand term, we take the exterior product on the forms and use the contraction between E and $K_X \otimes E^*$, with values in K_X. It follows immediately that for $\alpha,\ \beta \in A^{0,q}(E)$ we have

$$(\alpha, \beta)_{L^2} = \int_X \alpha \wedge *_E \beta.$$

Now consider the operator

$$\overline{\partial}_E^* : A^{0,q}(E) \to A^{0,q-1}(E)$$

defined by $\overline{\partial}_E^* = (-1)^q *_E^{-1} \circ \overline{\partial}_{K_X \otimes E^*} \circ *_E$.

Lemma 5.10 *The operator $\overline{\partial}_E^*$ is the formal adjoint of $\overline{\partial}_E$.*

Proof We want to show that

$$(\overline{\partial}_E \alpha, \beta)_{L^2} = (-1)^{d^0\beta} \big(\alpha, *_E^{-1} \circ \overline{\partial}_{K_X \otimes E^*} \circ *_E \beta\big)_{L^2},$$

i.e.

$$\int_X \overline{\partial}_E \alpha \wedge *_E \beta = (-1)^{d^0\beta} \int_X \alpha \wedge *_E *_E^{-1} \overline{\partial}_{K_X \otimes E^*} *_E \beta. \tag{5.3}$$

But we have $\int_X \overline{\partial}(\alpha \wedge *_E \beta) = 0$ and

$$\overline{\partial}(\alpha \wedge *_E \beta) = \overline{\partial}_E \alpha \wedge *_E \beta + (-1)^{d^0\alpha} \alpha \wedge \overline{\partial}_{K_X \otimes E^*} *_E \beta.$$

We thus obtain

$$\int_X \overline{\partial} \alpha \wedge *_E \beta = -\int_X (-1)^{d^0\alpha} \alpha \wedge \overline{\partial}_{K_X \otimes E^*} *_E \beta.$$

The equality (5.3) thus follows from the fact that $d^0\alpha + 1 = d^0\beta$. □

Remark 5.11 *In particular, we can take the bundle E to be one of the holomorphic bundles Ω_X^p equipped with its induced Hermitian metric. The operators $\overline{\partial}_E$ and $\overline{\partial}_E^*$ then only coincide with the operators $\overline{\partial}, \overline{\partial}^*$ (restricted to the forms of type (p, q)) up to a coefficient. For example, thanks to Leibniz' rule (lemma 2.28), we have $\overline{\partial}_E = (-1)^p \overline{\partial}$ on $A^{p,q}(X) = A^{0,q}(\Omega_X^p)$. Moreover, we have $\overline{\partial}^* = (-1)^p \frac{1}{2} \overline{\partial}_E^*$, where the coefficient 2 comes from the difference between the metrics used (cf. lemma 5.6).*

5.1.4 Laplacians

For a Riemannian manifold M, let Δ_d denote the operator $dd^* + d^*d$ which acts on the C^∞ differential forms of degree k for each k. This operator is called the Laplacian associated to d.

For a complex manifold X equipped with a Hermitian metric, we will write Δ_∂ and $\Delta_{\overline{\partial}}$ for the Laplacians associated to the operators ∂ and $\overline{\partial}$ respectively, i.e.

$$\Delta_\partial = \partial\partial^* + \partial^*\partial, \quad \Delta_{\overline{\partial}} = \overline{\partial}\,\overline{\partial}^* + \overline{\partial}^*\overline{\partial}.$$

Similarly, if E is a holomorphic vector bundle over X, where E and X are equipped with Hermitian metrics, we write Δ_E for the Laplacian associated to the operator $\overline{\partial}_E$, $\Delta_E = \overline{\partial}_E\overline{\partial}_E^* + \overline{\partial}_E^*\overline{\partial}_E$.

Lemma 5.12 *If X is compact, we have the equality*

$$(\alpha, \Delta_d\alpha)_{L^2} = (d\alpha, d\alpha)_{L^2} + (d^*\alpha, d^*\alpha)_{L^2}$$

and the analogous equalities for the other Laplacians introduced above.

Proof We have

$$(\alpha, \Delta_d\alpha)_{L^2} = (\alpha, dd^*\alpha)_{L^2} + (\alpha, d^*d\alpha)_{L^2},$$

and this is equal to $(d\alpha, d\alpha)_{L^2} + (d^*\alpha, d^*\alpha)_{L^2}$ by the adjunction property (5.2). □

Corollary 5.13 *On a compact manifold, we have* $\operatorname{Ker}\Delta_d = \operatorname{Ker} d \cap \operatorname{Ker} d^*$ *and the analogous equalities for the three other Laplacians.*

Proof If $\Delta_d\alpha = 0$ we have $(\alpha, \Delta_d\alpha)_{L^2} = (d\alpha, d\alpha)_{L^2} + (d^*\alpha, d^*\alpha)_{L^2} = 0$, and thus $(d\alpha, d\alpha)_{L^2} = 0$, $(d^*\alpha, d^*\alpha)_{L^2} = 0$. Thus $d\alpha = 0$, $d^*\alpha = 0$. The other inclusion is trivial. □

Definition 5.14 *A harmonic (or Δ_d-harmonic) form is a form which is annihilated by the Laplacian Δ_d, or equivalently, which is annihilated by d and d^*. Similarly, we can define $\Delta_{\overline{\partial}}$-harmonic forms or Δ_E-harmonic $(0, q)$-forms with coefficients in E.*

5.2 Elliptic differential operators

5.2.1 Symbols of differential operators

Let E and F be two (real or complex) C^∞ vector bundles over a manifold M. Let

$$P : \underline{C}^\infty(E) \to \underline{C}^\infty(F)$$

be an ($\mathbb{R}$ or $\mathbb{C}$)-linear morphism of sheaves.

Definition 5.15 *P is a differential operator of order k if, in the open sets U equipped with coordinates $x_1, \ldots, x_n$ and trivialisations*

$$E_{|U} \cong U \times \mathbb{R}^p, \quad F_{|U} \cong U \times \mathbb{R}^q,$$

we have $P((\alpha_1, \ldots, \alpha_p)) = (\beta_1, \ldots, \beta_q)$ with

$$\beta_i = \sum_{I,j} P_{I,i,j} \frac{\partial \alpha_j}{\partial x_I},$$

where the coefficients $P_{I,i,j}$ are C^∞, and zero for $|I| > k$ with at least one coefficient $P_{I,i,j}$ non-zero for $|I| = k$.

It is easily seen that this condition does not depend on the choice of coordinates and trivialisations.

Let P be a differential operator of order k, and in each open set U as above, let us define the matrix P^k_{ji} with coefficients in the space of differential operators by

$$P^k_{ij} = \sum_{|I|=k} P_{I,i,j} \frac{\partial}{\partial x_I}.$$

Applying the rule for differentiating compositions of functions and Leibniz' rule, we easily see that by a change of coordinates, the coefficients of the matrix P^k are transformed like the sections of the kth symmetric power of the tangent bundle T_U corresponding to them via

$$\frac{\partial}{\partial x_I} \mapsto \frac{\partial}{\partial x_{i_1}} \cdots \frac{\partial}{\partial x_{i_k}}.$$

Similarly, by a change of trivialisation of the bundles E and F, the matrix P^k_{ij} transforms like a section of $\mathrm{Hom}\,(E, F)$. In both cases, the point is that all the terms (differential operators) appearing in the new expression of P after a change of coordinates or trivialisations, using the derivatives of the Jacobian matrix or the derivatives of the transition matrix, are of order less than k.

Definition 5.16 *The section σ_P of* $\mathrm{Hom}(E, F) \otimes S^k T_X$ *given by the P_k in open sets of a trivialisation is called the symbol of the operator P.*

We can also view σ_P as giving, at each point $m \in M$, a homogeneous map $\sigma_{P,m}$ of degree k from $\Omega_{M,m}$ to $\mathrm{Hom}(E_m, F_m)$.

Definition 5.17 *We say that a differential operator is elliptic if for every $m \in M$ and $\alpha_m \neq 0$ in $\Omega_{M,m}$, the homomorphism $\sigma_{P,m}(\alpha) : E_m \to F_m$ is injective.*

5.2.2 Symbol of the Laplacian

Let (M, g) be a Riemannian manifold and let $\Delta = dd^* + d^*d$ be the Laplacian associated to g, acting on the differential forms. It is a differential operator of order 2, since d and d^* are differential operators of order 1.

Lemma 5.18 *The symbol σ_Δ of the Laplacian is described by*

$$\sigma_\Delta(\alpha)(\omega) = -||\alpha||^2\omega. \tag{5.4}$$

Here $||\alpha||^2$ is the function on M which takes value $||\alpha_x||^2 = (\alpha_x, \alpha_x)_x$ at the point x.

Proof It suffices to show this locally. Furthermore, if $m \in M$, the differential operator $d^* = \pm * d*$ is the sum of a differential operator of order 0 in whose expression the derivatives of the metric occur, and an operator of order 1, in which the metric appears only to the order 0. It follows immediately that the terms of order 2 in the expression of the Laplacian depend on the metric only to the order 0. Thus, it suffices to show (5.4) for the constant metric.

For the metric with constant coefficients, the forms $*dx_I$ have constant coefficients, and are thus annihilated by d. Let $\omega = \sum_I f_i dx_I$ be a differential form; we have $\Delta\omega = (-1)^q(d *^{-1} d * - *^{-1} d * d)\omega$, $q = d^0\omega$. Now,

$$d\omega = \sum_{i,I} \frac{\partial f_I}{\partial x_i} dx_i \wedge dx_I, \quad * d\omega = \sum_{i,I} \frac{\partial f_I}{\partial x_i} * (dx_i \wedge dx_I),$$

so

$$*^{-1} d * d\omega = \sum_{k,i,I} \frac{\partial^2 f_I}{\partial x_i \partial x_k} *^{-1} (dx_k \wedge *(dx_i \wedge dx_I)).$$

Similarly, we find that

$$d *^{-1} d * \omega = \sum_{k,i,I} \frac{\partial^2 f_I}{\partial x_i \partial x_k} dx_k \wedge *^{-1}(dx_i \wedge *dx_I).$$

We obtain

$$\Delta\omega = (-1)^q \left(\sum_{I,i,k} \frac{\partial^2 f_I}{\partial x_i \partial x_k} (*^{-1}(dx_k \wedge *(dx_i \wedge dx_I)) \right.$$
$$\left. - dx_k \wedge *^{-1}(dx_i \wedge *dx_I)) \right).$$

Suppose that the metric is the standard metric $\sum_i dx_i^2$. We easily check that $*^{-1}(dx_i \wedge *dx_I)$ is equal to $(-1)^{q+1}\mathrm{int}_{\frac{\partial}{\partial x_i}}(dx_I)$, where the interior product $\mathrm{int}_u(\alpha)$ for a tangent vector u and a differential k-form α is the $(k-1)$-form defined by

$$\mathrm{int}_u(\alpha)(v_1, \ldots, v_{k-1}) = \alpha(u, v_1, \ldots, v_{k-1}).$$

Now, the interior product by $\frac{\partial}{\partial x_i}$ anticommutes with the exterior product by dx_k when $i \neq k$, whereas for $k = i$ we have

$$\mathrm{int}_{\frac{\partial}{\partial x_i}} \circ (dx_i \wedge) + (dx_i \wedge) \circ \mathrm{int}_{\frac{\partial}{\partial x_i}} = \mathrm{Id}.$$

It follows immediately that for the standard metric, we have

$$\Delta\omega = -\sum_i \frac{\partial^2 f_I}{\partial x_i^2} dx_I,$$

which shows (5.4) since $-\sum_i (\frac{\partial}{\partial x_i})^2 \in S^2 T_{M,m}$ is exactly the degree 2 homogeneous map $\alpha \mapsto -||\alpha||^2$ on $\Omega_{M,m}$. □

We have similar results for the Laplacians Δ_∂, $\Delta_{\bar\partial}$, Δ_E introduced above. For example, we have the following lemma.

Lemma 5.19 *The symbols of Δ_∂ and $\Delta_{\bar\partial}$ are equal to*

$$\xi \mapsto \frac{-1}{2} ||\xi||^2 \,\mathrm{Id}$$

on Ω_X^k. The symbol of Δ_E is equal to

$$\xi \mapsto -||\xi||^2 \,\mathrm{Id}$$

on $\Omega^{0,q} \otimes E$.

Corollary 5.20 *The Laplacians Δ, Δ_∂, $\Delta_{\overline{\partial}}$, Δ_E are elliptic operators.*

5.2.3 The fundamental theorem

Let M be a manifold. Let E, F be two C^∞ bundles over M, and let

$$P : \underline{C}^\infty(E) \to \underline{C}^\infty(F)$$

be a differential operator. Suppose that E and F are equipped with metrics, and that M is compact, oriented and equipped with a volume form Vol. Then we can construct a formal adjoint P^* for P, which is a differential operator from $F \cong F^*$ to $E \cong E^*$ of the same order as P, and which satisfies the equality

$$(\alpha, P\beta)_{L^2} = (P^*\alpha, \beta)_{L^2}, \quad \forall \alpha \in C^\infty(F), \quad \beta \in C^\infty(E), \qquad (*)$$

where the L^2 metric is defined as in (5.1). Indeed, it suffices to construct such a P^* locally (in which case we require that the equality $(*)$ be satisfied by forms with compact support). By uniqueness, the operators defined locally glue together to form a global operator P^*.

Now, locally we may assume that E and F are trivial as vector bundles equipped with metrics. Moreover, by a theorem of Moser (1965), we may assume that the volume form is the Euclidean volume form. We are thus reduced to constructing a formal adjoint of $\phi \frac{\partial}{\partial x_I}$, where ϕ is a function of $x_1, \ldots, x_n$, acting on the functions of $\mathbb{R}^n$. But clearly $(-1)^{|I|} \frac{\partial}{\partial x_I} \circ \phi$ is such an adjoint, by repeated applications of Stokes' formula.

Remark 5.21 *The proof also shows that the symbol of P^* is equal to the adjoint of the symbol of P, i.e.*

$$\sigma_{P^*}(\alpha) = (\sigma_P(\alpha))^*.$$

In particular, if E and F are of equal rank, then P is elliptic if and only if P^* is.

The essential theorem on elliptic differential operators, which we will use without proof, is the following (see Demailly 1996).

Theorem 5.22 *Let $P : E \to F$ be an elliptic differential operator on a compact manifold. Assume that E and F are of the same rank, and are equipped with metrics. Then* $\operatorname{Ker} P \subset C^\infty(E)$ *is finite-dimensional,* $P(C^\infty(E)) \subset C^\infty(F)$ *is* *closed and of finite codimension, and we have a decomposition as an orthogonal direct sum (for the L^2 metric)*

$$\mathcal{C}^\infty(E) = \operatorname{Ker} P \oplus P^*(\mathcal{C}^\infty(F)).$$

Note that by the adjunction property, Ker P and Im P^* are certainly orthogonal. The main step in the proof of this theorem consists in showing that an equality in the sense of distributions $P^*\alpha = \beta$, where β is $\mathcal{C}^\infty$, implies that α is $\mathcal{C}^\infty$.

5.3 Applications

5.3.1 Cohomology and harmonic forms

Let us apply theorem 5.22 to the Laplacian Δ acting on the differential forms of degree k of a compact oriented Riemannian manifold (X, g). We obtain the following.

Theorem 5.23 *Let $\mathcal{H}^k$ be the vector space of Δ-harmonic differential forms of degree k. Then the natural map*

$$\mathcal{H}^k \to H^k(X, \mathbb{R}) \tag{5.5}$$

which to α associates the class of the closed form α in $H^k_{\mathrm{DR}}(X, \mathbb{R}) = H^k(X, \mathbb{R})$ is an isomorphism. Similarly, the natural map from the space of complex-valued harmonic forms to the cohomology group $H^k(X, \mathbb{C})$ in an isomorphism.

Proof The operator Δ is self-adjoint and elliptic. Thus, theorem 5.22 gives the decomposition

$$A^k(X) = \mathcal{H}^k \oplus \Delta(A^k(X)).$$

Let $\beta \in A^k(X)$ be a closed form, and write $\beta = \alpha + \Delta\gamma$ with α harmonic. Thus, we have $\beta = \alpha + dd^*\gamma + d^*d\gamma$. As β, α and $dd^*\gamma$ are closed, we deduce that the forme $d^*d\gamma$ is annihilated by d and lies in the image of d^*. So it is zero, and $\beta = \alpha$ modulo an exact form. The map (5.5) is thus surjective.

Now, let β be a harmonic form, and assume that β is exact. Then β is annihilated by d^* by corollary 5.13, and lies in the image of d. So β is zero. The map (5.5) is thus injective. □

Using the Dolbeault cohomology of a holomorphic vector bundle E over a complex manifold X, we have the analogous result for the cohomology groups $H^q(X, \mathcal{E})$ with values in the sheaf $\mathcal{E}$ of holomorphic sections of E, which we identified in corollary 4.38 with the groups

$$\frac{\operatorname{Ker}(\overline{\partial} : A^{0,q}(E) \to A^{0,q+1}(E))}{\operatorname{Im}(\overline{\partial} : A^{0,q-1}(E) \to A^{0,q}(E))}.$$

Theorem 5.24 *Let E be a Hermitian holomorphic vector bundle over a complex compact manifold X equipped with a Hermitian metric. Then if $\mathcal{H}^{0,q}(E)$ is the space of harmonic forms, i.e. forms annihilated by Δ_E, of type $(0, q)$ with coefficients in E, the natural map*

$$\mathcal{H}^{0,q}(E) \to H^q(X, \mathcal{E})$$

which to a harmonic form α associates the class of the $\overline{\partial}$-closed form α is an isomorphism.

A first consequence of these theorems is the following corollary.

Corollary 5.25

(a) If X is a compact manifold, then the cohomology groups $H^q(X, \mathbb{R})$ are finite-dimensional.

(b) If X is a compact complex manifold, the cohomology groups $H^q(X, E)$ are finite-dimensional for every holomorphic vector bundle E over X.

Indeed, introducing metrics, these cohomology groups are represented by spaces of harmonic forms, which are finite-dimensional by theorem 5.22. □

Remark 5.26 *The first statement has already been obtained by a much simpler argument (cf. remark 4.46). There is no easy proof of the second statement.*

5.3.2 Duality theorems

Cup-products

Let $\mathcal{F}$ and $\mathcal{G}$ be two sheaves of abelian groups over a topological space X. We will define a natural pairing

$$H^p(X, \mathcal{F}) \otimes H^q(X, \mathcal{G}) \to H^{p+q}(X, \mathcal{F} \otimes \mathcal{G}). \tag{5.6}$$

If $\mathcal{A}$ is a sheaf of rings, the map given by the product

$$\mathcal{A} \otimes \mathcal{A} \to \mathcal{A}$$

induces homomorphisms $H^l(\mathcal{A} \otimes \mathcal{A}) \to H^l(\mathcal{A})$ which, composed with the preceding maps, give the cup-product

$$H^p(X, \mathcal{A}) \otimes H^q(X, \mathcal{A}) \to H^{p+q}(X, \mathcal{A}),$$

which we will denote by $(\alpha, \beta) \to \alpha \cup \beta$.

If $\pi : X \to Y$ is continuous, and $\mathcal{F}$ is a sheaf over Y, then its pullback $\pi^{-1}\mathcal{F}$ is the sheaf over X defined by

$$\pi^{-1}\mathcal{F}(U) = \varinjlim_{\pi(U)\subset V} \mathcal{F}(V).$$

Similarly, if $\mathcal{F}$ is a sheaf over X, then $\pi_*\mathcal{F}$ is the sheaf over Y defined by

$$\pi_*\mathcal{F}(V) = \mathcal{F}(\pi^{-1}(V)).$$

Let $\mathcal{K} = \mathrm{pr}_1^{-1}\mathcal{F} \otimes \mathrm{pr}_2^{-1}\mathcal{G}$ on $X \times X$. If $\Delta_X \overset{j}{\hookrightarrow} X \times X$ is the diagonal map, we have a natural map of sheaves over $X \times X$:

$$\mathcal{K} \to j_*(\mathcal{F} \otimes \mathcal{G}). \tag{5.7}$$

The pairing (5.6) is obtained by composing the map induced by (5.7) in cohomology (noting that $H^{\cdot}(X, \mathcal{F} \otimes \mathcal{G}) = H^{\cdot}(X \times X, j_*(\mathcal{F} \otimes \mathcal{G})))$ with a natural map which we will now define:

$$H^p(X, \mathcal{F}) \otimes H^q(X, \mathcal{G}) \to H^{p+q}\big(X \times X, \mathrm{pr}_1^{-1}\mathcal{F} \otimes \mathrm{pr}_2^{-1}\mathcal{G}\big). \tag{5.8}$$

More generally, we will construct a map

$$H^p(X, \mathcal{F}) \otimes H^q(Y, \mathcal{G}) \to H^{p+q}\big(X \times Y, \mathrm{pr}_1^{-1}\mathcal{F} \otimes \mathrm{pr}_2^{-1}\mathcal{G}\big), \tag{5.9}$$

where $\mathcal{F}$ is a sheaf over X and $\mathcal{G}$ a sheaf over Y. Let us first introduce the following notion.

Definition 5.27 *The tensor product of two complexes $(M^{\cdot}, d_M)$, $(N^{\cdot}, d_N)$ is the simple complex $((M \otimes N)^{\cdot}, d)$ associated to the double complex*

$$(M \otimes N)^{p,q} := M^p \otimes N^q, \quad D_1 = d_M \otimes \mathrm{Id}, \quad D_2 = \mathrm{Id} \otimes d_N.$$

Similarly, we can define the tensor product of two complexes of sheaves of abelian groups over a topological space X.

Clearly, if $m \in M^p$ is d_M-closed and $n \in N^q$ is d_N-closed, then $m \otimes n \in (M \otimes N)^{p+q}$ is d-closed. Moreover, if m or n is exact, then this also holds for $m \otimes n$. We thus have a natural map

$$H^p(M^{\cdot}) \otimes H^q(N^{\cdot}) \to H^{p+q}((M \otimes N)^{\cdot}).$$

Now let $I^{\cdot}$, $J^{\cdot}$ be the Godement resolutions of $\mathcal{F}$, $\mathcal{G}$ respectively. We can show that the complex $\mathrm{pr}_1^{-1}(I^{\cdot}) \otimes \mathrm{pr}_2^{-1}(J^{\cdot})$ is a resolution of $\mathrm{pr}_1^{-1}\mathcal{F} \otimes \mathrm{pr}_2^{-1}\mathcal{G}$. Applying proposition 4.27, we thus obtain a natural map

$$H^p\big(\Gamma\big(X \times Y, \mathrm{pr}_1^{-1}(I^{\cdot}) \otimes \mathrm{pr}_2^{-1}(J^{\cdot})\big)\big) \to H^p\big(X \times Y, \mathrm{pr}_1^{-1}\mathcal{F} \otimes \mathrm{pr}_2^{-1}\mathcal{G}\big).$$

Moreover, the morphism of complexes

$$\Gamma(I^\cdot)\otimes\Gamma(J^\cdot)\to\Gamma\big(\mathrm{pr}_1^{-1}(I^\cdot)\otimes\mathrm{pr}_2^{-1}(J^\cdot)\big)$$

gives a map

$$\begin{aligned}H^p(X,\mathcal{F})\otimes H^q(Y,\mathcal{G})&\cong H^p(\Gamma(X,I^\cdot))\otimes H^q(\Gamma(Y,J^\cdot))\\&\to H^{p+q}\big(\Gamma\big(X\times Y,\mathrm{pr}_1^{-1}(I^\cdot)\otimes\mathrm{pr}_2^{-1}(J^\cdot)\big)\big)\end{aligned}$$

which, composed with the preceding one, gives the pairing (5.9).

Remark 5.28 *The pairing (5.6) was defined for the tensor product over* $\mathbb{Z}$. *If we consider sheaves of A-modules, where A is a commutative ring with unit, their cohomology groups naturally have the structure of an A-module, and via the natural map*

$$\mathcal{F}\otimes_{\mathbb{Z}}\mathcal{G}\to\mathcal{F}\otimes_A\mathcal{G},$$

the pairing (5.6) then gives a pairing

$$H^p(X,\mathcal{F})\otimes_A H^q(X,\mathcal{G})\to H^{p+q}(X,\mathcal{F}\otimes_A\mathcal{G}). \tag{5.10}$$

Indeed, this follows immediately from the definition of (5.6): it suffices to replace the tensor product over $\mathbb{Z}$ *by the tensor product over A everywhere to obtain (5.10).*

Now let A be a commutative ring with unit, and let $\mathcal{F}$, $\mathcal{G}$ be two sheaves of A-modules over X.

Suppose we have acyclic resolutions $\mathcal{F}^\cdot$, $\mathcal{G}^\cdot$ and $(\mathcal{F}\otimes_A\mathcal{G})^\cdot$ of $\mathcal{F}$, $\mathcal{G}$ and $\mathcal{F}\otimes_A\mathcal{G}$ respectively, by complexes of sheaves of A-modules. Suppose furthermore that there exists a morphism of complexes

$$\mathcal{F}^\cdot\otimes_A\mathcal{G}^\cdot\to(\mathcal{F}\otimes_A\mathcal{G})^\cdot \tag{5.11}$$

which makes the following diagram commute:

$$\begin{array}{ccc}\mathcal{F}\otimes_A\mathcal{G} & \to & \mathcal{F}\otimes_A\mathcal{G}\\ \downarrow & & \downarrow\\ \mathcal{F}^\cdot\otimes_A\mathcal{G}^\cdot & \to & (\mathcal{F}\otimes_A\mathcal{G})^\cdot.\end{array} \tag{5.12}$$

Taking the global sections in (5.11), we obtain a morphism of complexes of A-modules

$$\Gamma(X,\mathcal{F}^\cdot)\otimes_A\Gamma(X,\mathcal{G}^\cdot)\to\Gamma(X,(\mathcal{F}\otimes_A\mathcal{G})^\cdot),$$

and thus we obtain pairings

$$H^p(\Gamma(X, \mathcal{F}^\cdot)) \otimes_A H^q(\Gamma(X, \mathcal{G}^\cdot)) \to H^{p+q}(\Gamma(X, (\mathcal{F} \otimes_A \mathcal{G})^\cdot)),$$

i.e., by proposition 4.32,

$$H^p(X, \mathcal{F}) \otimes_A H^q(X, \mathcal{G}) \to H^{p+q}(X, \mathcal{F} \otimes_A \mathcal{G}). \tag{5.13}$$

We then have the following result.

Theorem 5.29 *The pairing (5.13) coincides with the pairing (5.10).*

Proof This theorem is proved using the notion of the hypercohomology of a complex of sheaves, which will be defined at the end of section 8.1.2. It then suffices to construct the cup-product (5.10) for the hypercohomology and to use the commutative diagram (5.12) to conclude. □

Duality

If X is an n-dimensional connected compact oriented manifold, integration gives an isomorphism $H^n(X, \mathbb{R}) \cong \mathbb{R}$, from which we deduce a pairing

$$H^p(X, \mathbb{R}) \otimes H^{n-p}(X, \mathbb{R}) \to \mathbb{R}. \tag{5.14}$$

It follows from theorem 5.29 that this pairing between $H^p(X, \mathbb{R}) \cong H^p_{\mathrm{DR}}(X, \mathbb{R})$ and $H^{n-p}(X, \mathbb{R}) \cong H^{n-p}_{\mathrm{DR}}(X, \mathbb{R})$ is given by

$$(\alpha, \beta) \mapsto \int_M \alpha \wedge \beta, \tag{5.15}$$

where α, β are closed forms of degree p and $n - p$ respectively. (Stokes' formula (1.5) shows that $\int_M \alpha \wedge \beta$ depends only on the class of α and β modulo the exact forms.)

Indeed, the de Rham resolution $\mathbb{R} \to \mathcal{A}^\cdot_X$ is a Γ-acyclic resolution. Moreover, the exterior product of the differential forms gives a morphism of complexes

$$\mathcal{A}^\cdot_X \otimes_{\mathbb{R}} \mathcal{A}^\cdot_X \to \mathcal{A}^\cdot_X$$

which extends the morphism given by the product

$$\mathbb{R} \otimes_{\mathbb{R}} \mathbb{R} \to \mathbb{R}.$$

Theorem 5.29 then says that if α and β are closed, the cup-product of the classes of α and β is represented by the form $\alpha \wedge \beta$.

Theorem 5.23 now allows us to prove Poincaré's duality theorem for the cohomology with real coefficients.

Theorem 5.30 *The pairing defined above between $H^p(X,\mathbb{R})$ and $H^{n-p}(X,\mathbb{R})$ is a perfect pairing, i.e. it induces an isomorphism*

$$H^p(X,\mathbb{R}) \cong H^{n-p}(X,\mathbb{R})^*.$$

Remark 5.31 *This pairing was already defined on the integral cohomology, so theorem 5.30 implies that the pairing is also perfect for the cohomology with rational coefficients. In fact, Poincaré duality (Spanier 1966) is a much stronger statement, which gives a canonical isomorphism (depending on the orientation)*

$$H_p(X,\mathbb{Z}) \cong H^{n-p}(X,\mathbb{Z}). \tag{5.16}$$

Proof Choosing a metric on X, we use the identifications

$$\mathcal{H}^p(X) \cong H^p(X,\mathbb{R}) \quad \text{and} \quad \mathcal{H}^{n-p}(X) \cong H^{n-p}(X,\mathbb{R}).$$

Note that the operator $*$ commutes with Δ, since for α of degree p, we have

$$d^*\alpha = (-1)^p *^{-1} d * \alpha = (-1)^{(p-1)n+1} * d * \alpha.$$

Thus, $\Delta\alpha = (-1)^{n(p-1)+1} d * d * \alpha + (-1)^{np+1} * d * d\alpha$, and for β of degree $n-p$, we have

$$\Delta * \beta = (-1)^{n(p-1)+1} d * d * *\beta + (-1)^{np+1} * d * d * \beta,$$

while

$$*\Delta\beta = (-1)^{n(n-p-1)+1} * d * d * \beta + (-1)^{n(n-p)+1} * *d * d\beta,$$

and the equality follows from $** = (-1)^{k(n-k)}\,\mathrm{Id}$ on the forms of degree k.

It follows that if α is harmonic, then $*\alpha$ is also harmonic, and as $** = (-1)^{p(n-p)}\,\mathrm{Id}$ on the forms of degree p, the map

$$* : \mathcal{H}^p(X) \to \mathcal{H}^{n-p}(X)$$

is an isomorphism.

But furthermore, if $\alpha \in \mathcal{H}^p(X)$, then

$$\int_X \alpha \wedge *\alpha = ||\alpha||^2_{L^2},$$

which immediately implies that the pairing (5.15) is non-degenerate. □

Similarly, if E is a holomorphic vector bundle over a complex manifold X, theorem 5.24 allows us to prove Serre's duality theorem: we have a natural pairing

$$\mathcal{E} \otimes \mathcal{E}^* \otimes K_X \to K_X.$$

Furthermore, integrating the forms gives an isomorphism $H^n(X, K_X) \cong \mathbb{C}$. Thus we have a pairing

$$H^q(X, \mathcal{E}) \otimes H^{n-q}(X, \mathcal{E}^* \otimes K_X) \to H^n(X, K_X) \cong \mathbb{C}. \tag{5.17}$$

By theorem 5.29, this pairing can be computed in the Dolbeault cohomology as follows: if $\alpha \in A^{0,q}(E)$, $\beta \in A^{0,n-q}(K_X \otimes E^*)$, we can define the pairing

$$\langle \alpha, \beta \rangle = \int_X \alpha_x \wedge \beta_x,$$

where the form $\alpha_x \wedge \beta_x$ of degree $2n$ is obtained by taking the exterior product on the forms, contracting E and $K_X \otimes E^*$ and noting that $A^{0,n}(K_X) = A^{2n}(X)$. One checks immediately by Stokes' formula that

$$\langle \alpha, \overline{\partial}\beta \rangle = (-1)^{q+1} \langle \overline{\partial}\alpha, \beta \rangle,$$

so that if α and β are $\overline{\partial}$-closed, $\langle \alpha, \beta \rangle$ depends only on their cohomology classes: it is the desired pairing (5.17). Indeed, we note that the pairing

$$\mathcal{E} \otimes \mathcal{E}^* \otimes K_X \to K_X$$

extends, thanks to the exterior product of the forms and the contraction, to a morphism of complexes

$$\mathcal{A}^{0,\cdot}(E) \otimes \mathcal{A}^{0,\cdot}(E^* \otimes K_X) \to \mathcal{A}^{0,\cdot}(K_X),$$

and as the $\mathcal{A}^{0,\cdot}(F)$ give acyclic resolutions of the corresponding sheaves $\mathcal{F}$, we can apply theorem 5.29.

Theorem 5.32 (Serre) *The pairing (5.17) between*

$$H^q(X, \mathcal{E}) \quad \textit{and} \quad H^{n-q}(X, \mathcal{E}^* \otimes K_X)$$

is perfect.

Proof The proof is the same as above. We equip X and E with Hermitian metrics, which allow us to represent the classes by Δ_E-harmonic forms. The operator $*_E$ commutes with Δ_E; thus it gives an isomorphism of the space of Δ_E-harmonic forms $\mathcal{H}^q(X, E)$ with the space of $\Delta_{E^*\otimes K_X}$-harmonic forms $\mathcal{H}^{n-q}(X, E^* \otimes K_X)$, which satisfies $\langle \alpha, *_E\alpha \rangle = ||\alpha||^2$. □

Exercises

1. Let us consider $\mathbb{R}^n$, endowed with the Euclidean metric.
   (a) Compute the Laplacian on the differential forms of $\mathbb{R}^n$. Deduce from this that a differential form of degree k

$$\alpha = \sum_I \alpha_I dx_I,$$

   where $I = \{i_1 < \cdots < i_k\}$, is harmonic if and only if its coefficients α_I are harmonic functions.
   Now consider a real torus

$$T = \mathbb{R}^n / \Gamma,$$

   where Γ is a lattice in $\mathbb{R}^n$. We put on T the metric which is induced by the Euclidean metric by passing to the quotient.
   (b) Show that the only harmonic functions on T are the constants. Deduce from this that the harmonic forms on T are induced by the differential forms with constant coefficients on $\mathbb{R}^n$.
   (c) Show that

$$H^k(T, \mathbb{R}) \cong \bigwedge^k H^1(T, \mathbb{R}) = \bigwedge^k (\mathbb{R}^n)^*.$$

2. *An application of Serre duality.* Let X be a connected compact complex manifold of dimension n and let L be a holomorphic line bundle on X. We assume that there exists an integer $N > 0$, such that

$$H^0(X, L^{\otimes N}) \neq 0.$$

   Show that if $H^n(X, L \otimes K_X) \neq 0$, the line bundle L is trivial.

6

The Case of Kähler Manifolds

When X is a complex manifold equipped with a Hermitian metric, the three Laplacians Δ_d, Δ_∂, $\Delta_{\overline{\partial}}$ all act on the complex differential forms of X. Here, the Laplacian $\Delta_{\overline{\partial}}$ acts on each space $A^{p,q}(X)$, which can be seen as the space of $(0, q)$-forms with values in the vector bundle Ω_X^p of holomorphic p-differential forms.

If X is compact, the results of the preceding chapter show that the Δ_d-harmonic k-forms are in bijection with the cohomology classes of degree k, and that the $\Delta_{\overline{\partial}}$-harmonic (p, q)-forms are in bijection with the cohomology classes $\alpha \in H^q(X, \Omega_X^p)$.

Theorem 6.1 *If the metric on X is Kähler, then we have the equalities*

$$\Delta_d = 2\Delta_\partial = 2\Delta_{\overline{\partial}}. \tag{6.1}$$

As the operator $\Delta_{\overline{\partial}}$ preserves the decomposition of the forms into types, the same holds for Δ_d, and we deduce the following result.

Corollary 6.2 *Under the same hypothesis, if α is a k-harmonic form, then its components of type (p, q) are also harmonic.*

The space $\mathcal{H}^k$ is thus the direct sum of the spaces $\mathcal{H}^{p,q}$ of harmonic forms of type (p, q). One of the major results in this book follows from this, namely the Hodge decomposition

$$H^k(X, \mathbb{C}) = \bigoplus\nolimits_{p+q=k} H^{p,q}(X),$$

where $H^{p,q}(X)$ is the set of classes representable by a closed form of type (p, q). We have also the isomorphisms given by the representation by harmonic forms and the equality (6.1):

$$H^{p,q}(X) \cong \mathcal{H}^{p,q} \cong H^q(X, \Omega_X^p).$$

In chapter 8, we will see that the composed isomorphism

$$H^{p,q}(X) \cong H^q(X, \Omega_X^p)$$

is in fact independent of the choice of Kähler metric.

The equality (6.1) is obtained using Kähler identities. These identities are commutation relations between the adjoint Λ of the operator L of exterior product with the Kähler form, and the operators ∂, $\overline{\partial}$ and their formal adjoints. For example, we have the relation

$$[\Lambda, \overline{\partial}] = -i\partial^*.$$

Other important but much easier relations are the commutation relations between the operators L and Λ. For $n = \dim X$, we have

$$[L, \Lambda] = (k - n)\mathrm{Id} \;\text{ on }\; A_X^k. \tag{6.2}$$

This allows us to prove the hard Lefschetz theorem and the Lefschetz decomposition on forms.

Theorem 6.3 *For $k \leq n$ the operator*

$$L^{n-k} : A_X^k \to A_X^{2n-k}$$

is an isomorphism, and writing $A^r_{X\,\mathrm{prim}} := \mathrm{Ker}\, L^{n-r+1} \subset A_X^r$ for $r \leq n$, we have the decomposition

$$A_X^k \cong \bigoplus_{k-2r\geq 0} L^{k-2r} A_X^{k-2r}{}_{\mathrm{prim}}.$$

These results in Hermitian geometry can be translated into analogous statements on the cohomology of a compact Kähler manifold, thanks to the fact that L commutes with the Laplacian. Applying the preceding statements to harmonic forms, we obtain the hard Lefschetz theorem and Lefschetz decomposition given in the following statement.

Theorem 6.4 *For $k \leq n$ the operator*

$$L^{n-k} : H^k(X, \mathbb{C}) \to H^{2n-k}(X, \mathbb{C})$$

is an isomorphism, and writing $H^r(X,\mathbb{C})_{\text{prim}} := \operatorname{Ker} L^{n-r+1} \subset H^r(X,\mathbb{C})$ *for* $r \le n$*, we have the decomposition*

$$H^k(X,\mathbb{C}) \cong \bigoplus_{k-2r\ge 0} L^r H^{k-2r}(X,\mathbb{C})_{\text{prim}}.$$

6.1 The Hodge decomposition

6.1.1 Kähler identities

Let X be a Kähler manifold with Kähler form ω. The exterior product with ω defines an operator (which is of order 0, i.e. C^∞-linear)

$$L : \mathcal{A}_X^k \to \mathcal{A}_X^{k+2}$$

called the Lefschetz operator.

We write

$$\Lambda : \mathcal{A}_X^k \to \mathcal{A}_X^{k-2}$$

for its adjoint relative to the metric $(,)_x$ induced by ω on each $\Omega_{X,x}$. For all $x \in X$, we have $(L\alpha, \beta)_x = (\alpha, \Lambda\beta)_x$. As we have

$$(L\alpha, \beta)\text{Vol} = L\alpha \wedge *\beta = \omega \wedge \alpha \wedge *\beta = \alpha \wedge \omega \wedge *\beta,$$

we obtain $\Lambda\beta = *^{-1}L* = (-1)^k(*L*)$.

The following relation between the operators ∂, $\overline{\partial}$, their formal adjoints ∂^*, $\overline{\partial}^*$ relative to the metric g induced by ω, and the operator Λ, is at the very root of the Hodge decomposition.

Proposition 6.5 *We have the identities*

$$[\Lambda, \overline{\partial}] = -i\partial^*, \quad [\Lambda, \partial] = i\overline{\partial}^*. \tag{6.3}$$

Proof Note that in the local expressions of the operators under consideration, only the coefficients of the metric to the first order occur. Indeed, L and Λ use the metric only to the order 0, and the formula $\partial^* = -*\overline{\partial}*$ shows that the expression of ∂^* uses only the metric and its first derivatives.

Using proposition 3.14, we deduce that it suffices to prove the identities (6.3) for the Kähler metric with constant coefficients on $\mathbb{C}^n$.

Note, moreover, that the two equalities are equivalent by passage to the complex conjugate. Indeed, we have $\overline{\partial}\alpha = \overline{\partial(\overline{\alpha})}$, and thus $\overline{\partial}^*\alpha = \overline{\partial^*(\overline{\alpha})}$, whereas

$$[\Lambda, \partial](\alpha) = \overline{[\Lambda, \overline{\partial}](\overline{\alpha})}$$

since Λ is real.

Moreover, note that $[\Lambda, \overline{\partial}]$ and $-i\partial^*$ are differential operators of order 1 which, in the case of the flat metric, annihilate the forms with constant coefficients. To show that they are equal, it thus suffices to show that they have the same symbol. The symbol of the differential operators is compatible with composition. Moreover, the symbol of an operator $E \to F$ of order 0 is equal to itself, considered as a section of $\mathrm{Hom}(S^0\Omega_X, \mathrm{Hom}\,(E, F))$. Thus, the symbol of $[\Lambda, \overline{\partial}]$ is equal to $[\Lambda, \sigma_{\overline{\partial}}]$, and the symbol of $-i\partial^* = i * \overline{\partial} *$ is equal to $i * \sigma_{\overline{\partial}} *$.

Finally, the symbol of $\overline{\partial} : \mathcal{A}_X^{p,q} \to \mathcal{A}_X^{p,q+1}$ is the map

$$\eta \mapsto \eta^{0,1} \wedge : \Omega_X^{p,q} \to \Omega_X^{p,q+1}.$$

Indeed,

$$\overline{\partial}\left(\sum_{I,J} f_{I,J} dz_I \wedge d\overline{z}_J\right) = \sum_{i,I,J} \frac{\partial f_{I,J}}{\partial \overline{z}_i} d\overline{z}_i \wedge dz_I \wedge d\overline{z}_J,$$

so that the corresponding section of $T_X \otimes \mathrm{Hom}\left(\Omega_X^{p,q}, \Omega_X^{p,q+1}\right)$ is equal to $\sum_i \frac{\partial}{\partial \overline{z}_i} \otimes (d\overline{z}_i \wedge)$.

It remains to show the following. □

Lemma 6.6 *Let η be a section of the bundle $\Omega_X^{0,1}$. We have the equality*

$$[\Lambda, (\eta\wedge)] = i * (\eta\wedge) * \in \mathrm{Hom}\left(\Omega_X^{p,q}, \Omega_X^{p-1,q}\right).$$

Proof We may assume that $\eta = d\overline{z}_1$. First, we have

$$* \, (d\overline{z}_1\wedge) * = 2\,\mathrm{int}\left(\frac{\partial}{\partial z_1}\right). \tag{6.4}$$

Indeed, we have $d\overline{z}_1 = dx_1 - i dy_1$, and $*^{-1}(dx_1\wedge)* = (-1)^{q+1}\,\mathrm{int}\left(\frac{\partial}{\partial x_1}\right)$ on the forms of degree q. Similarly, $*^{-1}(dy_1\wedge)* = (-1)^{q+1}\,\mathrm{int}\left(\frac{\partial}{\partial y_1}\right)$. As $*^2 = (-1)^{d^0}$, we find $*(d\overline{z}_1\wedge)* = \mathrm{int}\left(\frac{\partial}{\partial x_1} - i\frac{\partial}{\partial y_1}\right) = 2\,\mathrm{int}\left(\frac{\partial}{\partial z_1}\right)$.

Furthermore,

$$[\Lambda, (\eta\wedge)] = [*^{-1}L*, (\eta\wedge)] = *^{-1}L * \circ (\eta\wedge) - (\eta\wedge) \circ *^{-1}L * .$$

Now, by (6.4), we have $*(d\overline{z}_1\wedge) = 2\,\mathrm{int}\left(\frac{\partial}{\partial z_1}\right) *^{-1}$, and moreover, we have the relation

$$L \circ \mathrm{int}\left(\frac{\partial}{\partial z_1}\right) = \mathrm{int}\left(\frac{\partial}{\partial z_1}\right) \circ L - \frac{i}{2}(d\overline{z}_1\wedge),$$

since $\omega = \frac{i}{2}\sum_i dz_i \wedge d\overline{z}_i$. Thus,

$$L \circ * \circ (d\overline{z}_1 \wedge) = 2\,\mathrm{int}\left(\frac{\partial}{\partial z_1}\right) \circ L \circ *^{-1} - i(d\overline{z}_1\wedge) \circ *^{-1}$$
$$= *(d\overline{z}_1\wedge) * L *^{-1} - i(d\overline{z}_1\wedge) *^{-1}.$$

Now, $*L *^{-1} = *^{-1} L*$, because as the dimension is even, we have $*^{-1} = (-1)^{d^0} *$, and L preserves the parity of the degree. Thus, we find

$$[\Lambda, d\overline{z}_1\wedge] = *^{-1}L * (d\overline{z}_1\wedge) - (d\overline{z}_1\wedge) *^{-1} L *$$
$$= -i *^{-1} (d\overline{z}_1\wedge) *^{-1}.$$

As $*^{-1} = (-1)^{d^0} *$, the right-hand term is equal to $i * (d\overline{z}_1\wedge)*$. The lemma is thus proved, which also concludes the proof of proposition 6.5. □

6.1.2 Comparison of the Laplacians

Proposition 6.5 now gives the following result.

Theorem 6.7 *Let (X, ω) be a Kähler manifold, and let Δ_d, Δ_∂, $\Delta_{\overline{\partial}}$ be the Laplacians associated respectively to the operators d, ∂, $\overline{\partial}$. Then we have the relations*

$$\Delta_\partial = \Delta_{\overline{\partial}} = \frac{1}{2}\Delta_d.$$

Proof We have

$$\Delta_d = (\partial + \overline{\partial})(\partial^* + \overline{\partial}^*) + (\partial^* + \overline{\partial}^*)(\partial + \overline{\partial}).$$

By proposition 6.5, this is equal to

$$(\partial + \overline{\partial})(\partial^* - i[\Lambda, \partial]) + (\partial^* - i[\Lambda, \partial])(\partial + \overline{\partial})$$
$$= \partial\partial^* + \overline{\partial}\partial^* + i\overline{\partial}\partial\Lambda - i\overline{\partial}\Lambda\partial + \partial^*\partial + \partial^*\overline{\partial} - i\Lambda\partial\overline{\partial} + i\partial\Lambda\overline{\partial}.$$

Writing $\partial^* = i[\Lambda, \overline{\partial}]$, we obtain

$$\partial^*\overline{\partial} = -i\overline{\partial}\Lambda\overline{\partial} = -\overline{\partial}\partial^*.$$

Thus,

$$\Delta_d = \Delta_\partial + i\partial[\Lambda, \overline{\partial}] + i[\Lambda, \overline{\partial}]\partial = 2\Delta_\partial.$$

The other equality is proved similarly. □

Corollary 6.8 *If X is Kähler, the Laplacian Δ_d is bihomogeneous, i.e.*

$$\Delta_d(A^{p,q}(X)) \subset A^{p,q}(X).$$

Proof Δ_∂ is clearly bihomogeneous. □

Corollary 6.9 *If $\alpha \in A^k(X)$ is harmonic, its components $\alpha^{p,q}$ are harmonic.*

Proof $\Delta_d\alpha = 0 = \sum_{p+q=k} \Delta_d\alpha^{p,q}$ with $\Delta_d\alpha^{p,q}$ of type (p, q). □

Corollary 6.10 *We have the decomposition as a direct sum*

$$\mathcal{H}^k(X) = \bigoplus_{p+q=k} \mathcal{H}^{p,q},$$

where $\mathcal{H}^{p,q}$ is the set of forms of type (p, q) which are harmonic for Δ_d. By theorem 6.7, this is also the set of forms of type (p, q) which are harmonic for $\Delta_{\bar{\partial}}$.

6.1.3 Other applications

We showed (theorem 5.23) that if X is compact, then we have an isomorphism $\mathcal{H}^k(X) \cong H^k(X, \mathbb{C})$, where $\mathcal{H}^k(X)$ is the set of complex valued harmonic forms for the Laplacian associated to any metric on X. When the metric is Kähler, then by corollary 6.10 we have the decomposition of harmonic forms into harmonic forms of type (p, q). Thus, we have an induced decomposition

$$H^k(X, \mathbb{C}) = \bigoplus_{p+q=k} H^{p,q}.$$

This decomposition is called the Hodge decomposition of the cohomology of a compact Kähler manifold.

Proposition 6.11 *This decomposition does not depend on the choice of Kähler metric.*

Proof Let $K^{p,q} \subset H^k(X, \mathbb{C})$ be the subspace consisting of the (de Rham) cohomology classes which are representable by a closed form of type (p, q). Obviously, we have $H^{p,q} \subset K^{p,q}$. We will show the inverse inclusion: let ω be a closed form of type (p, q). In a unique way, we can write $\omega = \alpha + \Delta\beta$ with α harmonic. Taking the components of type (p, q) and recalling that Δ is bihomogeneous, we obtain $\omega = \alpha^{p,q} + \Delta\beta^{p,q}$, where $\alpha^{p,q}$ is harmonic. But then

$\Delta\beta^{p,q} = dd^*\beta^{p,q} + d^*d\beta^{p,q}$ must be closed, which implies that $d^*d\beta^{p,q} = 0$ and thus that $\omega = \alpha^{p,q} + dd^*\beta^{p,q}$. Therefore, ω and $\alpha^{p,q}$ are of the same class, and $[\omega] \in H^{p,q}(X)$. Thus we have $K^{p,q} = H^{p,q}$, and as $K^{p,q}$ does not depend on the choice of the metric, proposition 6.11 is proved. □

The proof of proposition 6.11 also implies the following.

Corollary 6.12 *We have $\overline{H^{p,q}} = H^{q,p}$, where complex conjugation acts naturally on $H^{p+q}(X, \mathbb{C}) = H^{p+q}(X, \mathbb{R}) \otimes \mathbb{C}$.*

Proof Indeed, it is obvious that we have $\overline{K^{p,q}} = K^{q,p}$, where the $K^{p,q}$ are as above. □

A first consequence of this corollary and the Hodge decomposition is the following result.

Corollary 6.13 *The odd Betti numbers $b_{2k+1} = \dim_{\mathbb{C}} H^{2k+1}(X, \mathbb{C})$ of a compact Kähler manifold are even.*

Thus, a Hopf surface, which is a compact quotient of $\mathbb{C}^2 - \{0\}$ by the free action of a group isomorphic to $\mathbb{Z}$ acting by biholomorphic transformations $(z_1, z_2) \mapsto (\lambda_1 z_1, \lambda_2 z_2)$ is not a Kähler surface, since because $\mathbb{C}^2 - \{0\}$ is simply connected, the fundamental group of such a surface X is equal to $\mathbb{Z}$, and thus by Hurewitz' theorem (Spanier 1966), we have $H^1(X, \mathbb{Z}) = \mathbb{Z}$ and $b_1(X) = 1$.

The identification $H^{p,q} = K^{p,q}$ given in the proof of proposition 6.11 and the Hodge decomposition also have the following consequences for a compact Kähler manifold X.

Corollary 6.14 *If a cohomology class on X is representable by a closed form of type (p, q) and also by a closed form of type (p', q') with $(p, q) \neq (p', q')$, it is zero.*

Corollary 6.15 *The cup-product $H^k(X, \mathbb{C}) \otimes H^l(X, \mathbb{C}) \to H^{k+l}(X, \mathbb{C})$ is bigraded for the bigraduation given by the Hodge decomposition.*

Proof Indeed, if α is a closed form of type (p, q) and β is a closed form of type (p', q'), then the exterior product $\alpha \wedge \beta$ is a closed form of type $(p + p', q + q')$. □

Remark 6.16 *We cannot use the decomposition of corollary 6.10 of harmonic forms directly to prove this result, since the cup-product of two harmonic forms is not in general harmonic.*

Finally, theorem 6.7 implies the following result, known as the "$\partial\overline{\partial}$ lemma".

Proposition 6.17 *Let X be a Kähler manifold, and let ω be a form which is both ∂ and $\overline{\partial}$-closed. Then if ω is d or ∂ or $\overline{\partial}$-exact, there exists a form χ such that $\omega = \partial\overline{\partial}\chi$.*

Proof We will give the proof in the case where ω is $\overline{\partial}$-exact. Let us write $\omega = \overline{\partial}\beta$, and let $\beta = \alpha + \Delta\gamma$ be the decomposition of β, with α harmonic. As $\Delta = 2\Delta_{\overline{\partial}}$, we have $\overline{\partial}\alpha = 0$.

Furthermore, we noted above the equality $\overline{\partial}\partial^* = -\partial^*\overline{\partial}$. Thus,

$$\omega = 2\overline{\partial}(\partial^*\partial + \partial\partial^*)\gamma = -2\partial^*(\overline{\partial}\partial\gamma) + 2\overline{\partial}\partial\partial^*\gamma.$$

As both ω and $2\overline{\partial}\partial\partial^*\gamma = -2\partial\overline{\partial}\partial^*\gamma$ are ∂-closed, it follows that $\partial^*(\overline{\partial}\partial\gamma)$ is also. As it also lies in the image of the adjoint ∂^*, it is necessarily zero, so we have $\omega = 2\overline{\partial}\partial(\partial^*\gamma)$. □

To conclude, note the following result, which we will see again in chapter 8.

Lemma 6.18 *$H^{p,q}(X)$ is canonically isomorphic to $H^q(X, \Omega_X^p)$.*

Proof This follows from Hodge theory. If we put a Kähler metric on X, the elements of $H^k(X, \mathbb{C})$ are represented by Δ_d-harmonic forms, and $H^{p,q}(X)$ can be identified with the harmonic forms of type (p, q). But as $\Delta_d = 2\Delta_{\overline{\partial}}$, the harmonic forms of type (p, q) are the $\Delta_{\overline{\partial}}$-harmonic forms of type (p, q), and by theorem 5.24, they are in bijection with $H^q(X, \Omega_X^p)$. The fact that this isomorphism is in fact canonical will follow from the results of chapter 8. □

6.2 Lefschetz decomposition

6.2.1 Commutators

Let X be a Kähler manifold of complex dimension equal to n, and let L, Λ be the Lefschetz operators, $\Lambda = *^{-1}L*$.

Lemma 6.19 *We have the commutation relation*

$$[L, \Lambda] = (k - n)\,\mathrm{Id} \text{ on } \mathcal{A}_X^k. \tag{6.5}$$

Proof This is a lemma of Hermitian geometry, since we are considering operators of order 0. Thus, we will assume that the metric is the standard flat metric. Recall that L is the exterior product with $\omega = \frac{i}{2}\sum_i dz_i \wedge d\overline{z}_i$. Let A_i be the operator given by the exterior product with $\frac{i}{2}dz_i \wedge d\overline{z}_i$. By formula (6.4), we have

$$*^{-1}(d\overline{z}_i\wedge)* = (-1)^{k+1}2\,\mathrm{int}\left(\frac{\partial}{\partial z_i}\right) \text{ on } \mathcal{A}^k_X,$$

and similarly,

$$*^{-1}(dz_i\wedge)* = (-1)^{k+1}2\mathrm{int}\left(\frac{\partial}{\partial \overline{z}_i}\right) \text{ on } \mathcal{A}^k_X.$$

Thus,

$$*^{-1}A_i* = -2i\ \mathrm{int}\left(\frac{\partial}{\partial z_i} \wedge \frac{\partial}{\partial \overline{z}_i}\right).$$

As $L = \sum_i A_i$ and $\Lambda = \sum_i *^{-1}A_i*$, we have

$$[L, \Lambda] = \sum_{i,j}[A_i, *^{-1}A_j*].$$

Now, by the above, A_i and $*^{-1}A_j*$ commute for $i \neq j$. We thus obtain

$$[L, \Lambda] = \sum_i \left[dz_i \wedge d\overline{z}_i, \mathrm{int}\left(\frac{\partial}{\partial z_i} \wedge \frac{\partial}{\partial \overline{z}_i}\right)\right].$$

For $M \subset \{1, \dots, n\}$, set $w_M = \bigwedge_{m\in M} dz_m \wedge d\overline{z}_m$. Every form ω of degree k can be written as a linear combination of the forms $\omega_{A,B,M} = dz_A \wedge d\overline{z}_B \wedge w_M$, where the subsets A, B and M of $\{1, \dots, n\}$ are disjoint, and $|A| + |B| + 2|M| = k$. If A, B and M are fixed, let $J = \{1, \dots, n\} - (A \cup B \cup M)$. Then we have $dz_i \wedge d\overline{z}_i \wedge \omega_{A,B,M} = 0$ if $i \notin J$, and $\mathrm{int}\left(\frac{\partial}{\partial z_i} \wedge \frac{\partial}{\partial \overline{z}_i}\right)(\omega_{A,B,M}) = 0$ if $i \notin M$. Moreover, if $i \in M$, we have

$$(dz_i \wedge d\overline{z}_i) \circ \mathrm{int}\left(\frac{\partial}{\partial z_i} \wedge \frac{\partial}{\partial \overline{z}_i}\right)(\omega_{A,B,M}) = \omega_{A,B,M}.$$

Finally, if $i \in J$, we have

$$\mathrm{int}\left(\frac{\partial}{\partial z_i} \wedge \frac{\partial}{\partial \overline{z}_i}\right) \circ (dz_i \wedge d\overline{z}_i)(\omega_{A,B,M}) = \omega_{A,B,M}.$$

We conclude that

$$[L, \Lambda](\omega_{A,B,M}) = (|M| - |J|)\omega_{A,B,M}.$$

Now, $|J| = n-|A|-|B|-|M|$ and thus $|M|-|J| = k-n$ and $[L, \Lambda](\omega_{A,B,M}) = (k-n)\omega_{A,B,M}$. This proves the equality (6.5). □

6.2.2 Lefschetz decomposition on forms

The commutation relation (6.5) has the following consequence.

Lemma 6.20 *The morphism of vector bundles*

$$L^{n-k} : \Omega^k_{X,\mathbb{R}} \to \Omega^{2n-k}_{X,\mathbb{R}},$$

or equivalently, the operator of order 0 $L^{n-k} : \mathcal{A}^k_X \to \mathcal{A}^{2n-k}_X$, *is an isomorphism.*

Proof As we are dealing with vector bundles of equal rank, it suffices to prove injectivity. We have $[L, \Lambda] = (k-n)\mathrm{Id}$ on $\Omega^k_{X,\mathbb{R}}$. For $r \geq 1$, it follows that

$$[L^r, \Lambda] = (r(k-n) + r(r-1))L^{r-1}. \tag{6.6}$$

Indeed, we have

$$[L^r, \Lambda] = L[L^{r-1}, \Lambda] + [L, \Lambda]L^{r-1},$$

and thus (6.6) is obtained by induction on r. If we have $L^r\alpha = 0$ with $d^0\alpha = k$, then we also have

$$\begin{aligned} L^r\Lambda\alpha - (r(k-n) + r(r-1))L^{r-1}\alpha = 0 \\ = L^{r-1}(L\Lambda - (r(k-n) + r(r-1))\mathrm{Id})(\alpha) = 0. \end{aligned}$$

Now, by induction on $r \leq n-k$, we can assume that L^{r-1} is injective on $\Omega^k_{X,\mathbb{R}}$, and this implies that $(L\Lambda - (r(k-n) + r(r-1))\mathrm{Id})(\alpha) = 0$. But as $r \leq n-k$, we have $k-n+(r-1) \neq 0$ and thus the last equality implies, in particular, that $\alpha = L\beta$, with $d^0\beta = k-2$ and $L^{r+1}\beta = 0$. Thus we can reason by induction on the degree of α, to conclude that the last equality implies that $\beta = 0$, and thus $\alpha = 0$. □

Let us introduce the following notion.

Definition 6.21 *We say that an element* $\alpha \in \Omega^k_{X,x,\mathbb{R}}$, $k \leq n$ *is primitive, if it satisfies the condition* $L^{n-k+1}\alpha = 0$.

Lemma 6.20 now implies the Lefschetz decomposition on the differential forms, as follows.

Proposition 6.22 *Every element $\alpha \in \Omega^k_{X,x,\mathbb{R}}$ admits a unique decomposition of the form $\alpha = \sum_r L^r\alpha_r$, where each α_r is of degree $k - 2r \leq \inf(2n - k, k)$ and primitive.*

Proof We first reduce to the case where $d^0\alpha \leq n$, by using lemma 6.20 to write $\alpha = L^{k-n}\beta$ if this is not the case. We then show uniqueness as follows. Suppose that $\sum_r L^r\alpha_r = 0$ with α_r primitive, and $k = 2r + d^0\alpha_r \leq n$. Then if the smallest integer appearing in this decomposition is not zero, we have

$$L\Big(\sum_r L^{r-1}\alpha_r\Big) = 0,$$

and by lemma 6.20, this implies that $\sum_r L^{r-1}\alpha_r = 0$. An induction hypothesis on the degree then allows us to conclude that $\alpha_r = 0$. If, on the other hand, this smallest integer is zero, we have $d^0\alpha_0 = k$, and as α_0 is primitive, we have $L^{n-k+1}\alpha_0 = 0$. But then

$$L^{n-k+1}\sum_{r>0} L^r\alpha_r = 0 = L^{n-k+2}\sum_{r>0} L^{r-1}\alpha_r.$$

Lemma 6.20 then implies that $\sum_{r>0} L^{r-1}\alpha_r = 0$, and an induction hypothesis on the degree allows us to conclude that $\alpha_r = 0$ for $r > 0$ and thus that $\alpha_0 = 0$.

For the existence, we can also assume that $k = d^0\alpha \leq n$. Consider

$$L^{n-k+1}\alpha \in \Omega^{2n-k+2}_{X,x,\mathbb{R}}.$$

Lemma 6.20 shows that there exists $\beta \in \Omega^{k-2}_{X,x,\mathbb{R}}$ such that $L^{n-k+2}\beta = L^{n-k+1}\alpha$. Thus, $\alpha_0 = \alpha - L\beta$ is primitive, and we have $\alpha = \alpha_0 + L\beta$. An induction hypothesis on the degree then ensures the existence of the Lefschetz decomposition for β, and thus for α. □

Remark 6.23 *The Lefschetz decomposition is also valid for the complexified forms $\alpha \in \Omega^k_{X,x,\mathbb{C}}$. It is then bihomogeneous, in the sense that if α_r denote the primitive components of $\alpha = \sum_{p+q=k} \alpha^{p,q}$, the components $\alpha_r^{p-r,q-r}$ of type $(p - r, q - r)$ of α_r are the primitive components of $\alpha^{p,q}$.*

The following result also gives the classical definition of a primitive element.

Lemma 6.24 *An element α of degree $k \leq n$ is primitive if and only if $\Lambda\alpha = 0$.*

Proof This follows from the commutation relation (6.5). Indeed, we have $[L, \Lambda]\alpha = (k - n)\alpha$, so that if $k = n$, L and Λ commute, and then since L

is injective on $\Omega^k_{X,\mathbb{R}}$, $k < n$, we have

$$\Lambda\alpha = 0 \Leftrightarrow L\Lambda\alpha = 0 \Leftrightarrow \Lambda L\alpha = 0,$$

and this is also equivalent to $L\alpha = 0$, since Λ is injective on $\Omega^{n+2}_{X,x}$ (it is the adjoint of L).

More generally, we have the commutation relation

$$[L^r, \Lambda] = (r(k-n) + r(r-1))L^{r-1},$$

and in particular for $r = n - k + 1$, $k = d^0\alpha$ we find $[L^r, \Lambda](\alpha) = 0$. The argument can then be concluded as above. □

In general, we can thus define a primitive element as an element which is annihilated by Λ. As Λ is injective on the forms of degree strictly greater than n, a primitive element is necessarily of degree at most n, and primitive in the sense of definition 6.21.

6.2.3 Lefschetz decomposition on the cohomology

The following lemma gives the Lefschetz decomposition on the cohomology classes. Recall that a cohomology class $\eta \in H^k(X)$ induces a cup-product operator by η:

$$\eta : H^l(X) \to H^{k+l}(X), \tag{6.7}$$

where the cohomology we are considering is the cohomology with values in any locally constant sheaf of rings. If we consider the cohomology with real coefficients, it can be identified with the de Rham cohomology, and the operator η is induced by the exterior product by any representative closed form η (cf. section 5.3.2). Now consider the case of a Kähler manifold X, of Kähler form ω. As ω is closed, of de Rham class $[\omega]$, we have an operator

$$L : H^k(X, \mathbb{R}) \to H^{k+2}(X, \mathbb{R})$$

which is induced by the operator L on the forms. Now, in the compact case, we have the following theorem, known as the hard Lefschetz theorem.

Theorem 6.25 *If X is a compact Kähler manifold of dimension n, then for every $k \leq n$,*

$$L^{n-k} : H^k(X, \mathbb{R}) \to H^{2n-k}(X, \mathbb{R})$$

is an isomorphism.

By the same argument as in the case of forms, this implies the following theorem, called the Lefschetz decomposition theorem.

Corollary 6.26 *Every cohomology class $\alpha \in H^k(X, \mathbb{R})$ admits a unique decomposition*

$$\alpha = \sum_r L^r \alpha_r,$$

where the α_r are of degree $k - 2r \leq \inf(n, 2n - k)$ and are primitive in the sense that $L^{n-k+2r+1}\alpha_r = 0$ in $H^{2n-k+2r+2}(X, \mathbb{R})$.

Remark 6.27 *The Lefschetz decomposition is also valid for the cohomology with complex coefficients. It is then compatible with the Hodge decomposition of the cohomology. Indeed, the operator L is of bidegree $(1, 1)$ for the bigraduation of the cohomology given by the Hodge decomposition. Thus, a class is primitive if and only if its components of type (p, q) are primitive.*

The proof of theorem 6.25 follows from lemma 6.20, theorem 5.23 and the following lemma.

Lemma 6.28 *The Laplacian Δ_d commutes with L.*

Proof We have $\Delta_d = 2\Delta_\partial = 2(\partial\partial^* + \partial^*\partial)$. Thus,

$$[\Delta_d, L] = 2([\partial\partial^*, L] + [\partial^*\partial, L]) = 2(\partial[\partial^*, L] + [\partial^*, L]\partial)$$

since L commutes with ∂ (the form ω is ∂-closed). Now, we have the identity $[\partial^*, L] = -i\bar{\partial}$, which anticommutes with ∂. Thus $[\Delta_d, L] = 0$. □

Proof of theorem 6.25 As the Laplacian commutes with the operator L on the forms, L^{n-k} sends harmonic forms to harmonic forms:

$$L^{n-k} : \mathcal{H}^k(X) \to \mathcal{H}^{2n-k}(X).$$

As X is compact, the spaces of harmonic forms can be identified with the corresponding cohomology groups, and as remarked above, the operator L on the harmonic forms induces the operator L on the cohomology classes. Now, by theorem 5.30, we know that $H^k(X, \mathbb{R})$ and $H^{2n-k}(X, \mathbb{R})$ have the same dimension. Finally, by lemma 6.20, L^{n-k} is injective on the forms of degree k, and in particular on the harmonic forms. Thus $L^{n-k} : \mathcal{H}^k(X) \to \mathcal{H}^{2n-k}(X)$ is injective, and thus it is an isomorphism. Then $L^{n-k} : H^k(X, \mathbb{R}) \to H^{2n-k}(X, \mathbb{R})$ is also an isomorphism, and theorem 6.25 is proved. □

6.3 The Hodge index theorem

6.3.1 Other Hermitian identities

In the preceding section, we defined the notion of a primitive element ω of $\Omega^k_{X,x,\mathbb{R}}$, relative to the Hermitian form on the n-dimensional complex vector space $T_{X,x}$. Such an ω must satisfy one of the following equivalent conditions:

- We have $\Lambda\omega = 0$ in $\Omega^{k-2}_{X,x}$.
- We have $k \leq n$ and $L^{n-k+1}\omega = 0$.

We now have the following third property of primitive elements.

Proposition 6.29 *Let $\omega \in \Omega^{p,q}_{X,x} \subset \Omega^k_{X,x} \otimes \mathbb{C}$ be a primitive element. Then we have*

$$* \, \omega = (-1)^{\frac{k(k+1)}{2}} i^{p-q} \frac{L^{n-k}}{(n-k)!}\omega. \tag{6.8}$$

Proof Let $dz_1, \ldots, dz_n$ be a basis of the $\mathbb{C}$-vector space $\Omega_{X,x}$, such that the hermitian metric h takes the form $h_x = \sum_i dz_i d\overline{z}_i$ at the point x. In a unique way, we can write

$$\omega = \sum_{A,B,M} \gamma_{A,B,M} dz_A \wedge d\overline{z}_B \wedge w_M,$$

where $A,\ B,\ M$ are subsets disjoint of $\{1, \ldots, n\}$. We have

$$\Lambda = -2i \sum_i \mathrm{int}\left(\frac{\partial}{\partial z_i} \wedge \frac{\partial}{\partial \overline{z}_i}\right)$$

and thus

$$\Lambda(dz_A \wedge d\overline{z}_B \wedge w_M) = -2i \sum_{i \in M} dz_A \wedge d\overline{z}_B \wedge w_{M-\{i\}}.$$

Thus, $\Lambda\omega = 0$ implies that $\Lambda(\omega_{A,B}) = 0$, where $\omega_{A,B} = \sum_M \gamma_{A,B,M} dz_A \wedge d\overline{z}_B \wedge w_M$. It thus suffices to prove the result for $\omega = \omega_{A,B}$. Let us write $\omega_{A,B} = dz_A \wedge d\overline{z}_B \wedge \sum_M \gamma_M w_M$. In this sum, only the subsets $M \subset K := \{1, \ldots, n\} - (A \cup B)$ and of cardinal $m = \frac{1}{2}(k - |A| - |B|)$ appear. By the above, the condition $\Lambda\omega = 0$ can be written

$$\forall N \subset K, \quad |N| = m-1, \quad \sum_{i \in K-N} \gamma_{N \cup \{i\}} = 0. \tag{6.9}$$

We have the following lemma.

Lemma 6.30 *The equality (6.9) implies that for every fixed $J \subset K$ of cardinal $|K| - m$, we have*

$$\sum_{N \subset J} \gamma_N = (-1)^m \gamma_{^cJ} \tag{6.10}$$

where in this sum we must of course have $|N| = m$, and the complement ${}^cJ$ is taken in K.

Admitting this lemma, the proof of (6.8) can be concluded as follows. Firstly, we easily see that

$$* (dz_A \wedge d\overline{z}_B \wedge w_M) = (-1)^{m+\frac{k(k+1)}{2}} \left(\frac{i}{2}\right)^{n-k} i^{p-q} dz_A \wedge d\overline{z}_B \wedge w_{^cM}, \tag{6.11}$$

where ${}^cM$ is the complement of M in K. Indeed, this follows from the equality $(\alpha, \beta)\mathrm{Vol} = \alpha \wedge \overline{*\beta}$, with $\mathrm{Vol} = (\frac{i}{2})^n dz_1 \wedge d\overline{z}_1 \wedge \cdots \wedge dz_n \wedge d\overline{z}_n$, and from the fact that the $\omega_{A,B,M}$ are orthogonal and of norm 2^k for the metric at each point. We thus obtain

$$*\omega_{A,B} = \sum_{M \subset K} (-1)^{m+\frac{k(k+1)}{2}} \left(\frac{i}{2}\right)^{n-k} i^{p-q} \gamma_M dz_A \wedge d\overline{z}_B \wedge w_{^cM}.$$

Furthermore, we have

$$(-1)^{\frac{k(k+1)}{2}} i^{p-q} \frac{L^{n-k}}{(n-k)!} \omega_{A,B} = (-1)^{\frac{k(k+1)}{2}} i^{p-q} \left(\frac{i}{2}\right)^{n-k} \sum_{M,N} \gamma_M dz_A \wedge d\overline{z}_B \wedge w_{M \cup N},$$

where in this sum, N runs through the subsets of cardinal $n - k$ contained in K and disjoint from M. For $J \subset K$ fixed, the coefficient of $dz_A \wedge d\overline{z}_B \wedge w_J$ in $(-1)^{\frac{k(k+1)}{2}} i^{p-q} \frac{L^{n-k}}{(n-k)!} \omega_{A,B}$ is thus equal to $\sum_{M \subset J} (-1)^{\frac{k(k+1)}{2}} i^{p-q} \left(\frac{i}{2}\right)^{n-k} \gamma_M$. Now, by lemma 6.30, this is equal to $(-1)^m (-1)^{\frac{k(k+1)}{2}} i^{p-q} \left(\frac{i}{2}\right)^{n-k} \gamma_{^cJ}$, i.e. to the coefficient of $dz_A \wedge d\overline{z}_B \wedge w_J$ in $*\omega_{A,B}$ by the equality (6.11). □

Proof of lemma 6.30 For every $r \geq m$, let $S_r = \sum_{|N \cap J| = r} \gamma_N$. Of course, we have $S_m = \sum_{N \subset J} \gamma_N$, and $S_0 = \gamma_{^cJ}$. The equalities (6.9), which are valid for all $N \subset K$ of cardinal $m - 1$ such that $|N \cap J| = r$, then imply that

$$(r+1) S_{r+1} = -(m-r) S_r,$$

and thus

$$S_m = (-1)^m \frac{1 \cdots (m-1)m}{m \cdots 2} S_0 = (-1)^m S_0.$$

□

6.3.2 The Hodge index theorem

Let X be a compact Kähler manifold of dimension n, with Kähler form ω. We have the pairing $\langle\,,\,\rangle$ described in section 5.3.2:

$$H^k(X,\mathbb{R})\otimes H^{2n-k}(X,\mathbb{R})\to\mathbb{R}.$$

L being the Lefschetz operator acting on the cohomology, let us define the following intersection form Q on $H^k(X,\mathbb{R})$, $k\le n$:

$$Q(\alpha,\beta)=\langle L^{n-k}\alpha,\beta\rangle=\int_X\omega^{n-k}\wedge\alpha\wedge\beta.$$

Clearly, this form is symmetric for k even, and alternating otherwise. Thus the sesquilinear form $H_k(\alpha,\beta)=i^kQ(\alpha,\overline{\beta})$ is a Hermitian form on $H^k(X,\mathbb{C})$. By proposition 6.29, we can describe the signature of this Hermitian form.

Lemma 6.31 *The Lefschetz decomposition of corollary 6.26*

$$H^k(X,\mathbb{C})=\bigoplus_{2r\le k}L^rH^{k-2r}(X,\mathbb{C})_{\text{prim}}$$

is an orthogonal decomposition for H_k. Moreover, on each primitive component $L^rH^{k-2r}(X,\mathbb{C})_{\text{prim}}$, H_r induces the form $(-1)^rH_{k-2r}$.

Proof If $\alpha=L^r\alpha'$ and $\beta=L^s\beta'$ with α', β' primitive and $r<s$, we have $L^{n-k}\alpha\wedge\beta=(L^{n-k+r+s}\alpha')\wedge\beta'$, where α' is primitive of degree $k-2r$. As $r+s>2r$, we have $L^{n-k+r+s}\alpha'=0$ and $H_k(\alpha,\beta)=0$. The second statement is obvious. □

Finally, we have the following result, which is due to Riemann in the case of the H^1 of a curve.

Theorem 6.32 *The subspaces $H^{p,q}(X)\subset H^k(X,\mathbb{C})$ form an orthogonal direct sum for H_k. Moreover, the form $(-1)^{\frac{k(k-1)}{2}}i^{p-q-k}H_k$ is positive definite on the complex subspace $H^{p,q}_{\text{prim}}:=H^k(X,\mathbb{C})_{\text{prim}}\cap H^{p,q}(X)$.*

Proof If $\alpha^{p,q}$, $\beta^{p',q'}\in H^k(X,\mathbb{C})$ with $(p,q)\neq(p',q')$, we certainly have $L^{n-k}\alpha^{p,q}\wedge\overline{\beta^{p',q'}}=0$, since it is a class of type $(n-k+p+q',n-k+p'+q)$ and $H^{2n}(X)$ is of type (n,n).

The second assertion follows from proposition 6.29. Indeed, by lemma 6.28, the operator L acts on the harmonic forms, and if we identify the cohomology classes with the harmonic forms, the operator L on the classes can be identified with the operator L on the forms. Thus, a cohomology class α is primitive if and only if its harmonic representative $\tilde{\alpha}$ is a primitive form, as well as its complex conjugate, and we then have

$$*\overline{\tilde{\alpha}} = (-1)^{\frac{k(k-1)}{2}} i^{p-q} \frac{L^{n-k}}{(n-k)!}\overline{\tilde{\alpha}}.$$

Thus,

$$\begin{aligned} H_k(\alpha) &= i^k \int_X \tilde{\alpha} \wedge L^{n-k}\overline{\tilde{\alpha}} \\ &= i^k (n-k)!(-1)^{\frac{k(k-1)}{2}} i^{q-p} \int_X \tilde{\alpha} \wedge *\overline{\tilde{\alpha}} \\ &= i^k (n-k)!(-1)^{\frac{k(k-1)}{2}} i^{q-p} ||\tilde{\alpha}||^2_{L^2}, \end{aligned}$$

and $i^{p-q-k}(-1)^{\frac{k(k-1)}{2}} H_k(\alpha) > 0$ for $\alpha \neq 0$ primitive of type (p, q) with $p + q = k$. □

The Hodge index theorem is an immediate consequence of theorem 6.32. This theorem describes the index (or rather, the signature) of the (symmetric) intersection form on $H^n(X, \mathbb{R})$, where $n = \dim_{\mathbb{C}} X$ is assumed to be even, and X is compact Kähler.

Theorem 6.33 *The signature of the intersection form $Q(\alpha, \beta) = \int_X \alpha \wedge \beta$ on $H^n(X, \mathbb{R})$ is equal to $\sum_{a,b}(-1)^a h^{a,b}(X)$, where $h^{a,b}(X) = \dim(H^{a,b}(X))$.*

Proof Indeed, the signature is also equal to the signature of the Hermitian form $H(\alpha, \beta) = \int_X \alpha \wedge \overline{\beta}$. Now, we have a decomposition of $H^n(X, \mathbb{C})$ as an orthogonal direct sum of the $L^r H^{a,b}_{\text{prim}}$, $a + b = n - 2r$, and the sign of H on $L^r H^{a,b}_{\text{prim}}$ is equal to $(-1)^a$ by theorem 6.32 and because n is even. We thus have

$$\text{sign}(Q) = \sum_{a+b=n-2r} (-1)^a h^{a,b}_{\text{prim}}.$$

But $h^{a,b}_{\text{prim}} = h^{a,b} - h^{a-1,b-1}$, so

$$\text{sign}(Q) = \sum_{a+b=n-2r} (-1)^a (h^{a,b} - h^{a-1,b-1}) \tag{6.12}$$

$$= \sum_{a+b \cong n (\text{mod } 2)} (-1)^a h^{a,b} \tag{6.13}$$

$$= \sum_{a+b \cong 0 (\text{mod } 2)} (-1)^a h^{a,b}. \tag{6.14}$$

Note that we used Poincaré duality here, in order to write

$$2\sum_{a+b=n-2r,r>0}(-1)^a h^{a,b} = \sum_{a+b\cong n(\mathrm{mod}2),a+b\neq n}(-1)^a h^{a,b}.$$

By applying complex conjugation, it is obvious that $\sum_{a+b\cong 1(\mathrm{mod}\,2)}(-1)^a h^{a,b} = 0$, so the theorem is proved. □

Exercises

1. Let $H_{\mathbb{R}}$ be a $\mathbb{R}$-vector space, and $H_{\mathbb{C}} := H_{\mathbb{R}} \otimes \mathbb{C}$.
   (a) Show that a decomposition

   $$H_{\mathbb{C}} = \bigoplus_{p+q=k} H^{p,q}, \quad H^{p,q} = \overline{H^{q,p}}$$

   determines a continuous action $\rho : \mathbb{C}^* \to \mathrm{Gl}(H_{\mathbb{C}})$ of $\mathbb{C}^*$ on $H_{\mathbb{C}}$, given by

   $$z \cdot \alpha^{p,q} = z^p \overline{z}^q \alpha^{p,q},$$

   for $\alpha^{p,q} \in H^{p,q}$. Show that this action satisfies

   $$\rho(\overline{z}) = \overline{\rho(z)}$$

   where the conjugacy on $\mathrm{Gl}(H_{\mathbb{C}})$ is defined by

   $$\overline{g}(u) = \overline{g(\overline{u})}.$$

   Show that one also has $\rho(t) = t^k \mathrm{Id}$ for $t \in \mathbb{R}^*$.
   Conversely, let $\rho : \mathbb{C}^* \to \mathrm{Gl}(H_{\mathbb{C}})$ be a continuous action of $\mathbb{C}^*$ on $H_{\mathbb{C}}$ satisfying $\rho(t) = t^k \mathrm{Id}$ for $t \in \mathbb{R}^*$, $\rho(\overline{z}) = \overline{\rho(z)}$ for $z \in \mathbb{C}^*$.
   (b) Applying the diagonalisation theorem for the actions of torsion abelian groups to the torsion points of $\mathbb{C}^*$, show that there exists a decomposition into a direct sum

   $$H = \bigoplus_{\chi} H_{\chi},$$

   where χ belongs to the set of characters of $\mathbb{C}^*$, and $\mathbb{C}^*$ acts by $z \mapsto \chi(z)\mathrm{Id}$ on H_{χ}.
   (c) Show that only the characters $\chi_{p,q} : z \mapsto z^p \overline{z}^q$ with $p + q = k$ appear in this decomposition.
   (d) Let $H^{p,q} := H_{\chi_{p,q}}$. Show that $H^{p,q} = \overline{H^{q,p}}$.
2. *The Hodge decomposition for curves.* Let X be a compact connected complex curve. We have the differential

   $$d : \mathcal{O}_X \to \Omega_X$$

between the sheaf of holomorphic functions and the sheaf of holomorphic differentials.

(a) Show that d is surjective with kernel equal to the constant sheaf $\mathbb{C}$. Hence we have an exact sequence

$$0 \to \mathbb{C} \to \mathcal{O}_X \xrightarrow{d} \Omega_X \to 0. \tag{6.15}$$

(b) Deduce from Serre duality that $H^1(X, \Omega_X) \cong \mathbb{C}$. Deduce from Poincaré duality that $H^2(X, \mathbb{C}) = \mathbb{C}$.

(c) Show that (6.15) induces a short exact sequence

$$0 \to H^0(X, \Omega_X) \to H^1(X, \mathbb{C}) \to H^1(X, \mathcal{O}_X) \to 0.$$

(d) Show that the map which to a holomorphic form α associates the class of $\overline{\alpha}$ in $H^1(X, \mathcal{O}_X)$ is injective.

(e) Deduce from Serre duality that it is also surjective and that we have the decomposition

$$H^1(X, \mathbb{C}) = H^0(X, \Omega_X) \oplus \overline{H^0(X, \Omega_X)},$$

with

$$\overline{H^0(X, \Omega_X)} \cong H^1(X, \mathcal{O}_X).$$

7

Hodge Structures and Polarisations

In this chapter, we give a synthesis of all the results proved up to now, and using it, we prove that the rational cohomology of a complex compact polarised manifold admits a decomposition as a direct sum of polarised Hodge structures.

To begin with, we define the integral and rational Hodge structures. These are the structures which lie naturally on the integral or rational cohomology of a compact Kähler manifold; they are given by the Hodge decomposition of the cohomology with complex coefficients. We study the case of the Hodge structure of weight 1; giving such a structure is equivalent to giving a complex torus. In the last section, we will study morphisms of Hodge structures, and the functoriality properties under direct or inverse image of the Hodge structure on the cohomology of Kähler manifolds relative to the holomorphic maps between two such manifolds. We also prove a very simple result on morphisms of Hodge structures, whose generalisation to mixed Hodge structures (which will be explained in the second volume of this work) has numerous applications.

Lemma 7.1 *The morphisms of Hodge structures are strict for the Hodge filtration.*

Polarisation is the major notion introduced in this chapter. The Lefschetz decomposition and the Hodge index theorem allow us to write the cohomology of a compact Kähler manifold as a direct sum of primitive components, compatible with the Hodge decomposition, on which the Hermitian intersection form given by the Lefschetz operator has signs defined on each component of type (p, q).

This Lefschetz decomposition is not a decomposition as a direct sum of rational sub-Hodge structures, except when the operator L preserves the rational cohomology. Thus, we are led to distinguish the class of Kähler manifolds X for which the Kähler form is of integral class. We prove Lefschetz' theorem on $(1, 1)$ classes, in the following form.

Theorem 7.2 *If α is a real closed form of type $(1, 1)$ on a compact Kähler manifold whose cohomology class is integral, then α is the Chern form of a holomorphic line bundle equipped with a Hermitian metric.*

We also show the following result, known as the Kodaira embedding theorem (the proof of the vanishing theorem will not be given here; we refer to Demailly (1996) for this).

Theorem 7.3 *Let L be a holomorphic line bundle over a compact complex manifold X, and let h be a Hermitian metric on L whose Chern form is a Kähler form. Then for sufficiently large N, the holomorphic sections of $L^{\otimes N}$ give a holomorphic embedding of X into a projective space $\mathbb{P}^M$.*

These two theorems show that the manifolds admitting a polarisation are in fact the smooth complex projective varieties.

7.1 Definitions, basic properties

7.1.1 Hodge structure

If X is a compact manifold, and R is a field of characteristic 0, we have a natural isomorphism

$$H^k(X, \mathbb{Z}) \otimes R \cong H^k(X, R),$$

given by the morphism of constant sheaves $\mathbb{Z} \to R$. We can see this by applying theorem 4.41, which enables us to identify $H^k(X, \mathbb{Z})$ (resp. $H^k(X, R)$) with the Čech cohomology group $\check{H}^k(\mathcal{U}, \mathbb{Z})$ (resp. $\check{H}^k(\mathcal{U}, R)$), where $\mathcal{U}$ is a finite covering by contractible open sets whose multi-intersections are contractible. Now, as the Čech complex $C^\cdot(\mathcal{U}, \mathbb{Z})$ consists of free abelian groups of finite rank, and furthermore we have $C^\cdot(\mathcal{U}, \mathbb{Z}) \otimes R = C^\cdot(\mathcal{U}, R)$, then since R is of characteristic 0, we also have

$$\check{H}^k(\mathcal{U}, \mathbb{Z}) \otimes R = \check{H}^k(\mathcal{U}, R).$$

In what follows, we will often use the same notation for the integral cohomology modulo torsion and the integral cohomology. By the above, applied to the case $R = \mathbb{R}$, the integral cohomology modulo torsion can be identified with the image of the integral cohomology in the real cohomology. It forms a lattice in the real cohomology.

Suppose now that X is a compact Kähler manifold. By proposition 6.11, we have a decomposition

$$H^k(X,\mathbb{C}) = \bigoplus_{p+q=k} H^{p,q}(X),$$

where $H^{p,q}(X)$ is a complex subspace. Recall (corollary 6.12) that it satisfies the Hodge symmetry

$$H^{p,q}(X) = \overline{H^{q,p}(X)},$$

where complex conjugation acts naturally on $H^k(X,\mathbb{C}) = H^k(X,\mathbb{R}) \otimes \mathbb{C}$. Thus, we are led to the following definition.

Definition 7.4 *An integral Hodge structure of weight k is given by a free abelian group $V_{\mathbb{Z}}$ of finite type, together with a decomposition*

$$V_{\mathbb{C}} := V_{\mathbb{Z}} \otimes \mathbb{C} = \bigoplus_{p+q=k} V^{p,q}$$

satisfying $V^{p,q} = \overline{V^{q,p}}$.

Given such a decomposition, we define the associated Hodge filtration $F^{\cdot}V$ by

$$F^p V_{\mathbb{C}} = \bigoplus_{r \geq p} V^{r,k-r}.$$

It is a decreasing filtration on $V_{\mathbb{C}}$, which satisfies

$$V_{\mathbb{C}} = F^p V_{\mathbb{C}} \oplus \overline{F^{k-p+1} V_{\mathbb{C}}}.$$

The Hodge filtration determines the Hodge decomposition by

$$V^{p,q} = F^p V_{\mathbb{C}} \cap \overline{F^q V_{\mathbb{C}}}.$$

When X is a compact Kähler manifold and $V_{\mathbb{Z}} = H^k(X,\mathbb{Z})$, we have the following result.

Proposition 7.5 *Let $F^p A^k(X)$ be the set of complex differential forms which are sums of forms of type $(r, k-r)$ with $r \geq p$ at every point. Then we have*

$$F^p H^k(X,\mathbb{C}) = \frac{\mathrm{Ker}\,(d : F^p A^k(X) \to F^p A^{k+1}(X))}{\mathrm{Im}\,(d : F^p A^{k-1}(X) \to F^p A^k(X))}.$$

Proof We have an obvious map from $\mathrm{Ker}\,(d : F^p A^k(X) \to F^p A^{k+1}(X))$ to $H^k(X,\mathbb{C})$; to a closed form in $F^p A^k(X)$, it associates its class. Its image contains $F^p H^k(X,\mathbb{C})$, which is generated by the classes representable by a closed form of type $(r, k-r)$, with $r \geq p$, since such a form lies in

$\mathrm{Ker}\,(d : F^pA^k(X) \to F^pA^{k+1}(X))$. Conversely, let α be a class represented by a closed form β in $F^pA^k(X)$. Given a Kähler metric on X, let us write $\beta = \gamma + \Delta_d\epsilon$, with γ harmonic. As Δ_d is bihomogeneous, and by the uniqueness of this expression, we see that γ belongs to $F^pA^k(X)$, and we can also assume that $\epsilon \in F^pA^k(X)$. As β and γ are closed, we must have $d^*d\epsilon = 0$, since $d^*d\epsilon$ is both closed and in the image of d^*, and so $\beta = \gamma + dd^*\epsilon$. Thus, β is cohomologous to γ. But γ is harmonic, so its components of type $(r, k-r)$ are harmonic, and they are zero for $r < p$. Thus $[\gamma^{r,k-r}] \in H^{r,k-r}(X) \subset F^pH^k(X, \mathbb{C})$, $\forall r$ and $\beta \in F^pH^k(X, \mathbb{C})$.

It remains to show that the kernel of this map is exactly

$$\mathrm{Im}\,(d : F^pA^{k-1}(X) \to F^pA^k(X)).$$

We will use decreasing induction on p. If $p = k$, a closed form of type $(k, 0)$ is holomorphic, and both $\overline{\partial}$ and ∂-closed. If it is exact, it is equal to $\overline{\partial}\partial\gamma$ by the $\partial\overline{\partial}$-lemma 6.17, and thus it is zero for reasons of type.

Assume now that the property is satisfied for $p+1$, and let $\alpha \in F^pA^k(X)$ be a closed form of class zero. Then its harmonic representative β is zero for any Kähler metric. Thus, we have $\alpha = \Delta_d\gamma$. By the bihomogeneity of Δ_d, we also have $\alpha^{p,q} = \Delta_d\gamma^{p,q}$, and by theorem 6.7, we obtain $\alpha^{p,q} = 2\Delta_{\overline{\partial}}\gamma^{p,q}$. As the component $\alpha^{p,q}$ of α is $\overline{\partial}$-closed, by the fact that α has no component of type (k, l) with $l > q$, we deduce that that $\overline{\partial}^*\overline{\partial}\gamma^{p,q} = 0$, and thus that

$$\alpha^{p,q} = 2\overline{\partial}\,\overline{\partial}^*\gamma^{p,q}.$$

Then the form $\alpha' = \alpha - 2d\overline{\partial}^*\gamma^{p,q}$ is in $F^{p+1}A^k(X)$ and of class zero. We can thus apply the induction hypothesis to it to conclude that $\alpha' = d\beta'$, $\beta' \in F^{p+1}A^{k-1}(X)$. As $\overline{\partial}^*\gamma^{p,q}$ lies in $F^pA^{k-1}(X)$ and

$$\alpha = 2d\overline{\partial}^*\gamma^{p,q} + d\beta',$$

the result is also shown for p. □

Corollary 7.6 *For every $p \leq n$, $H^{p,0}(X)$ is isomorphic to the space of holomorphic forms of degree p on X.*

Proof Indeed, the preceding result shows that $H^{p,0}(X)$ is isomorphic to the space of closed forms of degree p and of type $(p, 0)$ on X. Such a form is holomorphic since it is $\overline{\partial}$-closed. Thus, it suffices to show that the holomorphic forms are closed, i.e. ∂-closed. But this follows from lemma 6.17, since if α is holomorphic, then $\partial\alpha$ is both $\overline{\partial}$-closed and ∂-exact. Thus it is $\partial\overline{\partial}$-exact and thus zero for reasons of type. □

7.1.2 Polarisation

If X is an n-dimensional compact Kähler manifold of Kähler form ω, the cup-product

$$L : H^k(X, \mathbb{R}) \to H^{k+2}(X, \mathbb{R})$$

with the class $[\omega] \in H^2(X, \mathbb{R})$ of ω gives the Lefschetz decomposition

$$H^k(X, \mathbb{R}) = \bigoplus_r L^r H^{k-2r}_{\text{prim}},$$

where each component admits an induced Hodge decomposition, since the operator L is of bidegree $(1, 1)$ for the bigraduation given by the Hodge decomposition.

Moreover, L gives an intersection form on $H^k(X, \mathbb{R})$ for $k \leq n$:

$$Q(\alpha, \beta) = \int_X \omega^{n-k} \wedge \alpha \wedge \beta = \langle L^{n-k}\alpha, \beta \rangle,$$

where in the middle term, α, β are representative forms of the cohomology classes under consideration. Q is alternating if k is odd, symmetric otherwise. The induced Hermitian form

$$H(\alpha, \beta) = i^k Q(\alpha, \overline{\beta})$$

on $H^k(X, \mathbb{C})$ satisfies the following properties (cf. section 6.3.2):

(i) The Hodge decomposition is orthogonal for H.

Moreover, we know that the Lefschetz decomposition is orthogonal for this form, and that on the primitive component $H^k(X)_{\text{prim}}$, we have

(ii) $i^{p-q-k}(-1)^{\frac{k(k-1)}{2}} H(\alpha) > 0$ for α non-zero of type (p, q).

When the class $[\omega]$ is integral, i.e. belongs to $H^2(X, \mathbb{Z}) \subset H^2(X, \mathbb{R})$, the operator L acts on the integral cohomology and the primitive component

$$H^k(X)_{\text{prim}} = \operatorname{Ker} L^{n-k+1}$$

is defined on $\mathbb{Z}$. Moreover, the intersection form Q is integral, i.e. takes integral values on the integral classes. The structure obtained on the primitive part of the cohomology of such a Kähler manifold (X, ω), using the intersection form Q defined by the operator L, is then the following.

Definition 7.7 *An integral polarised Hodge structure of weight k is given by a Hodge structure $(V_{\mathbb{Z}}, F^p V_{\mathbb{C}})$ of weight k, together with an intersection form Q on $V_{\mathbb{Z}}$, which is symmetric if k is even, alternating otherwise, and satisfies conditions (i) and (ii) above.*

We can weaken this definition a little, by considering rational polarised Hodge structures: we then simply require V to have a rational structure, for which Q is rational.

7.1.3 Polarised varieties

To associate polarised Hodge structures to a Kähler manifold X, we need to choose a class $[\omega] \in H^2(X, \mathbb{R})$, which is the cohomology class of a Kähler form ω, and which is integral.

Definition 7.8 *A polarised manifold is a pair* $(X, [\omega])$*, where* X *is a compact complex manifold, and* $[\omega]$ *is an integral Kähler class on* X.

In this section, we propose to show that such a manifold X is then necessarily projective, i.e. admits a holomorphic embedding into $\mathbb{P}^N$ for some sufficiently large N. Let X be a complex manifold. We have a natural map

$$H^2(X, \mathbb{C}) \to H^2(X, \mathcal{O}_X) \tag{7.1}$$

given by the inclusion of sheaves $\mathbb{C} \subset \mathcal{O}_X$. As this morphism extends to a morphism between the de Rham resolution of $\mathbb{C}$ and the Dolbeault resolution of $\mathcal{O}_X$

$$(\mathcal{A}_X^{\cdot}, d) \to \left(\mathcal{A}_X^{0,\cdot}, \overline{\partial}\right),$$

which to a form of degree k associates its component of bidegree $(0, k)$, the morphism (7.1) associates to a de Rham class $[\eta]$ the Dolbeault class of $\eta^{0,2}$. The fact that the Laplacians Δ_d and $\Delta_{\overline{\partial}}$ coincide up to a factor of 2 implies that for η harmonic, the component $\eta^{0,2}$ is $\Delta_{\overline{\partial}}$-harmonic, so that the map (7.1) can be identified with the projection

$$H^2(X, \mathbb{C}) \to H^{0,2}(X)$$

given by the Hodge decomposition. The classes in the kernel of (7.1) are thus those which are representable (given a Kähler metric on X) by harmonic forms in $F^1A^2(X)$, and the real classes are those which are representable by real harmonic forms of type $(1, 1)$.

Now, consider the exponential exact sequence

$$0 \to \mathbb{Z} \stackrel{2i\pi}{\to} \mathcal{O}_X \stackrel{\exp}{\to} \mathcal{O}_X^* \to 0.$$

By the associated long exact sequence, it gives a morphism

$$c_1 : H^1(X, \mathcal{O}_X^*) \to H^2(X, \mathbb{Z})$$

whose image is, by the above, exactly the set of integral classes of degree 2 representable by a real closed form of type (1, 1).

Remark 7.9 *The map $H^2(X, \mathbb{Z}) \to H^2(X, \mathcal{O}_X)$ obviously annihilates the torsion of $H^2(X, \mathbb{Z})$. The exponential exact sequence shows that its kernel is equal to the image of c_1. In particular, the torsion classes are in the image of c_1.*

Recall also (cf. theorem 4.49) that the group $H^1(X, \mathcal{O}_X^*)$ can be identified with the Picard group of isomorphism classes of holomorphic line bundles. The class $c_1(L)$ is called the first Chern class of L. It is a topological invariant (Euler class) of the underlying complex vector bundle of rank 1. Indeed, by its construction, the map c_1 factors through the natural map

$$H^1(X, \mathcal{O}_X^*) \to H^1(X, (\underline{\mathcal{C}}^0)^*)$$

which to a holomorphic line bundle associates the underlying topological line bundle.

Theorem 7.10 *Let L be a holomorphic line bundle over X, and let h be a Hermitian metric on L. Then the class of the Chern form $\omega_{L,h}$ (cf. section 3.3.1) is equal to the image of $c_1(L)$ in $H^2(X, \mathbb{R})$. Moreover, for every real form ω of type $(1, 1)$ whose class is equal to the image of $c_1(L)$ in $H^2(X, \mathbb{R})$, there exists a metric h on L such that $\omega_{L,h} = \omega$.*

Proof Let $(U_i)_{i \in I}$ be a covering of X by open sets on which L is trivialised by everywhere non-zero holomorphic sections σ_i. Then

$$\omega_{L,h|U_i} = \frac{1}{2i\pi} \partial\overline{\partial} \log h_i, \quad h_i = h(\sigma_i).$$

Thus, we also have

$$\omega_i := \omega_{L,h|U_i} = d\beta_i, \quad \beta_i = \frac{1}{2i\pi} \overline{\partial} \log h_i$$

and

$$\beta_i - \beta_j = \frac{1}{2i\pi} \overline{\partial}(\log h_i - \log h_j)$$

on $U_i \cap U_j$. Recalling that $\sigma_i = g_{ij}\sigma_j$ on U_{ij}, where the g_{ij} are holomorphic and invertible and, up to refining the cover, can be assumed of the form $\exp 2i\pi f_{ij}$, we also have $h_i = |g_{ij}|^2 h_j$ and

$$\beta_i - \beta_j = -d\overline{f_{ij}}.$$

Finally, we have $g_{ij}g_{jk}g_{ki} = 1$ on U_{ijk}, and thus $f_{ij} + f_{jk} + f_{ki} \in \mathbb{Z}$ on U_{ijk} (which is assumed to be connected).

Now, considering the resolution of the sheaf $\mathbb{C}$ by the simple complex (cf. definition 4.42) $(\mathcal{K}, D)$ associated to the double complex

$$\mathcal{K}^{p,q} = \underline{\mathcal{C}}^q(\mathcal{A}^p_{\mathbb{C}}), \quad D_1 = d, \quad D_2 = \delta,$$

where δ is the Čech differential and $D = d + (-1)^p\delta$, we first show that the class $[\omega_{L,h}] \in H^2(X, \mathbb{R})$ admits the cocycle $a_{ijk} = f_{ij} + f_{jk} + f_{ki} \in \mathbb{Z} \subset \mathbb{R}$ as a representative in the Čech cohomology.

Indeed, if $(K^\cdot, D)$ is the complex of the global sections of the complex of sheaves $\mathcal{K}$, then since $(\mathcal{K}, D)$ is a resolution of the sheaf $\mathbb{C}$, we have a natural map $H^k(K^\cdot) \to H^k(X, \mathbb{C})$, and furthermore the de Rham complex of X and the Čech complex of X associated to the covering $\mathcal{U} = (U_i)$ are naturally subcomplexes of $K^\cdot$ such that the composition maps

$$H^k_{\mathrm{DR}}(X) \to H^k(K^\cdot) \to H^k(X, \mathbb{C}),$$

$$\check{H}^k(\mathcal{U}, \mathbb{C}) \to H^k(K^\cdot) \to H^k(X, \mathbb{C})$$

are the natural maps. Now, the equalities written above can be translated as

$$(\omega_i) - D(\beta_i) = \delta(\beta_i), \quad \delta(\beta_i) + D(\overline{f_{ij}}) = \delta(\overline{f_{ij}}) = \delta(f_{ij})$$

in $K^\cdot$, where the last equality follows from the fact that $a_{ijk} = \delta(\overline{f_{ij}})$ is a cocycle with real (in fact integral) coefficients. Thus, $\omega = (\omega_i)$ is cohomologous in $K^\cdot$ to (a_{ijk}).

Moreover, the element of $H^1(\mathcal{O}_X^*)$ corresponding to L is described by the Čech cocycle $g_{ij} \in \mathcal{O}_{U_{ij}}^*$, and its image in $H^2(X, \mathbb{Z})$ is obtained precisely by lifting g_{ij} in $\mathcal{O}_{U_{ij}}$ and applying $\frac{1}{2i\pi}\delta$, where δ is the Čech differential, to the cochain $\log g_{ij}$ thus obtained. So this element is also represented in Čech cohomology by (a_{ijk}). This proves the first statement.

As for the second assertion, let H be a metric on L, and let $\omega_{L,H}$ be the corresponding Chern form. Now let ω be a real closed form of type $(1, 1)$, of the same class as $\omega_{L,H}$. As $\omega - \omega_{L,H}$ is ∂ and $\overline{\partial}$-closed, and is exact, the

$\partial\overline{\partial}$-lemma 6.17 shows that there exists a function ϕ, which we can clearly assume to be real, such that

$$\frac{1}{2i\pi}\partial\overline{\partial}\phi = \omega - \omega_{L,H}.$$

Then the Chern form of the metric $h = e^{\phi} H$ is equal to ω. □

Now, let $(X, [\omega])$ be a polarised manifold. The above shows that there exists a holomorphic line bundle L on X and a metric h on L such that $\omega_{L,h} = \omega$ is a positive form. We say that L is positive, and we have the following result, known as the Kodaira embedding theorem.

Theorem 7.11 *Let X, L be as above, with L positive. Then for every sufficiently large integer N, there exists a holomorphic embedding $\phi : X \to \mathbb{P}^r$ such that $\phi^*(\mathcal{O}_{\mathbb{P}^r}(1)) = \mathcal{L}^{\otimes N}$ where $\mathcal{L}$ is the sheaf of holomorphic sections of L.*

Remark 7.12 *The sheaf $\phi^*(\mathcal{O}_{\mathbb{P}^r}(1))$ is the sheaf of holomorphic sections of the vector bundle $\phi^* S^*$, where S is the tautological subbundle of $\mathbb{P}^r$. Its relation with the pullback $\phi^{-1}\mathcal{O}_{\mathbb{P}^r}(1)$, as defined in section 5.3.2, is given by*

$$\phi^*(\mathcal{O}_{\mathbb{P}^r}(1)) = \phi^{-1}\mathcal{O}_{\mathbb{P}^r}(1) \otimes_{\phi^{-1}\mathcal{O}_{\mathbb{P}^r}} \mathcal{O}_X.$$

This theorem shows that a polarised manifold is a projective variety. The theorem is essentially based on the following special case of the Kodaira–Akizuki–Nakano vanishing theorem.

Theorem 7.13 *Let L be a positive holomorphic line bundle over a compact complex manifold. Then for every $q > 0$, we have $H^q(X, K_X \otimes \mathcal{L}) = 0$.*

This theorem is obtained by applying theorem 5.24, and proving the Kodaira–Bochner–Nakano identities which compare the Laplacians Δ_L and Δ'_L, where Δ_L was defined in 5.1.4 and Δ'_L is the Laplacian associated to the operator $\nabla^{1,0}$, the part of type $(1, 0)$ of the Chern connection on L. We refer to Demailly (1996) for a complete proof.

Proof of theorem 7.11 Let $x \in X$, and let $X_x \xrightarrow{\tau} X$ be the blowup of X at x. We first note that there exists a holomorphic section $\sigma \in H^0(X, \mathcal{L})$ which does not vanish at x if and only if there exists a holomorphic section of $\tau^* L$ which does not vanish along the exceptional divisor $E = \pi^{-1}(x)$, i.e. if and only if the restriction morphism

$$H^0(X_x, \tilde{\mathcal{L}}) \to H^0(E, \tilde{\mathcal{L}}_{|E}) \tag{7.2}$$

is surjective, where $\tilde{\mathcal{L}}$ is the sheaf of holomorphic sections of the bundle $\tau^* L$, and

$$\tilde{\mathcal{L}}_{|E} = \tilde{\mathcal{L}}/\mathcal{I}_E\tilde{\mathcal{L}}$$

is the sheaf of holomorphic sections of $\tau^* L_{|E}$. Here, $\mathcal{I}_E \subset \mathcal{O}_{X_x}$ is the sheaf of ideals of E, which to each open set U associates the set of holomorphic functions on U that vanish on $E \cap U$.

Indeed, the holomorphic vector bundle $\tau^* L$ is trivial along E, and its holomorphic sections along E can be identified with the sections of L at x, i.e. with L_x. Moreover, the holomorphic sections of $\tau^* L$ can be identified with the holomorphic sections of L by Hartogs' theorem 1.25.

Now, as E is a hypersurface, E is defined locally by a single equation, and $\mathcal{I}_E$ is thus a sheaf of free $\mathcal{O}_{X_x}$-modules of rank 1. Furthermore, the surjectivity of the map (7.2) is implied by the vanishing of the group $H^1(X_x, \mathcal{I}_E\tilde{\mathcal{L}})$, by the long exact sequence associated to the short exact sequence of sheaves

$$0 \to \mathcal{I}_E\tilde{\mathcal{L}} \to \tilde{\mathcal{L}} \to \tilde{\mathcal{L}}_{|E} \to 0.$$

But we have $\tau^* K_X = K_{X_x} \otimes \mathcal{I}_E^{\otimes n-1}$. Indeed, if ω is a section generating the canonical bundle $K_X = \bigwedge^n \Omega_X$ in the neighbourhood of x, then $\tau^*\omega$ is a holomorphic n-form on X_x which vanishes to order $n-1$ along the exceptional divisor E. Moreover, X_x and X are isomorphic outside x, and thus $\tau^* K_X$ and K_{X_x} are isomorphic outside E. We thus obtain

$$\mathcal{I}_E\tilde{\mathcal{L}} \cong K_{X_x} \otimes \left(\tau^* K_X^{-1} \otimes \tilde{\mathcal{L}} \otimes \mathcal{I}_E^{\otimes n}\right).$$

Now, we can put a metric on the line bundle corresponding to $\mathcal{I}_E^{\otimes n}$ whose Chern form ω_{nE} is positive on $E \cong \mathbb{P}^{n-1}$. Indeed, the restriction of this bundle to E is isomorphic to $H^{\otimes n}$, where H is the dual of the tautological subbundle S over $E_x \cong \mathbb{P}^{n-1}$ (cf. the proof of lemma 3.25). As L admits a metric h with positive Chern form and the Chern form of $L^{\otimes N} \otimes K_X^{-1}$ equipped with the metric $h^{\otimes n} \otimes h_X^{-1}$, where h_X is a metric on K_X, is equal to $N\omega_{L,h} - \omega_{K_X,h_X}$, the line bundle corresponding to

$$\tau^* K_X^{-1} \otimes \tilde{\mathcal{L}}^{\otimes N} \otimes \mathcal{I}_E^{\otimes n}$$

admits a metric of Chern form equal to $\omega_{nE} + \tau^*(N\omega_{L,h} - \omega_{K_X,h_X})$. Now, using the fact that $\omega_{L,h}$ is positive and $\omega_{nE|E_E}$ is positive, we easily see that $\omega_{nE} + \tau^*(N\omega_{L,h} - \omega_{K_X,h_X})$ is positive on X_x for sufficiently large N.

We can then apply theorem 7.13 to this bundle, to conclude that

$$H^1(X_x, \mathcal{I}_E \tilde{\mathcal{L}}^{\otimes N}) = 0$$

for sufficiently large N. Thus, for sufficiently large N, there exists a holomorphic section of $L^{\otimes N}$ which does not vanish at x. This still holds in a neighbourhood of x, and by a compactness argument, we conclude that for sufficiently large N and for every $x \in X$, there exists a holomorphic section of $L^{\otimes N}$ which does not vanish at x.

Similarly, we show that for sufficiently large N, for every pair of points $x, y \in X$, there exists a holomorphic section σ of $L^{\otimes N}$ which vanishes at x but not at y. Finally, by the same type of argument, we show that for sufficiently large N, for every point $x \in X$ and every non-zero tangent vector $u \in T_{X,x}$ of type $(1,0)$, there exists a holomorphic section σ of $L^{\otimes N}$ which vanishes at x but is such that $d\sigma(u) \neq 0$. (Note that the differential of a holomorphic section of L is not defined, since L is not equipped with a connection. But the differential of a section σ is defined at a point x where σ vanishes. Then it is an element of $\Omega_{X,x} \otimes L_x$. It can be computed by locally trivialising L and differentiating the function corresponding to σ via this trivialisation.)

To conclude, it suffices to recall the following construction. Let L be a holomorphic line bundle over a complex manifold. Suppose that there exists a finite-dimensional space V of holomorphic sections of L on X, such that for every $x \in X$ there exists a section $\sigma \in V$ which does not vanish at x. (Note that if X is compact, the space of sections of L is finite-dimensional by corollary 5.25, and thus if such a V exists, we can take $V = H^0(X, L)$. We say that L is base-point-free.) Let

$$\phi_{V,L} : X \to \mathbb{P}(V^*)$$

be the map which associates to x the hyperplane of V consisting of the sections $\sigma \in V$ which vanish at x.

Let $\sigma_0, \ldots, \sigma_r$ be a basis of V, and let $x \in X$ be such that $\sigma_0 \neq 0$. Then σ_0 does not vanish on a neighbourhood U of x in X, and if σ_i^* is the dual basis of V^*, we see immediately that on U, the map ϕ is described by

$$\phi(x) = \left\langle \left(1, \frac{\sigma_1(x)}{\sigma_0(x)}, \ldots, \frac{\sigma_r(x)}{\sigma_0(x)}\right) \right\rangle \subset V^*,$$

where the symbol $\langle u \rangle$ means "line generated by u".

As the functions $\frac{\sigma_i}{\sigma_0}$ are holomorphic, ϕ is holomorphic. Moreover, when V separates the points of X as above, ϕ is a holomorphic embedding. Finally, it is not difficult to see that $\phi^*(\mathcal{O}_{\mathbb{P}^r}(1)) \cong \mathcal{L}$. Theorem 7.11 is thus proved. □

7.2 Examples

7.2.1 Projective space

The integral cohomology of projective space is easy to compute.

Theorem 7.14 *We have $H^k(\mathbb{P}^n, \mathbb{Z}) = 0$ for k odd, and $H^{2k}(\mathbb{P}^n, \mathbb{Z}) = \mathbb{Z}H^k$, $k \leq n$, where $H = c_1(\mathcal{O}_{\mathbb{P}^n}(1))$.*

Proof By induction on n; we have $\mathbb{P}^n = \mathbb{C}^n \cup \mathbb{P}^{n-1}$. As $\mathbb{C}^n$ is contractible, its cohomology is zero in positive degree. We have the exact sequence of relative (singular) cohomology for the pair $(\mathbb{P}^n, \mathbb{C}^n)$ (Spanier 1966):

$$\to H^{k-1}(\mathbb{P}^n, \mathbb{Z}) \to H^{k-1}(\mathbb{C}^n, \mathbb{Z}) \to H^k(\mathbb{P}^n, \mathbb{C}^n, \mathbb{Z}) \to\to H^k(\mathbb{P}^n, \mathbb{Z}) \to \cdots. \tag{7.3}$$

Recall that the relative singular cohomology of the pair (X, U), where U is open in X, is the cohomology of the complex of relative singular cochains

$$C^{\cdot}_{\text{sing}}(X, U) = \operatorname{Ker}(C^{\cdot}_{\text{sing}}(X) \to C^{\cdot}_{\text{sing}}(U)).$$

The exact sequence (7.3) is the long exact sequence associated to the exact sequence of complexes

$$0 \to C^{\cdot}_{\text{sing}}(X, U) \to C^{\cdot}_{\text{sing}}(X) \to C^{\cdot}_{\text{sing}}(U) \to 0.$$

We now have an isomorphism

$$H^k(\mathbb{P}^n, \mathbb{C}^n, \mathbb{Z}) \cong H^{k-2}(\mathbb{P}^{n-1}, \mathbb{Z})$$

obtained by combining the excision isomorphism

$$H^k(\mathbb{P}^n, \mathbb{C}^n, \mathbb{Z}) \cong H^k(T, T - \mathbb{P}^{n-1}, \mathbb{Z}),$$

where T is a tubular neighbourhood of $\mathbb{P}^{n-1}$ in $\mathbb{P}^n$, diffeomorphic to the normal bundle $N \to \mathbb{P}^{n-1}$ of $\mathbb{P}^{n-1}$ in $\mathbb{P}^n$, and the Thom isomorphism

$$H^k(N, N - 0_N, \mathbb{Z}) \cong H^{k-2}(\mathbb{P}^{n-1}, \mathbb{Z}), \tag{7.4}$$

where $0_N \cong \mathbb{P}^{n-1}$ is the zero section of the vector bundle N over $\mathbb{P}^{n-1}$.

(In general, the Thom isomorphism is an isomorphism

$$H^k(E, E - 0_E) \cong H^{k-r}(Z),$$

where $E \to Z$ is an oriented vector bundle of (real) rank r. In the case where Z is a point, E is a vector space, and $E - 0$ has the homotopy type of a sphere, so the result follows from the long exact sequence (7.3). The general result is then an immediate application of the Leray–Hirsch theorem 7.33, whose proof is given in the following chapter (section 8.1.3).)

The exact sequence (7.3) then shows that $H^k(\mathbb{P}^n, \mathbb{Z}) \cong H^{k-2}(\mathbb{P}^{n-1}, \mathbb{Z})$ for $k \geq 1$. By induction on n, we deduce that $H^{2k}(\mathbb{P}^n, \mathbb{Z})$ is free of rank 1 for every $k \leq n$. To see that $H^{2k}(\mathbb{P}^n, \mathbb{Z})$ is generated by H^k, we note that H^k is non-zero for $k \leq n$, since $H^n = 1$ in $H^{2n}(\mathbb{P}^n, \mathbb{Z}) \cong \mathbb{Z}$. Thus, every $\alpha \in H^{2k}(\mathbb{P}^n, \mathbb{Z})$ is a rational multiple βH^k of H^k. But as $\langle \alpha, H^{n-k} \rangle \in \mathbb{Z}$ and $H^n = 1$, we find that $\beta \in \mathbb{Z}$, and thus α is in fact an integral multiple of H^k. □

The Hodge structure on the cohomology groups of $\mathbb{P}^n$ is thus trivial, i.e. all the cohomology classes are of type (p, p).

7.2.2 Hodge structures of weight 1 and abelian varieties

Let X be a Kähler manifold. The Hodge structure on $H^1(X)$ is described by the decomposition

$$H^1(X, \mathbb{C}) = H^{1,0}(X) \oplus H^{0,1}(X).$$

Note that by corollary 7.6, $H^{1,0}(X)$ is represented by the holomorphic forms on X.

We know that $H^{0,1} = \overline{H^{1,0}}$, and thus we have an isomorphism of real vector spaces

$$H^1(X, \mathbb{R}) \subset H^1(X, \mathbb{C}) \to H^{0,1}(X),$$

where the last map is the projection given by the Hodge decomposition. The lattice $H^1(X, \mathbb{Z}) \subset H^1(X, \mathbb{R})$ thus projects onto a lattice in the complex vector space $H^{0,1}(X)$. Thus, we have a complex torus $T = H^{0,1}(X)/H^1(X, \mathbb{Z})$ associated to the Hodge structure on $H^1(X)$. This torus is the Picard variety $\mathrm{Pic}^0(X)$ of X, which parametrises the isomorphism classes of holomorphic line bundles L over X, of Chern class $c_1(L)$ zero in $H^2(X, \mathbb{Z})$. Indeed, we have a natural isomorphism $H^{0,1}(X) \cong H^1(X, \mathcal{O}_X)$, which to the class of a closed form of type $(0, 1)$ associates the Dolbeault class of the corresponding $\overline{\partial}$-closed form. Writing the long exact sequence associated to the exponential exact sequence

$$0 \to \mathbb{Z} \to \mathcal{O}_X \to \mathcal{O}_X^* \to 0,$$

we find that the quotient $T = H^1(X, \mathcal{O}_X)/H^1(X, \mathbb{Z})$ can be identified with

$$\mathrm{Pic}^0(X) := \mathrm{Ker}\,(H^1(X, \mathcal{O}_X^*) \xrightarrow{c_1} H^2(X, \mathbb{Z})),$$

i.e. by theorem 4.49, with the set of isomorphism classes of holomorphic line bundles of first Chern class equal to zero.

Note that the torus T is itself a Kähler manifold, and thus admits a Hodge structure on its group H^1. The relation with the Hodge structure on $H^1(X)$ is the following: for a torus $T = V/\Gamma$, where V is a complex vector space, we have a natural identification of Γ with the singular homology group $H_1(T, \mathbb{Z})$ and of V^* with $H^1(T, \mathbb{R})$. Indeed, an element γ of Γ can be identified with the corresponding segment $[0, \gamma]$ of V. Now, the two endpoints of this segment project onto the same point in T, which gives a closed singular chain $\overline{\gamma}$ in T. The fact that this map is an isomorphism follows from the fact that it gives an isomorphism $\pi_1(T) \cong \Gamma$ since V is simply connected, and from Hurewitz' theorem (see Spanier 1966). The second isomorphism follows from this, but can also be proved directly by noting that if we put a metric with constant coefficients on T, the harmonic 1-forms are the forms with constant coefficients, i.e. the elements of $\mathrm{Hom}\,(V, \mathbb{R})$.

Furthermore, the holomorphic cotangent bundle of T is trivial, as its global sections are given by the complex linear forms on V, considered as holomorphic forms on V invariant under Γ. Thus, $H^{1,0}(T) = V^*$. Thus we have shown that the Hodge structure on $H^1(T)$ is dual to that of $H^1(X)$, i.e. $H^1(T, \mathbb{Z}) = H^1(X, \mathbb{Z})^*$ and $H^{1,0}(T) = H^{0,1}(X)^*$.

Note that T is determined by the Hodge structure on $H^1(T)$. Indeed, by the above, the corresponding Hodge structure on the dual $\Gamma = H^1(T, \mathbb{Z})^*$, $\Gamma^{1,0} = H^{0,1}(T)^* \subset \Gamma \otimes \mathbb{C}$, $\Gamma^{0,1} = \overline{\Gamma^{1,0}}$ satisfies

$$\Gamma^{0,1}/\Gamma \cong T.$$

Suppose now that X is a polarised manifold, and let L be the Lefschetz operator acting on the integral cohomology of X. Obviously, the cohomology of degree 1 is primitive, and thus the alternating intersection form

$$Q(\alpha, \beta) = \langle L^{n-1}\alpha, \beta\rangle, \quad n = \dim X$$

defined on $H^1(X)$ and with integral values on $H^1(X, \mathbb{Z})$ satisfies the property that the Hermitian form $H(\alpha, \beta) = i\,Q(\alpha, \overline{\beta})$ is positive definite on $H^{1,0}(X)$, which is orthogonal to $H^{0,1}(X)$ for H. This can be reinterpreted as follows: the form $Q \in \bigwedge^2 (H^1(X, \mathbb{Z}))^*$ can be considered as an element ω of

$$\bigwedge^2 (H_1(T, \mathbb{Z})^*) = \bigwedge^2 (H^1(T, \mathbb{Z})) = H^2(T, \mathbb{Z}),$$

where the second isomorphism is given by the cup-product. In fact, the de Rham class of ω is simply the class of the constant (and thus closed) 2-form Ω on T obtained by extending Q by $\mathbb{R}$-linearity. (Here, we are using the identification of $H_1(T, \mathbb{Z}) \otimes \mathbb{R}$ with the real tangent space of T at each of its points.) If we identify $H_1(T, \mathbb{Z})$ with $H^1(X, \mathbb{Z})$, and thus $H_1(T, \mathbb{Z}) \otimes \mathbb{R}$ with $H^1(X, \mathbb{R})$, this differential form Ω is equal to Q on $H^1(X, \mathbb{R})$.

The properties of Q then imply the following result.

Lemma 7.15 *The form Ω is a Kähler form on T.*

Proof First of all, we must check that $\Omega = Q$ is of type $(1, 1)$ at each point of T. This means that the $\mathbb{C}$-bilinear extension of Ω to a 2-form on $T_T \otimes \mathbb{C}$ vanishes on $\bigwedge^2 T_T^{1,0}$. But this bilinear extension is the form Q on $H^1(X, \mathbb{C})$, and by definition, $T_T^{1,0} = H^{0,1}(X) \subset H^1(X, \mathbb{C})$, since the complex structure of T_T is given by the identification

$$T_{T,\mathbb{R}} \cong H^{0,1}(X).$$

Thus, the fact that $H^{0,1}(X)$ is totally isotropic for Q implies that Ω is of type $(1, 1)$ over T. Finally the positivity of Ω follows immediately from the positivity of H on $H^{1,0}$. Indeed, if $u \in T_{T,\mathbb{R}}$ is a tangent vector, u can be seen as an element of $H^1(X, \mathbb{R})$, and since the complex structure I on $T_{T,\mathbb{R}}$ is given by the identification $H^{0,1}(X) \cong T_{T,\mathbb{R}}$, we have

$$\Omega(u, Iu) = \Omega(u_1 + \overline{u_1}, -iu_1 + i\overline{u_1}),$$

where $u = 2\Re\, u_1$, $u_1 \in H^{1,0}(X)$, and in the right-hand term Ω can be identified with Q. But this is equal to $2iQ(u_1, \overline{u_1}) = 2H(u_1)$. □

As the Kähler form thus defined on T is of integral class, Kodaira's theorem 7.11 implies that the torus T is in fact an algebraic projective variety. Such a torus is called an abelian variety. We have shown the following.

Proposition 7.16 *The Picard variety* $\mathrm{Pic}^0(X)$ *of a projective smooth variety is an abelian variety.*

In chapter 12, we will establish a direct relation between the manifold X and the dual torus of T, which is called the Albanese variety of X; this relation will show clearly that the Albanese variety is algebraic whenever X is.

7.2.3 Hodge structures of weight 2

We now turn to polarised Hodge structures of weight 2. Such a Hodge structure is given by a lattice $V_{\mathbb{Z}}$ equipped with a symmetric intersection form Q, and a

Hodge decomposition

$$V_{\mathbb{C}} = V^{2,0} \oplus V^{1,1} \oplus V^{0,2}, \quad V^{0,2} = \overline{V^{2,0}}, \quad \overline{V^{1,1}} = V^{1,1}.$$

Moreover, we must have the following:

- This decomposition is orthogonal for the Hermitian form $H(\alpha, \beta) = Q(\alpha, \overline{\beta})$. Equivalently, the orthogonal complement of $V^{2,0}$ for Q is equal to $V^{2,0} \oplus V^{1,1}$.
- The Hermitian form H is negative definite on $V^{1,1}$ and positive definite on $V^{2,0}$ and $V^{0,2}$. Equivalently, the form Q is negative definite on $V^{1,1} \cap V_{\mathbb{R}}$ and positive definite on $(V^{2,0} \oplus V^{0,2}) \cap V_{\mathbb{R}}$.

Lemma 7.17 *A Hodge decomposition of weight 2 polarised by Q is determined by the complex subspace $V^{2,0} \subset V_{\mathbb{C}}$ of rank $h^{2,0}$. This subspace must satisfy:*

- *$V^{2,0}$ is totally isotropic for Q and H is positive definite on $V^{2,0}$. The intersection form Q must of course be of signature $(2h^{2,0}, h^{1,1})$.*

Proof If we are given $V^{2,0}$ satisfying the above conditions, it is clear that $V^{2,0} \cap V_{\mathbb{R}} = 0$, since otherwise $V^{2,0}$ would contain a real isotropic element for Q on which H is positive. But Q and H coincide on the real elements.

If we define $V^{0,2} = \overline{V^{2,0}}$, we have $V^{2,0} \cap V^{0,2} = 0$. We then set

$$V^{1,1} = (V^{2,0} \oplus V^{0,2})^{\perp}.$$

As the form H is positive definite on $V^{2,0}$ and on $V^{0,2}$, which are orthogonal, we have an orthogonal decomposition for H:

$$V_{\mathbb{C}} = V^{2,0} \oplus V^{1,1} \oplus V^{0,2},$$

and H must be negative definite on $V^{1,1}$, since H is positive definite on $V^{2,0} \oplus V^{0,2}$ and of signature $(2h^{2,0}, h^{1,1})$. □

If we take the special case of Hodge structures of weight 2 satisfying $h^{2,0} = 1$, we obtain the following result.

Theorem 7.18 *Let Q be a symmetric non-degenerate intersection form of signature $(2, \dim V - 2)$ on a lattice V. Then the Hodge structures of weight 2 on V satisfying $h^{2,0} = 1$ and polarised by Q are parametrised by the complex manifold*

$$\mathcal{D} = \{\omega \in \mathbb{P}(V_{\mathbb{C}}) \mid Q(\omega, \omega) = 0, \ Q(\omega, \overline{\omega}) > 0\}.$$

The Hodge structures described above are said to be of $K3$ type. The manifold $\mathcal{D}$ is called the "period domain". In the case we are considering here, it is an open set in a projective quadric. We will see later on that the set of Hodge structures (polarised or not) with given Hodge numbers is always a complex manifold.

In the projective space $\mathbb{P}^3(\mathbb{C})$, consider a hypersurface S defined by a homogeneous equation F of degree 4

$$S = \{(z_0, \dots, z_3) \mid F(z_0, \dots, z_3) = 0\}.$$

Remark 7.19 *F is not a function on $\mathbb{P}^3$. It is only a function on $\mathbb{C}^4$, but by homogeneity, the condition $F(z) = 0$ does not depend on the choice of representative $(z_0, \dots, z_3)$. In fact, F must be considered as a section of the sheaf $\mathcal{O}_{\mathbb{P}^3}(4) := \mathcal{O}_{\mathbb{P}^3}(1)^{\otimes 4}$ on $\mathbb{P}^3$.*

If the equation F is chosen generically, i.e. if it belongs to a Zariski open subset of the complex vector space $H^0(\mathbb{P}^3, \mathcal{O}_{\mathbb{P}^3}(4))$, the surface S is smooth. Moreover, we know that $H^{2,0}(S)$ can be identified with the space of holomorphic forms of degree 2 on S, by corollary 7.6. In other words, it is the space of holomorphic sections of the bundle $\bigwedge^2 \Omega_S$ of rank 1.

Lemma 7.20 *The canonical bundle $K_S = \bigwedge^2 \Omega_S$ of such a surface S is trivial.*

Proof Consider the meromorphic differential form on $\mathbb{C}^4$ given by

$$\omega = \sum_i (-1)^i z_i \frac{dz_0 \wedge \cdots \hat{dz_i} \cdots \wedge dz_3}{F(z_0, \dots, z_3)}.$$

We show that it comes from a meromorphic differential form of type (3, 0) on $\mathbb{P}^3$, which we will also call ω. Indeed, if $E = \sum_i z_i \frac{\partial}{\partial z_i}$ is the Euler vector field tangent to the fibres of the quotient map $\pi : \mathbb{C}^4 - \{0\} \to \mathbb{P}^3$, we have

$$\omega = \frac{\operatorname{int}(E)(dz_0 \wedge \cdots \wedge dz_3)}{F(z_0, \dots, z_3)}.$$

This shows that ω is a meromorphic section of $\pi^* \Omega^3_{\mathbb{P}^3}$, and as it is invariant under the action of $\mathbb{C}^*$ because the homogeneity degree of F is 4, ω must be the pullback of a meromorphic section ω of $\Omega^3_{\mathbb{P}^3}$. The form ω admits a pole of order 1 along S. This means that along S, in local coordinates x_1, x_2, x_3 which are holomorphic for $\mathbb{P}^3$, and for a local equation f for S, ω can be written

$$\omega = \frac{g}{f} dx_1 \wedge dx_2 \wedge dx_3.$$

This is clear, since in the open set U_3 where $z_3 \neq 0$, we can take for example local coordinates $z_0, z_1, z_2, z_3 = 1$ (i.e. we identify U_3 with the affine hyperplane

$z_3 = 1$ of $\mathbb{C}^4$) and we can take $f = F(z_0, \dots, 1)$ as an equation for S. Then ω can be written $-dz_0 \wedge dz_1 \wedge dz_2/f$ in these coordinates. This also shows that g does not vanish along S. (This is clearly independent of the choice of coordinates.)

We can then define the residue $\alpha = \mathrm{Res}_S\omega$ of such a form on S; α is a holomorphic 2-form on S, which is in fact everywhere non-zero when the local coefficient g is everywhere non-zero along S. The 2-form α is defined as follows: as S is smooth, in the neighbourhood of S the differential df is non-zero, and in the neighbourhood of each point of S, we can choose a system of holomorphic coordinates x_1, x_2, x_3 on $\mathbb{P}^3$ such that $f = x_3$. In these coordinates, ω can be written $\omega = g(x_1, x_2, f)dx_1 \wedge dx_2 \wedge \frac{df}{f}$, where g is a holomorphic function and the x_i, $i = 1, 2$ give local coordinates for S. We then set $\alpha = 2i\pi g(x_1, x_2, 0)dx_1 \wedge dx_2$. We check that this holomorphic 2-form on S depends only on ω and not on the choice of coordinates or of a local equation for S.

Since the form α is everywhere non-zero, the canonical bundle K_S admits a holomorphic section which is everywhere non-zero, i.e. which is trivial. □

This lemma implies that $\dim H^{2,0}(S) = 1$, since as S is compact, we have $H^0(\mathcal{O}_S) = \mathbb{C}$. Moreover, all the surfaces obtained in this way are diffeomorphic. Indeed, the smooth surfaces S as above are parametrised by a Zariski open set (i.e. the complement of an analytic or algebraic closed set) of the projective space of homogeneous polynomials of degree 4. Such an open set is connected. Thus, two such surfaces are the fibres of a family of smooth surfaces parametrised by a connected manifold, so they are diffeomorphic by Ehresmann's theorem 9.3. In particular, they are homeomorphic and have cohomology groups which are isomorphic, although not canonically. The primitive cohomology of such a surface, relative to the polarisation given by $H = c_1(\mathcal{O}_S(1))$ (it is the Kähler class of the restriction of the Fubini–Study metric), is a lattice $V_{\mathbb{Z}}$ of rank 21 (see Beauville and Bourguignon 1985). In fact, it is the orthogonal complement of H for the intersection form $\langle , \rangle$ on $H^2(S, \mathbb{Z})$, and up to isomorphism, this lattice $(V_{\mathbb{Z}}, \langle , \rangle)$ does not depend on the choice of S.

The corresponding period domain $\mathcal{D}$ is thus of complex dimension 19 (it is an open set of a quadric in $\mathbb{P}^{20}$). Now, the space of effective parameters for S is also of dimension 19. Indeed, the projective space U of homogeneous polynomials of degree 4 over $\mathbb{P}^3$ (or rather, the open set parametrising the smooth surfaces) is of dimension 34, and the group $\mathrm{PGl}(4)$, which acts on this space in the obvious way, transforms an equation into another equation defining an isomorphic surface. As $\dim \mathrm{PGl}(4) = 15$, the quotient $U/\mathrm{PGl}(4)$ is of dimension 19. We have the following result (see Beauville and Bourguignon 1985).

Theorem 7.21 *(Piateckii–Shapiro–Shafarevitch) The map which to a surface S as above, equipped with an isomorphism of polarised lattices*

$$\gamma : V_{\mathbb{Z}} \cong H^2(S, \mathbb{Z})_{\text{prim}},$$

associates the Hodge structure on $H^2(S)_{\text{prim}}$ is an isomorphism of $\tilde{U}/\mathrm{PGl}(4)$ with an open set of $\mathcal{D}$.

Here $\tilde{U}$ is the covering of U parametrising the pairs (S, γ). (Such a pair is called a "marked" surface.) The surfaces considered here are called $K3$ surfaces. They are studied in (Beauville and Bourguignon 1985) and in Barth *et al.* (1984). They occupy a privileged position in the classification of Kähler surfaces. Indeed, one can show that every Kähler surface having a trivial canonical bundle and satisfying the condition $b_1 = 0$ can be obtained as a deformation (cf. following chapter) of a surface of the type described above. These surfaces are quartic $K3$ surfaces, whereas a generic $K3$ surface (i.e. a $K3$ surface parametrised by a generic point of the basis of the universal family of deformations) cannot be embedded into projective space, since its Picard group is 0.

7.3 Functoriality

7.3.1 Morphisms of Hodge structures

Let $(V_{\mathbb{Z}}, F^p V_{\mathbb{C}})$ and $(W_{\mathbb{Z}}, F^p W_{\mathbb{C}})$ be Hodge structures of weight n and $m = n + 2r$, $r \in \mathbb{Z}$ respectively.

Definition 7.22 *A morphism of groups $\phi : V_{\mathbb{Z}} \to W_{\mathbb{Z}}$ is a morphism of Hodge structures if the morphism $\phi : V_{\mathbb{C}} \to W_{\mathbb{C}}$ obtained by $\mathbb{C}$-linear extension satisfies $\phi(F^p V_{\mathbb{C}}) \subset F^{p+r} W_{\mathbb{C}}$, or equivalently, $\phi(V^{p,q}) \subset W^{p+r,q+r}$.*

We say that ϕ is a morphism of Hodge structures of type (r, r).

Lemma 7.23 *If ϕ is a morphism of Hodge structures, then ϕ is strict for the Hodge filtration, i.e $\mathrm{Im}\,\phi \cap F^{k+r} W_{\mathbb{C}} = \phi(F^k V_{\mathbb{C}})$.*

Proof Let $\alpha = \phi(\beta)$, $\alpha \in F^{k+r} W_{\mathbb{C}}$. Let us write $\beta = \sum_{p+q=n} \beta^{p,q}$. Then $\alpha = \sum_{p,q} \phi(\beta^{p,q})$, where $\phi(\beta^{p,q})$ is of type $(p + r, q + r)$. Thus, $\alpha \in F^{k+r} W_{\mathbb{C}}$ if and only if $\phi(\beta^{p,q}) = 0$ for $p < k$. But then $\alpha = \phi(\sum_{p \geq k} \beta^{p,q})$ with $\sum_{p \geq k} \beta^{p,q} \in F^k V_{\mathbb{C}}$. □

Corollary 7.24 *The quotient filtration induced by $F^{\cdot}V_{\mathbb{C}}$ on $\operatorname{Im}\phi$ coincides with the filtration induced by $F^{\cdot+r}W_{\mathbb{C}}$ on the vector subspace $\operatorname{Im}\phi$. This filtration defines a Hodge structure on $\operatorname{Im}\phi$.*

Proof Only the second point still needs to be proved. We need to see that if

$$(\operatorname{Im}\phi)^{p+r,q+r} = \operatorname{Im}\phi \cap F^{p+r}W_{\mathbb{C}} \cap \overline{F^{q+r}W_{\mathbb{C}}} = \operatorname{Im}\phi \cap W^{p+r,q+r},\ p+q=n,$$

then we have

$$\operatorname{Im}\phi = \bigoplus_{p+q=n} (\operatorname{Im}\phi)^{p+r,q+r}. \tag{7.5}$$

But the argument used above shows that $(\operatorname{Im}\phi)^{p+r,q+r} = \phi(V^{p,q})$, which proves (7.5). □

With the morphism of Hodge structures ϕ as above, we also have an induced Hodge structure on $\operatorname{Ker}\phi$.

Lemma 7.25 *Let $F^pK_{\mathbb{C}} = K_{\mathbb{C}} \cap F^pV_{\mathbb{C}}$, where $K_{\mathbb{Z}} = \operatorname{Ker}\phi \subset V_{\mathbb{Z}}$, $K_{\mathbb{C}} = \operatorname{Ker}\phi \subset V_{\mathbb{C}}$, so that $K_{\mathbb{C}} = K_{\mathbb{Z}} \otimes \mathbb{C}$. Then $(K_{\mathbb{Z}}, F^pK_{\mathbb{C}})$ is a Hodge structure.*

Proof It suffices to see that if $K^{p,q} = F^pK_{\mathbb{C}} \cap \overline{F^qK_{\mathbb{C}}}$, $p+q=n$, we have $K_{\mathbb{C}} = \bigoplus_{p+q=n} K^{p,q}$. But if $\alpha \in K_{\mathbb{C}}$, let $\alpha = \sum_{p+q=n} \alpha^{p,q}$ be its Hodge decomposition in $V_{\mathbb{C}}$. Then $\phi(\alpha) = 0 = \sum_{p+q=n} \phi(\alpha^{p,q})$, with $\phi(\alpha^{p,q}) \in W^{p+r,q+r}$. Thus, $\phi(\alpha^{p,q}) = 0$ and $\alpha^{p,q} \in K^{p,q}$. □

One can show similarly that the induced filtration on the cokernel of ϕ (modulo torsion) defines a Hodge structure on $\operatorname{Coker}\phi$. If we consider the category of rational Hodge structures of given weight, where the morphisms are the morphisms of rational Hodge structures of type (0, 0), the preceding results show that this category is an abelian category, where exact sums are defined in the obvious way.

The polarised Hodge structures, together with morphisms which are the morphisms of Hodge structures, form an even more rigid category.

Lemma 7.26 *Let $V_{\mathbb{Q}} \subset W_{\mathbb{Q}}$ be a rational sub-Hodge structure. Then if the Hodge structure on W is polarised, the same holds for the Hodge structure on V, and we have a decomposition as a direct sum*

$$W_{\mathbb{Q}} = V_{\mathbb{Q}} \oplus V'_{\mathbb{Q}},$$

where $V'_{\mathbb{Q}}$ is also a sub-Hodge structure of $W_{\mathbb{Q}}$.

A sub-Hodge structure is, of course, simply a $\mathbb{Q}$-vector subspace which, when tensored with $\mathbb{C}$, has a Hodge decomposition induced by that of W.

Proof Let Q be the form giving the polarisation of W. Let us first show that $Q_{|V}$ is non-degenerate on V. It suffices to show that the associated Hermitian form H

$$H(\alpha, \beta) = i^k Q(\alpha, \overline{\beta}),$$

where k is the weight of W, is non-degenerate on $V_{\mathbb{C}}$. But $V_{\mathbb{C}} = \bigoplus_{p+q=k} V^{p,q}$ with $V^{p,q} = W^{p,q} \cap V_{\mathbb{C}}$. Now, on $W^{p,q}$, the form H is of a definite sign, and the $W^{p,q}$ are orthogonal for H. Thus, H remains definite on $V^{p,q}$, and the $V^{p,q}$ are orthogonal for H. So $H_{|V_{\mathbb{C}}}$ is non-degenerate.

Now set $V'_{\mathbb{Q}} = V_{\mathbb{Q}}^{\perp}$, where the orthogonal complement is taken with respect to the rational intersection form Q. It is easy to check that $V'_{\mathbb{C}} = \oplus V'^{p,q}$, where $V'^{p,q} \subset W^{p,q}$ is the orthogonal complement of $V^{p,q}$ in $W^{p,q}$ with respect to H. Thus, V' is indeed a sub-Hodge structure of W, and an orthogonal complement to V. □

7.3.2 The pullback and the Gysin morphism

From the geometric point of view, the most natural morphisms of Hodge structure are the morphisms ϕ^* and ϕ_* (respectively the pullback and Gysin morphisms) induced by a holomorphic map $\phi : X \to Y$ between two compact Kähler manifolds.

The morphism

$$\phi^* : H^k(Y, \mathbb{Z}) \to H^k(X, \mathbb{Z}),$$

where ϕ is a continuous map between two topological spaces, is simply induced by the natural morphism of sheaves

$$\mathbb{Z}_Y \to \phi_* \mathbb{Z}_X. \tag{7.6}$$

This induces a map $H^k(Y, \mathbb{Z}) \to H^k(Y, \phi_*\mathbb{Z})$. Moreover, there exists a natural map

$$H^k(Y, \phi_*\mathbb{Z}) \to H^k(X, \mathbb{Z}) \tag{7.7}$$

obtained as follows. If $\mathbb{Z} \subset I^\cdot$ is a flasque resolution of $\mathbb{Z}_X$, we know that $H^k(X, \mathbb{Z})$ is the kth cohomology group of the complex $\Gamma(X, I^\cdot) = \Gamma(Y, \phi_* I^\cdot)$. As the $\phi_* I^\cdot$ are flasque, this is also the hypercohomology (cf. following chapter) of the complex $\phi_* I^{\cdot}$, and the map (7.7) is simply the map in (hyper)cohomology induced by the morphism of complexes of sheaves

$$\phi_* \mathbb{Z} \to \phi_* I^{\cdot}.$$

Remark 7.27 *We can also define ϕ^* by using the identification of the cohomology $H^k(X, \mathbb{Z})$ with the singular cohomology (section 4.3.2). Then ϕ^* is the morphism induced on cohomology by the morphism ϕ^* between the complexes of singular cochains of Y and X. If α is a singular cochain, and $\psi : \Delta \to X$ is a singular chain of X, then $\phi^*(\alpha)(\psi) = \alpha(\phi \circ \psi)$.*

When ϕ is a differentiable map between differentiable manifolds, the corresponding morphism $\phi^* : H^k(Y, \mathbb{R}) \to H^k(X, \mathbb{R})$ can be better understood as the morphism induced by the pullback of the differential forms: if α is a closed (resp. exact) differential form on Y, $\phi^*\alpha$ is a closed (resp. exact) differential form on X, which gives a map

$$H^k_{DR}(Y) \stackrel{\phi^*}{\to} H^k_{DR}(X).$$

To see that these morphisms are actually the same, using the isomorphisms

$$H^k(X, \mathbb{R}) \cong H^k_{DR}(X), \quad H^k(Y, \mathbb{R}) \cong H^k_{DR}(Y),$$

it suffices to note that we have a morphism

$$\phi^* : \mathcal{A}^{\cdot}_Y \to \phi_* \mathcal{A}^{\cdot}_X$$

between the de Rham complex of Y and the direct image of that of X, which extends the morphism (7.6).

This gives a simple proof in the differentiable case of the fact that the morphism ϕ^* is compatible with the cup-product:

$$\phi^*(\alpha \cup \beta) = \phi^*(\alpha) \cup \phi^*(\beta).$$

When ϕ is a holomorphic map between Kähler manifolds, the morphism ϕ^* is a morphism of Hodge structures. Indeed, $H^{p,q}(Y)$ is the set of classes representable by a closed form of type (p, q), and clearly the pullback of such a form is still of type (p, q), so its class is in $H^{p,q}(X)$.

The following point is an important one.

Lemma 7.28 *Let $\phi : X \to Y$ be a surjective holomorphic map between two compact complex manifolds, with X Kähler. Then the map $\phi^* : H^k(Y, \mathbb{Q}) \to H^k(X, \mathbb{Q})$ is injective for every k.*

Proof It suffices to show the injectivity on the cohomology with real coefficients. Moreover, we may assume that Y is connected. First let $n = \dim Y$, and let $\eta \in H^{2n}(Y, \mathbb{R}) \cong \mathbb{R}$ be a non-zero element. We may assume that η is represented by a positive form. Then if $r = \dim X - \dim Y$, and ω is a Kähler form on X, the form of maximal degree $\phi^*\eta \wedge \omega^r$ on X is positive or zero, and strictly positive exactly where ϕ is a submersive map, i.e. at least on an open set of X. Thus, $\int_X \phi^*\eta \wedge \omega^r \neq 0$ in $H^{2n+2k}(X)$ and so $\phi^*\eta \neq 0$ in $H^{2n}(X)$.

Finally, let $0 \neq \alpha \in H^k(Y)$. By the Poincaré duality theorem 5.30, there exists $\beta \in H^{2n-k}(Y)$ such that $\alpha \cup \beta \neq 0 \in H^{2n}(Y)$, where $\cup$ is the cup-product. But then, $\phi^*(\alpha \cup \beta) \neq 0$ in $H^{2n}(X)$ by the above. This is equal to $\phi^*\alpha \cup \phi^*\beta$, so $\phi^*\alpha \neq 0$ in $H^k(X)$. □

Remark 7.29 *In the case where* $\dim Y = \dim X$, *we do not need the Kähler hypothesis. When the morphism* ϕ *has finite generic fibre of cardinal* d, *it is of degree* $d > 0$ *(see Milnor 1965), since its differential is* $\mathbb{C}$*-linear and thus preserves the orientation. In this case we have the following formula:*

$$\phi_* \circ \phi^* = d\mathrm{Id} : H^k(Y, \mathbb{Z}) \to H^k(Y, \mathbb{Z}),$$

where $\phi_* : H^l(X, \mathbb{Z}) \to H^l(Y, \mathbb{Z})$ *is defined below for a differentiable map between compact oriented manifolds of the same dimension.*

In particular, we find that if $\phi : X \to Y$ *is a differentiable map of degree* $d \neq 0$ *between compact differentiable manifolds, the map* $\phi^* : H^l(Y, \mathbb{Z}) \to H^l(X, \mathbb{Z})$ *is injective for every* l.

If $\phi : X \to Y$ is a morphism between two Kähler manifolds of dimension n and m respectively, with $m = r + n$, the Gysin morphism

$$\phi_* : H^k(X, \mathbb{Z}) \to H^{k+2r}(Y, \mathbb{Z})$$

is defined using Poincaré duality for X and Y as the morphism

$$\phi_* : H_{2n-k}(X, \mathbb{Z}) \to H_{2n-k}(Y, \mathbb{Z}),$$

where H_{2n-k} is the singular homology (cf. section 4.3.2), and ϕ_* is defined on the singular chains $\psi : \Delta_l \to X$ by $\phi_*(\psi) = \phi \circ \psi$.

We can also define it (on the cohomology modulo torsion) as the composition

$$H^k(X, \mathbb{Z}) \overset{\mathrm{PD}}{\cong} H^{2n-k}(X, \mathbb{Z})^* \overset{{}^t(\phi^*)}{\to} H^{2n-k}(Y, \mathbb{Z})^* \overset{\mathrm{PD}}{\cong} H^{k+2r}(Y, \mathbb{Z}),$$

where PD is the Poincaré duality morphism (theorem 5.30) and ${}^t(\phi^*)$ is the adjoint of ϕ^*.

It is easy to see that ϕ_* is a morphism of Hodge structures of bidegree (r, r). Indeed, if α is a class of type (p, q) on X, we must check that $\phi_*(\alpha)$ is of type $(p + r, q + r)$. We use the following lemma.

Lemma 7.30 *We have*

$$H^{k,l}(Y) = \left(\bigoplus_{\substack{p'+q'=2m-k-l \\ (p',q')\neq(m-k,m-l)}} H^{p',q'}(Y) \right)^{\perp},$$

where the orthogonality is relative to the Poincaré duality on Y.

Now, by the last definition of $\phi_*(\alpha)$, we have

$$(\phi_*\alpha, \beta)_Y = (\alpha, \phi^*\beta)_X.$$

If $\beta \in \bigoplus H^{p',q'}(Y)$, $(p', q') \neq (m - p - r, m - q - r)$, then

$$\phi^*\beta \in \bigoplus H^{p',q'}(X), \quad (p', q') \neq (m - p - r, m - q - r),$$

so $(\alpha, \phi^*\beta)_X = 0$. □

Proof of lemma 7.30 The classes in $H^{k,l}(Y)$ are representable by closed forms of type (k, l). Thus, if $\alpha \in H^{k,l}$ is represented by a closed form $\tilde{\alpha}$ of type (k, l) and $\beta \in H^{p',q'}$ is represented by a closed form $\tilde{\beta}$ of type (p', q'), with $(p', q') \neq (m - k, m - l)$, we have $\tilde{\alpha} \wedge \tilde{\beta} = 0$, since the forms of degree $2m$ on Y are necessarily of bidegree (m, m), and thus

$$\langle \alpha, \beta \rangle = \int_Y \tilde{\alpha} \wedge \tilde{\beta} = 0.$$

So we have the inclusion

$$H^{k,l}(Y) \subset \left(\bigoplus_{\substack{p'+q'=2m-k-l \\ (p',q')\neq(m-k,m-l)}} H^{p',q'}(Y) \right)^{\perp}.$$

But also, the right-hand term is of dimension equal to

$$h^{m-k,m-l}(Y) := \dim H^{m-k,m-l}(Y),$$

which is the codimension of

$$\bigoplus_{\substack{p'+q'=2m-k-l \\ (p',q')\neq(m-k,m-l)}} H^{p',q'}(Y)$$

in $H^{2m-k-l}(Y)$. Now, we have $h^{k,l}(Y) = h^{m-k,m-l}(Y)$, since by Hodge symmetry, we have $h^{k,l}(Y) = h^{l,k}(Y)$, and by Lefschetz' theorem 6.25, we have $h^{l,k}(Y) = h^{m-k,m-l}(Y)$, since the Lefschetz isomorphism

$$L^{m-k-l} : H^{k+l}(Y) \cong H^{2m-k-l}(Y)$$

is bigraded of bidegree $(m-k-l, m-k-l)$ and thus sends $H^{l,k}(Y)$ to $H^{m-k,m-l}(Y)$. Thus, the two spaces have the same dimension, and the inclusion implies the equality. □

7.3.3 Hodge structure of a blowup

Let X be a Kähler manifold, and let $Z \subset X$ be a submanifold. By proposition 3.24, the blowup $\tilde{X}_Z \stackrel{\tau}{\to} X$ of X along Z is still a Kähler manifold. Let $E = \tau^{-1}(Z)$ be the exceptional divisor. E is a projective bundle of rank $r-1$, $r = \operatorname{codim} Z$. Moreover, $E \stackrel{j}{\hookrightarrow} \tilde{X}_Z$ is a smooth hypersurface. The Hodge structure on $H^k(\tilde{X}_Z, \mathbb{Z})$ is described as follows.

Theorem 7.31 *Let $h = c_1(\mathcal{O}_E(1)) \in H^2(E, \mathbb{Z})$, where the line bundle $\mathcal{O}_E(1)$ was described in section 3.3.2. Then we have an isomorphism of Hodge structures*

$$H^k(X, \mathbb{Z}) \oplus \left(\bigoplus_{i=0}^{r-2} H^{k-2i-2}(Z, \mathbb{Z})\right) \stackrel{\tau^* + \sum_i j_* \circ h^i \circ \tau_{|E}^*}{\longrightarrow} H^k(\tilde{X}_Z, \mathbb{Z}). \quad (7.8)$$

Here, h^i is the morphism of Hodge structures given by the cup-product by $h^i \in H^{2i}(E, \mathbb{Z})$. On the components $H^{k-2i-2}(Z, \mathbb{Z})$ of the left-hand term, we put the Hodge structure of Z, but shifted by $(i+1, i+1)$ in bidegree, so as to obtain a Hodge structure of weight k.

Proof By the results of the preceding section, the morphism (7.8) is a morphism of Hodge structures. It thus suffices to prove that it is an isomorphism of $\mathbb{Z}$-modules. Let $U \subset X$ be the open set $X - Z$. Then U is also isomorphic to the open set $\tilde{X}_Z - E$ of $\tilde{X}_Z$. As τ gives a morphism between the pair $(\tilde{X}_Z, U)$ and the pair (X, Z), we have a morphism τ^* between the long exact sequences of cohomology relative to these pairs (where we are considering cohomology groups with integral coefficients):

$$\begin{array}{ccccccc} H^{k-1}(U) & \to & H^{k-1}(X, U) & \to & H^k(X) & \to & H^k(U) \\ \downarrow \tau_U^* & & \downarrow \tau_{X,U}^* & & \downarrow \tau_X^* & & \downarrow \tau_U^* \\ H^{k-1}(U) & \to & H^{k-1}(\tilde{X}_Z, U) & \to & H^k(\tilde{X}_Z) & \to & H^k(U) \end{array} \quad (7.9)$$

The first and last maps are of course the identity. Furthermore, by excision and by the Thom isomorphism, we have

$$H^{k-1}(X, U) \cong H^{k-2r}(Z), \quad H^{k-1}(\tilde{X}_Z, U) \cong H^{k-2}(E).$$

Moreover the morphism $H^{k-1}(\tilde{X}_Z, U) \cong H^{k-2}(E) \to H^k(\tilde{X}_Z)$ can be identified with j_*, and the morphism $H^{k-1}(X, U) \cong H^{k-2r}(Z) \to H^k(X)$ can be identified with j_{Z*}, where j_Z is the inclusion of Z in X.

Lemma 7.32 *The cohomology $H^{\cdot}(E, \mathbb{Z})$ of the projective bundle $E \to Z$ is a free module over the ring $H^{\cdot}(Z, \mathbb{Z})$, with basis $1, h, \ldots, h^{r-1}$.*

Temporarily admitting this lemma, we conclude as follows: lemma 7.28, or even better, remark 7.29 implies that the map $\tau_X^* : H^k(X) \to H^k(\tilde{X}_Z)$ is injective. Then, lemma 7.32 implies that

$$\tau_{X,U}^* : H^{k-1}(X, U) \to H^{k-1}(\tilde{X}_Z, U)$$

is injective: more precisely, we can consider $\tau_{X,U}^*$ as a morphism which we denote by

$$\alpha : H^{k-2r}(Z) \to H^{k-2}(E) = \bigoplus_{i=0}^{i=r-1} h^i \tau^* H^{k-2-2i}(Z).$$

It is not difficult to see that the $(r-1)$th component α_{r-1} of α is equal to $h^{r-1}\tau^*$, and thus $\tau_{X,U}^*$ is indeed injective.

The commutativity of the diagram of long exact sequences of relative cohomology (7.9) then implies that the natural map

$$(\tau^*, j_*) : H^k(X) \oplus H^{k-2}(E) \to H^k(\tilde{X}_Z)$$

is surjective, and the injectivity of $\tau_{X,U}^*$ in degree $k-1$ shows that the kernel of this map is

$$\mathrm{Im}(j_{Z*}, -\alpha) : H^{k-2r}(Z) \to H^k(X) \oplus H^{k-2}(E).$$

Lemma 7.32 and the fact that $\alpha_{r-1} = -h^{r-1}\tau^*$ then show that (7.8) is an isomorphism. □

Proof of lemma 7.32 This follows from the following theorem (7.33), due to Leray and Hirsch. Indeed, by theorem 7.14, the classes $h^i \in H^{2i}(E, \mathbb{Z})$, $i = 0, \ldots, r-1$, restricted to each fibre E_x of $\tau : E \to Z$, form a basis of $H^{\cdot}(E_x, \mathbb{Z})$. □

Let X be a locally contractible topological space, and let $\phi : Y \to X$ be a fibration. This means that Y is locally homeomorphic, above X, to a product: for every $x \in X$, there exists a neighbourhood U_x of x such that

$$\phi^{-1}(U_x) \cong Y_x \times U_x, \quad Y_x = \phi^{-1}(x).$$

Assume that the cohomology groups $H^*(Y_x, \mathbb{Z})$ are torsion free.

Theorem 7.33 (Leray–Hirsch) *Suppose there exist cohomology classes*

$$\alpha_1, \ldots, \alpha_N \in H^*(Y, \mathbb{Z})$$

such that for every $x \in X$, the subgroup A of $H^*(Y, \mathbb{Z})$ generated by the α_i is isomorphic by restriction to the group $H^*(Y_x, \mathbb{Z})$. Then $H^*(Y, \mathbb{Z})$ is isomorphic, via the morphism ϕ^* and the cup-product, to $A \otimes_{\mathbb{Z}} H^*(X, \mathbb{Z})$.

This theorem is an immediate consequence of the theory of hypercohomology, developed in the following chapter. It will be proved in section 8.1.3.

Exercises

1. Let X be a compact Kähler manifold.
   (a) Show that the Kähler form is harmonic.
   (b) Deduce from this that if $H^2(X, \mathcal{O}_X) = 0$, the set of Kähler classes is open in $H^2(X, \mathbb{R})$.
   (c) Under the previous hypothesis, show that X is projective.
2. *Serre's vanishing theorem for vector bundles.* Let E be a holomorphic vector bundle of rank r over a complex manifold X. Let

$$\pi : \mathbb{P}(E^*) \to X$$

   be the associated projective bundle. Let $H := \mathcal{O}_{\mathbb{P}(E^*)}(1)$ (cf. section 3.3.2).
   (a) Show with the help of local trivializations that the natural evaluation morphism

$$E \to R^0\pi_* H$$

   is an isomorphism.
   (b) Deduce from this, using exercise 2 of chapter 3, that for any line bundle L on X, there is an evaluation isomorphism

$$E \otimes L \to R^0\pi_*(H \otimes \pi^* L)$$

   and hence an isomorphism

$$H^0(\mathbb{P}(E^*), H \otimes \pi^* L) \cong H^0(X, E \otimes L).$$

We shall admit that there is more generally an isomorphism for any $i \geq 0$:

$$H^i(\mathbb{P}(E^*), H \otimes \pi^* L) \cong H^i(X, E \otimes L).$$

(c) Show that the canonical line bundle of $\mathbb{P}(E^*)$ is isomorphic to $\pi^* K_X \otimes H^{-r} \otimes \pi^* \det E$.

(d) We assume now that X is compact and that L is positive, that is, can be endowed with a metric of positive curvature. Show that for N large enough,

$$K^{-1}_{\mathbb{P}(E^*)} \otimes H \otimes \pi^* L^{\otimes N}$$

can be endowed with a metric of positive curvature.

(e) Deduce from Kodaira's vanishing theorem that for N large enough,

$$H^i(X, E \otimes L^{\otimes N}) = 0, \quad \forall i > 0.$$

8

Holomorphic de Rham Complexes and Spectral Sequences

In this chapter, we develop the results obtained in the preceding chapters in various directions. First of all, we propose another definition of the Hodge filtration on the cohomology of a compact Kähler manifold, which makes it possible to generalise this filtration to the cohomology of a complex manifold without the Kähler hypothesis. For this, we introduce the holomorphic de Rham complex $\Omega_X^\cdot$, and we show (via the holomorphic Poincaré lemma) that this complex is a resolution of the constant sheaf. This resolution is not a resolution by acyclic sheaves, but by sheaves of free $\mathcal{O}_X$-modules. We also introduce the logarithmic holomorphic de Rham complex $\Omega_X^\cdot(\log D)$ of an open manifold, and more precisely, of an open subset $j : U \hookrightarrow X$ of a complex manifold whose complement is a normal crossing divisor D, and we show that the cohomology of this complex in degree k is equal to $R^k j_*\mathbb{C}$.

After this, we introduce the notion of derived functors $R^i F(M^\cdot)$ for a complex $M^\cdot$ in an abelian category $\mathcal{A}$, and for a left-exact functor $F : \mathcal{A} \to \mathcal{B}$. The important point is the fact that a morphism of complexes $\phi^\cdot : M^\cdot \to N^\cdot$ induces a canonical morphism $R^i F(\phi) : R^i F(M^\cdot) \to R^i F(N^\cdot)$, which is an isomorphism if $\phi^\cdot$ is a quasi-isomorphism. The objects $R^i F(M^\cdot)$ are defined using a quasi-isomorphism $i : M^\cdot \to I^\cdot$, where $I^\cdot$ is a complex of injective objects, and the canonical isomorphism above means that given quasi-isomorphisms $i_M : M^\cdot \to I^\cdot$ and $i_N : N^\cdot \to J^\cdot$ as above, there exists a canonical morphism $R^i F(\phi) : R^i F(M^\cdot) \to R^i F(N^\cdot)$ where the derived objects are computed using i_M and i_N.

Applying this to the case where $\mathcal{A}$ is the category of sheaves of abelian groups on X, and F is the global section functor, we obtain the hypercohomology $\mathbb{H}^i(X, \mathcal{F}^\cdot)$ of a complex of sheaves on X.

The holomorphic de Rham resolutions introduced above then give the following result.

Theorem 8.1 *We have canonical isomorphisms*

$$H^k(X, \mathbb{C}) = \mathbb{H}^k(X, \Omega_X^{\cdot}),$$

and in the case of an open subset $U = X - D$, we have

$$H^k(U, \mathbb{C}) = \mathbb{H}^k(X, \Omega_X^{\cdot}(\log D)).$$

This enables us to give the following definition of the Hodge filtration on the cohomology of a complex manifold.

Definition 8.2 *Set*

$$F^p H^k(X, \mathbb{C}) = \mathrm{Im}\,(\mathbb{H}^k(X, F^p\Omega_X^{\cdot}) \to \mathbb{H}^k(X, \Omega_X^{\cdot})),$$

where the complex $F^p\Omega_X\cdot$ is the truncated holomorphic de Rham complex

$$0 \to \Omega_X^p \to \Omega_X^{p+1} \to \cdots.$$

We show that if X is Kähler, the Hodge filtration thus defined coincides with the Hodge filtration deduced from the Hodge decomposition. In the case of an open set $U = X - D$ of a Kähler manifold, the above definition applied to the logarithmic complex enables us to equip $H^k(U, \mathbb{C})$ with a Hodge filtration (which is not the same as that obtained by truncating the de Rham complex of U). We sketch the proof of the existence of a mixed Hodge structure on the cohomology of U, whose Hodge filtration is defined as above.

A good part of this chapter is devoted to spectral sequences, which serve as an instrument to compute the cohomology of a filtered complex $(M^{\cdot}, F)$ by successive approximations. This means that we have complexes $(E_r^{p,q}, d_r)$, such that $E_r^{p,q}$ is equal to the cohomology of the complex E_{r-1} in bidegree (p, q), and moreover, for sufficiently large r, $E_r^{p,q} = \mathrm{Gr}_F^p H^{p+q}(M^{\cdot})$. The spectral sequence of hypercohomology associated to the filtration F^p on the holomorphic de Rham complex is called the Frölicher spectral sequence. This allows us to reformulate the Hodge decomposition theorem in the following nearly equivalent form.

Theorem 8.3 *The Frölicher spectral sequence of a compact Kähler manifold degenerates at E_1, i.e. the differentials d_r are zero for $r \geq 1$.*

In fact, this last statement is weaker, and does not imply either Hodge decomposition or Hodge symmetry. However, it does imply that if F is the filtration defined above on the cohomology of X, then we have an isomorphism

$$\mathrm{Gr}_F^p H^{p+q}(X,\mathbb{C}) \cong H^q(X,\Omega_X^p).$$

8.1 Hypercohomology

8.1.1 Resolutions of complexes

Let $\mathcal{A}$ and $\mathcal{B}$ be two abelian categories, and let F be a functor from $\mathcal{A}$ to $\mathcal{B}$. Assume that $\mathcal{A}$ has sufficiently many injective objects, and that F is left-exact. We will define the derived functors $R^iF(M^\cdot)$, for every left-bounded complex $M^\cdot$ (a left-bounded complex is a complex such that $M^i = 0,\ i \le r_0$) of the category $\mathcal{A}$.

We first show the following result.

Proposition 8.4 *For every $M^\cdot$ as above, there exists a complex $I^\cdot$ in the category $\mathcal{A}$, which is left-bounded and each of whose terms I^k is an injective object of $\mathcal{A}$, and a morphism $\phi^\cdot : M^\cdot \to I^\cdot$ of complexes, which is a quasi-isomorphism, and which furthermore satisfies the property that for every k, $\phi^k : M^k \to I^k$ is injective.*

Proof We first construct a double complex $I^{k,l}$, (D_1, D_2) satisfying the following conditions.

- Each $I^{k,l}$ is injective.
- Each $(I^{k,\cdot}, D_2)$ is a resolution of M^k.
- The inclusion $(M^\cdot, d_M) \stackrel{i^\cdot}{\to} (I^{\cdot,0}, D_1)$ given by these resolutions is a morphism of complexes.

For this, it suffices to know how to construct the first line $(I^{k,0}, D_1)$, and the injection of complexes

$$i^\cdot : (M^\cdot, d_M) \to (I^{\cdot,0}, D_1).$$

Then we merely have to apply this construction to the quotient complex $\mathrm{Coker}\, i^\cdot$, and so on.

We may assume that the complex $M^\cdot$ is zero in negative degrees. We first choose an inclusion $M^0 \stackrel{i^0}{\to} I^{0,0}$ with $I^{0,0}$ injective. We thus have an injective map

$$M^0 \stackrel{(i^0,-d_M)}{\to} I^{0,0} \oplus M^1.$$

Now let η: Coker $(i_0, -d_M) \to I^{1,0}$ be an injective morphism with $I^{1,0}$ injective. Set

$$i^1 = \eta \circ \pi \circ j : M^1 \to I^{1,0}, \quad D_1 = \eta \circ \pi \circ k : I^{0,0} \to I^{1,0},$$

where $\pi : I^{0,0} \oplus M^1 \to \operatorname{Coker}(i^0, -d_M)$ is the quotient morphism, and $j : M^1 \to I^{0,0} \oplus M^1$, $k : I^{0,0} \to I^{0,0} \oplus M^1$ are the natural injections. Then i^1 is injective, since i^0 is injective. Moreover, we clearly have $D_1 \circ i^0 = i^1 \circ d_M$, since $\eta \circ \pi$ vanishes on the image of $(i^0, -d_M)$.

The construction of $M^2 \overset{i^2}{\to} I^{2,0}$ is now done as follows. We have the diagram

$$\begin{array}{ccccc} M^0 & \overset{d_M}{\to} & M^1 & \overset{d_M}{\to} & M^2 \\ \downarrow^{i^0} & & \downarrow^{i^1} & & \\ I^{0,0} & \overset{D_1}{\to} & I^{1,0}. & & \end{array}$$

Consider the map

$$(\bar{i}^1, -d_M) : M^1 \to \operatorname{Coker} D_1 \oplus M_2,$$

where $\bar{i}^1 : M^1 \to \operatorname{Coker} D_1$ is the composition of $i^1 : M^1 \to I^{1,0}$ with the quotient map $I^{1,0} \to \operatorname{Coker} D_1$. Let us take an inclusion

$$\eta : \operatorname{Coker}(\bar{i}^1, -d_M) \to I^{2,0},$$

where $I^{2,0}$ is injective. Then we easily check that the morphism $i^2 = \eta \circ \pi \circ j : M^2 \to I^{2,0}$ is injective, where $j : M^2 \to \operatorname{Coker} D_1 \oplus M^2$ is the inclusion, and π is the quotient morphism

$$\operatorname{Coker} D_1 \oplus M^2 \to \operatorname{Coker}(\bar{i}^1, -d_M).$$

Furthermore, we have the morphism

$$D_1 = \eta \circ k : I^{1,0} \to I^{2,0},$$

where k is the composition of the quotient morphism $I^{1,0} \to \operatorname{Coker} D_1$ with the natural injection $\operatorname{Coker} D_1 \to \operatorname{Coker} D_1 \oplus M^2$ and with the quotient morphism π. Clearly the differentials $D_1 : I^{0,0} \to I^{1,0}$ and $D_1 : I^{1,0} \to I^{2,0}$ satisfy $D_1 \circ D_1 = 0$, and $D_1 \circ i^1 = i^2 \circ d_M$ on M^1, since $\eta \circ \pi$ vanishes on $\operatorname{Im}(\bar{i}^1, -d_M)$.

Generally, once the commutative diagram

$$\begin{array}{ccccc} M^{k-2} & \overset{d_M}{\to} & M^{k-1} & \overset{d_M}{\to} & M^k \\ \downarrow^{i^{k-2}} & & \downarrow^{i^{k-1}} & & \\ I^{k-2,0} & \overset{D_1}{\to} & I^{k-1,0}. & & \end{array}$$

is constructed, we complete the last square by choosing an injection of

$$\operatorname{Coker}((\bar{i}^{k-1}, -d_M) : M^{k-1} \to \operatorname{Coker} D_1 \oplus M^k)$$

into an injective object.

Having constructed the double complex $I^{\cdot\cdot}$, let

$$(I^{\cdot}, D),\ I^k = \bigoplus_{p+q=k} I^{p,q}, \quad D = D_1 + (-1)^p D_2$$

be the associated simple complex. Naturally, we have a morphism of complexes $i^{\cdot} : M^{\cdot} \to I^{\cdot}$, since by definition $D_2 \circ i = 0$ and $D_1 \circ i = i \circ d_M$. To conclude the proof of proposition 8.4, it thus remains to see the following.

Lemma 8.5 *Let $(I^{\cdot}, D)$ be the simple complex associated to a double complex $(I^{p,q}, D_1, D_2)$. Suppose that for every p, the complex $(I^{p,\cdot}, D_2)$ is a resolution of M^p via the injective morphism $M^p \xrightarrow{i^p} I^{p,0}$. Then the morphism of complexes $i^{\cdot}$ induces an isomorphism $H^k(M^{\cdot}, d_M) \cong H^k(I^{\cdot}, D)$ for every k.*

Proof We will give the proof for the category of abelian groups, since it is particularly revealing in this case. Suppose, to simplify, that $M^{\cdot}$ is zero in negative degrees. Let $\alpha = (\alpha_{p,q})_{p+q=k} \in \bigoplus_{p+q=k} I^{p,q} = I^k$ be such that $D\alpha = 0$. Then $D_2\alpha_{0,k} = 0$ and $D_1\alpha_{p,q} + (-1)^{p+1} D_2\alpha_{p+1,q-1} = 0$ for $q \geq 1$. If $k > 0$, the exactness of $(I^{0,\cdot}, D_2)$ in strictly positive degrees then implies that $\alpha_{0,k} = D_2\beta_{0,k-1}$. But then, if $\alpha' = \alpha - D\beta_{0,k-1}$, we have $\alpha' \in \bigoplus_{p+q=k, p\geq 1} I^{p,q}$, and of course α' is cohomologous to α in $I^{\cdot}$. We then have $D_2\alpha'_{1,k-1} = 0$, and we can use the preceding reasoning again if $k - 1 > 0$. Finally, iterating this argument, we conclude that α is cohomologous, in I, to an element β of $I^{k,0}$. But then $D\beta = 0$ implies that $D_2\beta = 0$ and $D_1\beta = 0$. The first condition says exactly that β is in $i^k(M^k)$, i.e. $\beta = i^k(\beta')$, and the second says that β' is d_M-closed. Thus, we have shown that the map

$$H^k(i^{\cdot}) : H^k(M^{\cdot}, d_M) \to H^k(I^{\cdot}, D)$$

induced by $i^{\cdot}$ is surjective.

Similarly, we show that $H^k(i^{\cdot})$ is injective. Indeed, let $\alpha \in M^k$ be a d_M-closed element, and assume that $i^k(\alpha) = D\beta$ in $I^{\cdot}$ with $\beta = \sum_{p+q=k-1} \beta_{p,q}$. Then if $k-1 = 0$, we have $D_2\beta_{0,0} = 0$, and thus $\beta \in i^{k-1}(M^{k-1})$. If $k-1 > 0$, we have $D_2(\beta_{0,k-1}) = 0$ and the exactness of D_2 in positive degrees implies that $\beta_{0,k-1} = D_2\gamma_{0,k-2}$. If we set $\beta' = \beta - D\gamma$, we then have $\beta' \in \bigoplus_{p+q=k-1, p\geq 1} I^{p,q}$. As $D\beta = D\beta'$ has all its components of type $(p, q), q > 0$ equal to zero, we can continue the reasoning and conclude that $i^k(\alpha) = D\beta$ with β in $I^{k-1,0}$. But then $D_2\beta = 0$, so β is in $i^{k-1}(M^{k-1})$, i.e. $\beta = i^{k-1}(\beta')$, $\beta' \in M^{k-1}$. The second condition then says that $\alpha = d_M\beta'$, so α is cohomologous to 0 in $M^{\cdot}$. Thus, $H^k(i^{\cdot})$ is injective. This concludes the proof of lemma 8.5, and of proposition 8.4. □

8.1.2 Derived functors

If $\mathcal{A}$ and $\mathcal{B}$ are two abelian categories, where $\mathcal{A}$ has sufficiently many injective objects, and F is a left-exact functor from $\mathcal{A}$ to $\mathcal{B}$, for a left-bounded complex $M^\cdot$ of $\mathcal{A}$, we define the ith derived object $R^iF(M^\cdot)$ as follows. Let $i^\cdot : M^\cdot \to I^\cdot$ be a quasi-isomorphism, with i^k injective for every k, where $I^\cdot$ is an injective left-bounded complex. We then set

$$R^iF(M^\cdot) := H^i(F(I^\cdot)).$$

Note that if $A^\cdot$ is zero in degree different from 0, and equal to $A \in \operatorname{Ob}\mathcal{A}$ in degree 0, we have $R^iF(A^\cdot) = R^iF(A)$, thanks to the following proposition.

Proposition 8.6 *$R^iF(M^\cdot)$ is well-defined up to canonical isomorphism, independently of the choice of the quasi-isomorphism $i^\cdot$.*

By this we mean that for another choice of quasi-isomorphism $j^\cdot : M^\cdot \to J^\cdot$, we have a canonical isomorphism

$$H^i(F(I^\cdot)) \cong H^i(F(J^\cdot)).$$

Proof Let $M^\cdot \stackrel{i^\cdot}{\to} I^\cdot$ and $M^\cdot \stackrel{j^\cdot}{\to} J^\cdot$ be two quasi-isomorphisms. Assume that each $i^k : M^k \to I^k$ is injective.

Lemma 8.7 *If $J^\cdot$ is injective, there exists a morphism of complexes*

$$\phi^\cdot : I^\cdot \to J^\cdot$$

such that $\phi^\cdot \circ i^\cdot = j^\cdot$. Moreover $\phi^\cdot$ is well defined up to homotopy.

Proof As usual, we can assume that $M^\cdot$ is zero in negative degrees, as well as $I^\cdot$ and $J^\cdot$. As i^0 is injective and J^0 is injective, there exists a morphism $\phi^0 : I^0 \to J^0$ extending $j^0 : M^0 \to J^0$. Now, consider the morphisms $\alpha = d_J \circ \phi^0 : I^0 \to J^1$ and $j^1 : M^1 \to J^1$. We will construct $\phi^1 : I^1 \to J^1$ such that $\phi^1 \circ i^1 = j^1$ and $\phi^1 \circ d_I = d_J \circ \phi^0$. For this, consider the morphism

$$d_I + i^1 : I^0 \oplus M^1 \to I^1.$$

As $H^1(M^\cdot) \to H^1(I^\cdot)$ is injective, $H^0(M^\cdot) \to H^0(I^\cdot)$ is surjective, and i^2 is injective, we see that the kernel of this morphism is exactly the image of

$$M^0 \stackrel{(i^0, -d_M)}{\to} I^0 \oplus M^1.$$

Now, the morphism $\beta : I^0 \oplus M^1 \to J^1$ which equals $d_J \circ \phi^0$ on I^0 and j^1 on M^1 vanishes on $(i^0, -d_M)(M^0)$. Thus, this morphism factors through Im $(d_I + i^1) \hookrightarrow I^1$ and extends to a morphism $\phi^1 : I^1 \to J^1$ since J^1 is injective. Similarly, we construct $\phi^k : I^k \to J^k$.

The proof of uniqueness up to homotopy is similar. □

End of the proof of proposition 8.6 If $I^\cdot$ is also injective, as well as $j^\cdot$, we can also construct a morphism of complexes

$$\psi^\cdot : J^\cdot \to I^\cdot$$

such that $\psi^\cdot \circ j^\cdot = i^\cdot$ is well-defined up to homotopy. As the morphism of injective complexes

$$1^\cdot_I - \psi^\cdot \circ \phi^\cdot : I^\cdot \to I^\cdot$$

is zero on the subcomplex $M^\cdot$, it induces 0 in cohomology and is even homotopic to 0, as is

$$1^\cdot_J - \phi^\cdot \circ \psi^\cdot : J^\cdot \to J^\cdot,$$

which can easily be seen by an argument similar to the one given in the proof of proposition 4.27. As there exists a homotopy equivalence between $I^\cdot$ and $J^\cdot$, we also obtain a homotopy equivalence between $F(I^\cdot)$ and $F(J^\cdot)$, by applying the functor F. Thus, these two complexes have canonically isomorphic cohomologies. □

We have thus shown that $R^i F(M^\cdot) \in \mathcal{B}$ is determined by the choice of an injective quasi-isomorphism $i_M : M^\cdot \cong I^\cdot$, with $I^\cdot$ a complex of injective objects, and that given two such quasi-isomorphisms i_M and i'_M, we have a canonical isomorphism

$$R^i F(M^\cdot)_{i_M} \cong R^i F(M^\cdot)_{i'_M}.$$

We will need the following more flexible definition of $R^i F(M^\cdot)$.

Proposition 8.8 *Let $\alpha^\cdot : M^\cdot \to I^\cdot$ be a quasi-isomorphism of left-bounded complexes in the category $\mathcal{A}$, with I^k injective for every k. Then we have an isomorphism induced by $\alpha^\cdot$:*

$$R^i F(M^\cdot) \cong H^i(F(I^\cdot)).$$

(The only difference with proposition 8.6 is that we no longer assume that each α^k is an injective morphism.)

Corollary 8.9 *Let $\alpha^\cdot : M^\cdot \to N^\cdot$ be a quasi-isomorphism of left-bounded complexes in the category $\mathcal{A}$. Then we have a canonical isomorphism induced by $\alpha^\cdot$:*

$$R^i F(\alpha^\cdot) : R^i F(M^\cdot) \cong R^i F(N^\cdot).$$

This statement (like that of proposition 8.8) is slightly ambiguous, because of the fact that the $R^i F(M^\cdot)$ are defined using an injective quasi-isomorphism $i_M : M^\cdot \to I^\cdot$ where $I^\cdot$ is a complex of injective objects. Its precise meaning is as follows: for every injective quasi-isomorphism $i_M : M^\cdot \to I^\cdot$ and $i_N : N^\cdot \to J^\cdot$, there exists a canonical isomorphism

$$R^i F(\alpha^\cdot) : R^i F(M^\cdot)_{i_M} \cong R^i F(N^\cdot)_{i_N},$$

where the derived objects are computed using i_M and i_N respectively. Moreover, these isomorphisms are compatible with the canonical isomorphisms

$$R^i F(M^\cdot)_{i_M} \cong R^i F(M^\cdot)_{i'_M}, \quad R^i F(N^\cdot)_{i_N} \cong R^i F(N^\cdot)_{i'_N}$$

given by proposition 8.6.

Proof If we have a quasi-isomorphism $i^\cdot : N^\cdot \to I^\cdot$, with $I^\cdot$ injective, then we also have a composed quasi-isomorphism $i^\cdot \circ \alpha^\cdot : M^\cdot \to I^\cdot$. Thus, by proposition 8.8, $H^k(F(I^\cdot))$ is canonically isomorphic to $R^k F(M^\cdot)$ and to $R^k F(N^\cdot)$. □

Proof of proposition 8.8 For this proof, we will use the notion of the cone of a morphism of complexes. Given a morphism of complexes $\phi^\cdot : A^\cdot \to B^\cdot$, its cone $C(\phi^\cdot)$ is the complex

$$C^k = A^k \oplus B^{k-1}, \quad d_C^k = \begin{pmatrix} d_A^k & (-1)^k \phi^k \\ 0 & d_B^{k-1} \end{pmatrix}.$$

Now let $i^\cdot : M^\cdot \to J^\cdot$ be a quasi-isomorphism, with i^k injective for every k. Then by lemma 8.7, there exists a morphism of complexes

$$\phi^\cdot : J^\cdot \to I^\cdot$$

such that $\phi^\cdot \circ i^\cdot = \alpha^\cdot$. This morphism $\phi^\cdot$ is also a quasi-isomorphism. Thus, we need to show that if two injective complexes $I^\cdot$ and $J^\cdot$ are quasi-isomorphic, the complexes $F(I^\cdot)$ and $F(J^\cdot)$ are quasi-isomorphic. But this follows from the following facts:

(i) If $M^\cdot$ is an exact injective left-bounded complex and F is a left-exact functor, the complex $F(M^\cdot)$ is exact.

(ii) If $\phi^\cdot : A^\cdot \to B^\cdot$ is a quasi-isomorphism between two complexes, the cone of $\phi^\cdot$ is acyclic. If moreover each of the complexes is injective, the cone is injective.

(i) follows from the fact that an exact injective left-bounded complex $(I^\cdot, d)$ is homotopically trivial, i.e. there exists a homotopy equivalence between this complex and the trivial complex, or in other words, there exists a homotopy

$$H^\cdot : I^\cdot \to I^{\cdot -1}$$

satisfying $H^{k+1} \circ d^k + d^{k-1} \circ H^k = \mathrm{Id}$ for every k. This homotopy is constructed using the universal property of injective objects and the injectivity of the maps $d^k : I^k/\mathrm{Im}\, d^{k-1} \to I^{k+1}$. Such a homotopy also induces a homotopy of $F(H^\cdot)$ to $F(I^\cdot)$, showing that the latter is also exact.

(ii) follows from the fact that we have a short exact sequence, split in the category $\mathcal{A}$ (cf. proof of proposition 4.30)

$$0 \to B^\cdot[1] \to C^\cdot \to A^\cdot \to 0 \tag{8.1}$$

and that the connection morphism $\delta : H^k(A^\cdot) \to H^{k+1}(B^\cdot[1]) = H^k(B^\cdot)$ appearing in the associated long exact sequence is equal to $H^k(\alpha^\cdot)$. As $H^k(\alpha^\cdot)$ is an isomorphism for every k, the long exact sequence of cohomology shows that $H^k(C^\cdot) = 0$ for every k. Finally, the cone is clearly injective if each complex is.

Let us apply this to the quasi-isomorphism $\phi^\cdot$. The cone $C^\cdot$ of $\phi^\cdot$ is then acyclic and injective. Thus, if we apply the functor F to it, we obtain an acyclic complex $F(C^\cdot)$. Furthermore, as the exact sequence (8.1) is split (with A replaced by J and B by I), it remains exact and even split in the category $\mathcal{B}$, after applying the functor F. We thus have a long exact sequence

$$H^k(F(C^\cdot)) \to H^k(F(J^\cdot)) \stackrel{H^k(F(\phi^\cdot))}{\to} H^k(F^\cdot) \to H^{k+1}(F(C^\cdot)),$$

and as $F(C^\cdot)$ is exact, $H^k(F(\phi^\cdot))$ is an isomorphism for every k. □

These "derived functors" $R^i F$ depend on the choice of a quasi-isomorphism $i_M : M^\cdot \cong I^\cdot$ with I^k injective. However, they have the following functorial property, which generalises corollary 8.9 and is proved in the same way.

Lemma 8.10 *Let $\phi^{\cdot} : A^{\cdot} \to B^{\cdot}$ be a morphism of complexes, and let $i^{\cdot} : A^{\cdot} \to I^{\cdot}$, $j^{\cdot} : B^{\cdot} \to J^{\cdot}$ be quasi-isomorphisms with I^k and J^k injective for all k. Then there exists a canonical morphism $R^i F(\phi^{\cdot}) : R^i F(A^{\cdot}) \to R^i F(B^{\cdot})$, where the derived functors are computed respectively using the quasi-isomorphisms $i^{\cdot}$ and $j^{\cdot}$.*

For this, we first show the following result.

Lemma 8.11 *Let $\phi^{\cdot} : A^{\cdot} \to B^{\cdot}$ be a morphism of complexes, and let $i^{\cdot} : A^{\cdot} \to I^{\cdot}$, $j^{\cdot} : B^{\cdot} \to J^{\cdot}$ be quasi-isomorphisms. Then if each i^k is injective, and $J^{\cdot}$ is injective, there exists a morphism of complexes $\psi^{\cdot} : I^{\cdot} \to J^{\cdot}$ which makes the following diagram commutative.*

$$\begin{array}{ccc} A^{\cdot} & \stackrel{\phi^{\cdot}}{\to} & B^{\cdot} \\ \downarrow^{i^{\cdot}} & & \downarrow^{j^{\cdot}} \\ I^{\cdot} & \stackrel{\psi^{\cdot}}{\to} & J^{\cdot} \end{array}$$

Moreover, $\psi^{\cdot}$ is unique up to homotopy. □

We now have the following analogue of proposition 4.32.

Proposition 8.12 *Under the hypotheses considered above, let $M^{\cdot} \stackrel{\alpha^{\cdot}}{\to} N^{\cdot}$ be a quasi-isomorphism, with N^k acyclic for the functor F, for every k. Then $\alpha^{\cdot}$ induces an isomorphism*

$$R^i F(M^{\cdot}) \cong H^i(F(N^{\cdot})).$$

Proof By corollary 8.9, $\alpha^{\cdot}$ induces an isomorphism

$$R^i F(M^{\cdot}) \cong R^i F(N^{\cdot}).$$

It thus suffices to show that for a complex $N^{\cdot}$ of F-acyclic objects, we have

$$R^i F(N^{\cdot}) \cong H^i(F(N^{\cdot})).$$

Let $i^{\cdot} : N^{\cdot} \to J^{\cdot}$ be a quasi-isomorphism, with i^k injective for all k. Let $Q^{\cdot}$ be the quotient complex. The exact sequence

$$0 \to N^{\cdot} \stackrel{i^{\cdot}}{\to} I^{\cdot} \to Q^{\cdot} \to 0,$$

where $i^{\cdot}$ is a quasi-isomorphism, shows that $Q^{\cdot}$ is an exact complex. Moreover, each Q^k is acyclic for F, as a quotient of two objects which are acyclic for F.

Finally, as each N^k is acyclic, the short exact sequence

$$0 \to N^k \xrightarrow{i^k} I^k \to Q^k \to 0$$

also induces a short exact sequence

$$0 \to F(N^k) \xrightarrow{F(i^k)} F(I^k) \to F(Q^k) \to 0.$$

We thus have an exact sequence of complexes

$$0 \to F(N^\cdot) \xrightarrow{F(i^\cdot)} F(I^\cdot) \to F(Q^\cdot) \to 0 \qquad (8.2)$$

in the category $\mathcal{B}$. We then conclude by using the following fact, which is easy to show:

Let $Q^\cdot$ be an exact left-bounded F-acyclic complex. Then $F(Q^\cdot)$ is an exact complex.

The exact sequence (8.2) with $F(Q^\cdot)$ exact then shows that $F(I^\cdot)$ and $F(N^\cdot)$ are quasi-isomorphic, so we have the desired isomorphism

$$R^i F(N^\cdot) = H^i(F(I^\cdot)) = H^i(F(N^\cdot)).$$

□

In the case where $\mathcal{A}$ is the category of sheaves of abelian groups over a topological space X, and F is the functor $\Gamma : \mathcal{F} \mapsto \mathcal{F}(X) = \Gamma(X, \mathcal{F})$ of global sections, we write

$$\mathbb{H}^k(X, \mathcal{F}^\cdot) = R^k\Gamma(\mathcal{F}^\cdot)$$

for the kth derived functor of Γ applied to the complex of sheaves $\mathcal{F}^\cdot$. The group $\mathbb{H}^k(X, \mathcal{F}^\cdot)$ is called the hypercohomology of the complex $\mathcal{F}^\cdot$. It is canonically determined by the choice of a quasi-isomorphism

$$\mathcal{F}^\cdot \cong \mathcal{K}^\cdot$$

where $\mathcal{K}^\cdot$ is a Γ-acyclic complex of sheaves.

8.1.3 Composed functors

The definition of the derived functors of a complex is particularly useful to compute the derived functors of a composed functor $G \circ F$. We assume here that $F : \mathcal{A} \to \mathcal{B}$ is a left-exact functor, where the category $\mathcal{A}$ has sufficiently many injective objects, and $G : \mathcal{B} \to \mathcal{C}$ is a left-exact functor, where the category $\mathcal{B}$ has sufficiently many injective objects. More precisely, suppose that the functor F transforms the injective objects of $\mathcal{A}$ (or the $G \circ F$-acyclic objects) into G-acyclic objects. Then let A be an object of $\mathcal{A}$, and let $A \to I^{\cdot}$ be an injective resolution of A (or more generally, a $G \circ F$-acyclic resolution). Then by definition we have

$$R^k(G \circ F)(A) \cong H^k(G \circ F(I^{\cdot})).$$

But furthermore, as the $F(I^{\cdot})$ are G-acyclic, by proposition 8.12 we also have

$$H^k(G \circ F(I^{\cdot})) = R^k G(F(I^{\cdot})).$$

We can thus compute $R^k(G \circ F)(A)$ as the kth derived functor of G applied to the complex $F(I^{\cdot})$. Moreover, by corollary 8.9, $R^k G(F(I^{\cdot}))$ depends on the complex $F(I^{\cdot})$ only up to quasi-isomorphism.

One important application concerns sheaf cohomology: if $\phi : X \to Y$ is a continuous map and $\mathcal{F}$ is a sheaf over X, we have

$$\Gamma(X, \mathcal{F}) = \Gamma(Y, \phi_* \mathcal{F}),$$

so we can see the functor of global sections on X as the composition of the functor ϕ_* (between sheaves over X and sheaves over Y) and the functor of global sections over Y. As the functor ϕ_* obviously transforms flasque (so Γ-acyclic) sheaves over X into flasque sheaves over Y, the observation above applies in this situation. Thus, for every flasque resolution $\mathcal{F}^{\cdot}$ of $\mathcal{F}$, we obtain

$$H^p(X, \mathcal{F}) = \mathbb{H}^p(Y, \phi_*(\mathcal{F}^{\cdot})).$$

Application: Proof of the Leray–Hirsch theorem 7.33

Recall that we have assumed that the graded group $A^{\cdot}$ has no torsion. Let $\mathcal{F}^{\cdot}$ be a flasque resolution of the sheaf $\mathbb{Z}$ over Y. Then $H^p(Y, \mathbb{Z})$ is equal to $H^p(\Gamma(Y, \mathcal{F}^{\cdot}))$. Let $\beta_i \in \Gamma(Y, \mathcal{F}^{k_i})$ be closed elements representing the classes α_i, with $k_i = d^0\alpha_i$. The β_i generate a graded group isomorphic to $A^{\cdot}$, and over X they give the inclusion of the complex of constant sheaves $A^{\cdot}$, equipped with the zero differential, in $\phi_* \mathcal{F}^{\cdot}$. Now, under the local triviality hypotheses on the map ϕ, the condition that $A^{\cdot}$ maps isomorphically to the cohomology of the fibre Y_x is equivalent to the fact that this inclusion is a quasi-isomorphism. Thus we deduce a graded isomorphism

$$\mathbb{H}^*(X, A^{\cdot}) \cong \mathbb{H}^*(X, \phi_* \mathcal{F}^{\cdot}).$$

Now, the right-hand term is equal to $H^*(Y, \mathbb{Z})$ by the above, while the left-hand term is equal, as a graded group, to $A^{\cdot} \otimes_{\mathbb{Z}} H^{\cdot}(X, \mathbb{Z})$, since $A^{\cdot}$ is a free $\mathbb{Z}$-module of finite rank. □

8.2 Holomorphic de Rham complexes

8.2.1 Holomorphic de Rham resolutions

Let $\mathcal{F}$ be a sheaf over X, and let $\mathcal{F} \overset{i}{\hookrightarrow} \mathcal{K}^{\cdot}$ be a resolution of $\mathcal{F}$. This exactly means that the complex of sheaves $\mathcal{K}^{\cdot}$ is quasi-isomorphic, via i, to the sheaf $\mathcal{F}$, considered as a complex supported in degree 0. Then, by corollary 8.9, we have

$$H^k(X, \mathcal{F}) = \mathbb{H}^k(X, \mathcal{K}^{\cdot}).$$

Suppose now that X is a complex manifold. Let $\mathbb{C}$ denote the locally constant sheaf of stalk $\mathbb{C}$ over X. Writing Ω_X^q for the sheaf of holomorphic differential forms of degree q, the exterior differentiation operator $d = \partial : \Omega_X^q \to \Omega_X^{q+1}$ satisfies $\partial \circ \partial = 0$. We thus have a complex, called the holomorphic de Rham complex:

$$0 \to \mathcal{O}_X \xrightarrow{\partial} \Omega_X \xrightarrow{\partial} \cdots \xrightarrow{\partial} \Omega_X^n \to 0,$$

with $n = \dim X$. Moreover, we have the inclusion $i : \mathbb{C} \hookrightarrow \mathcal{O}_X$ of the locally constant functions into the holomorphic functions.

Lemma 8.13 *Via i, the holomorphic de Rham complex is a resolution of $\mathbb{C}$.*

Proof We want to show that the sheaves of cohomology $\mathcal{H}^k = \mathcal{H}^k(\Omega_X^{\cdot})$ satisfy $\mathcal{H}^0 = i(\mathbb{C})$ and $\mathcal{H}^k = 0$ for $k > 0$.

Now, we have a inclusion of the holomorphic de Rham complex into the de Rham complex

$$(\Omega_X, \partial) \to \left(\mathcal{A}_X^k, d\right),$$

since d and ∂ coincide on holomorphic forms. Moreover, we can see the usual de Rham complex $\mathcal{A}_X^{\cdot}$ as the simple complex associated to the double complex

$$(\mathcal{A}^{p,q}, \partial, (-1)^p \overline{\partial}).$$

Each column $(\mathcal{A}_X^{p,q}, (-1)^p \overline{\partial})$ of this double complex is exact in positive degree by proposition 2.36 and gives a resolution of Ω_X^p. Thus, the de Rham complex is quasi-isomorphic to the holomorphic de Rham complex by lemma 8.5. Like the usual de Rham complex, it is exact in positive degree, and its cohomology is given by the locally constant sheaf $\mathbb{C}$ in degree 0, so this also holds for the holomorphic de Rham complex. ∎

Corollary 8.14 *By corollary 8.9, we have*

$$H^k(X, \mathbb{C}) = \mathbb{H}^k(X, \Omega_X^\cdot).$$

8.2.2 The logarithmic case

Let X be a complex manifold, and let $D \subset X$ be a hypersurface, i.e. D is locally defined by the vanishing of a holomorphic equation.

Definition 8.15 *We say that D is a normal crossing divisor if locally there exist coordinates $z_1, \ldots, z_n$ on X such that D is defined by the equation $z_1 \cdots z_r = 0$ for an integer r which naturally depends on the considered open set.*

Given a pair (X, D), where D is a normal crossing divisor in X, we will define the holomorphic de Rham complex with logarithmic singularities along D.

Let $\Omega_X^k(\log D)$ be the subsheaf of the sheaf $\Omega_X^k(*D)$ of meromorphic forms on X, holomorphic on $X - D$, defined by the condition:

- If α is a meromorphic differential form on U, holomorphic on $U - D \cap U$, $\alpha \in \Omega_X^k(\log D)_{|U}$ if α admits a pole of order at most 1 along (each component of) D, and the same holds for $d\alpha$.

Lemma 8.16 *Let $z_1, \ldots, z_n$ be local coordinates on an open set U of X, in which $D \cap U$ is defined by the equation $z_1 \cdots z_r = 0$. Then $\Omega_X^k(\log D)_{|U}$ is a sheaf of free $\mathcal{O}_U$-modules, for which $\frac{dz_{i_1}}{z_{i_1}} \wedge \cdots \wedge \frac{dz_{i_l}}{z_{i_l}} \wedge dz_{j_1} \wedge \cdots \wedge dz_{j_m}$ with $i_s \leq r,\ j_s > r$ and $l + m = k$ form a basis.*

Proof Let α be a section of $\Omega_X^k(\log D)$ on $V \subset U$. As α admits a pole of order at most 1 along D, we can write $\alpha = \frac{\beta}{z_1 \cdots z_r}$, with β a holomorphic k-form on V. As $d\alpha$ admits a pole of order at most 1 along D, we find that $\sum_{i \leq r} z_1 \cdots \hat{z}_i \cdots z_r dz_i \wedge \beta$ must vanish along D. It follows immediately that if $\beta = \sum_{I,J} \beta_{I,J} dz_I \wedge dz_J$ with $I \subset \{1, \ldots, r\}$, $J \subset \{r+1, \ldots, n\}$, the function $\beta_{I,J}$ must vanish on the hyperplanes of equation z_i, $i \in I' := \{1, \ldots, r\} - I$, and thus must be divisible by $z_{I'} = \prod_{i \in I'} z_i$. □

Corollary 8.17 *The sheaves $\Omega_X^k(\log D)$ are sheaves of free $\mathcal{O}_X$-modules.*

Furthermore, by definition, if α is a section of $\Omega_X^k(\log D)$ on $V \subset X$, then $d\alpha = \partial\alpha$ is in $\Omega_X^{k+1}(\log D)$. Indeed $d\alpha$ is meromorphic, with a pole of order at most 1 along D, and closed. Thus, $(\Omega_X^\cdot(\log D), \partial)$ is a complex of sheaves over X. This complex is called the logarithmic de Rham complex.

8.2.3 Cohomology of the logarithmic complex

The logarithmic de Rham complex is not locally exact in positive degrees. Indeed, if X is a complex curve and D is a point x defined by the equation $z = 0$, the form $\frac{dz}{z} \in \Omega_X(\log D)$ is not exact in the neighbourhood of x, since its integral over a small positively oriented circle around x is equal to $2i\pi$.

Let $U = X - D$, and let j be the inclusion of U into X. We have a natural inclusion

$$\Omega_X^k(\log D) \subset j_*\Omega_U^k,$$

which is compatible with differentials. We also have the inclusion of complexes $\Omega_U^{\cdot} \subset \mathcal{A}_U^{\cdot}$, and thus we have a morphism of complexes

$$\Omega_X^{\cdot}(\log D) \to j_*\mathcal{A}_U^{\cdot}. \tag{8.3}$$

The following result can be found in Griffiths (1969) and Deligne (1971).

Proposition 8.18 *The morphism (8.3) is a quasi-isomorphism.*

Corollary 8.19 *There is a canonical isomorphism*

$$H^k(U, \mathbb{C}) = \mathbb{H}^k(X, \Omega_X^{\cdot}(\log D)).$$

Proof By corollary 8.9 and proposition 8.18, we have a canonical isomorphism

$$\mathbb{H}^k(X, \Omega_X^{\cdot}(\log D)) = \mathbb{H}^k(X, j_*\mathcal{A}_U^{\cdot}).$$

But as $\mathcal{A}_U^{\cdot}$ is a sheaf of $\underline{\mathcal{C}}_U^\infty$-modules which is a resolution of $\mathbb{C}$ over U, the cohomology of its global sections is equal to the cohomology of U with values in $\mathbb{C}$. Furthermore, $j_*\mathcal{A}_U^{\cdot}$ is a sheaf of $\underline{\mathcal{C}}_X^\infty$-modules, so it is acyclic, and thus its hypercohomology is equal to the cohomology of the complex of its global sections by proposition 4.32. Thus, we have

$$\mathbb{H}^k(X, j_*\mathcal{A}_U^{\cdot}) = H^k(\Gamma(X, j_*\mathcal{A}_U^{\cdot})) = H^k(\Gamma(U, \mathcal{A}_U^{\cdot})) = H^k(U, \mathbb{C}).$$

□

Proof of proposition 8.18 The statement is obviously local. We may thus assume that X is a polydisk of dimension n and $D = \{(z_1, \dots, z_n) \in D_1 \times \dots \times D_n \mid z_1 \dots z_r = 0\}$. Then $U = X - D = D_1^* \times \dots \times D_r^* \times D_{r+1} \times \dots \times D_n$ retracts onto the product of circles $(S^1)^r = \prod_{i \le r} \partial D_i$. But the cohomology of such a product of circles (or torus) T_r is given by

$$H^1(T_r, \mathbb{C}) \cong \mathbb{C}^r, \quad \bigwedge^k H^1(T_r, \mathbb{C}) \cong H^k(T_r, \mathbb{C}).$$

The first equality was already noted in section 7.2.2. The second equality follows from Künneth's formula (theorem 11.38). The second isomorphism is given by the cup-product.

This immediately shows that the morphism (8.3) induces a surjective map in cohomology. Indeed, consider the closed forms $\omega_i = \frac{dz_i}{z_i} \in \Omega^1_U(\log D)$. Their integrals over the circles ∂D_i (where the other variables are fixed) satisfy

$$\int_{\partial D_j} \omega_i = 2i\pi\delta_{ij},$$

and thus the classes of these forms generate $H^1(T, \mathbb{C}) = \mathrm{Hom}(H_1(T, \mathbb{Z}), \mathbb{C})$. Thus, the exterior products $\omega_I = \bigwedge_{i\in I} \omega_i \in \Omega^k_U(\log D)$, $k = |I|$, also give a basis of the cohomology of U in degree k, and thus the map

$$H^k\big(\Gamma\big(U, \Omega^k_U(\log D)\big)\big) \to H^k(U, \mathbb{C}) = H^k(\Gamma(j_*(\mathcal{A}^{\cdot}_U)))$$

is surjective.

It remains to prove the injectivity. For this, it suffices to see that locally in the neighbourhood of 0, a holomorphic form α with logarithmic singularities along $D = \{z \in X \mid z_1 \cdots z_r = 0\}$ is cohomologous in $\Omega^{\cdot}_X(\log D)$ to a combination with constant coefficients of the $\frac{dz_I}{z_I}$, $I \subset \{1, \ldots, r\}$. We will proceed by induction on r. The case $r = 0$, i.e. without singularities, is the statement of lemma 8.13. Assume that the result is shown for $r - 1$. Let α be a holomorphic closed section of $\Omega^{\cdot}_X(\log D)$ in the neighbourhood of 0. Write $\alpha = \frac{dz_r}{z_r} \wedge \beta + \gamma$, where dz_r does not occur in β, and the coefficients of β are independent of z_r, while γ is holomorphic in z_r (i.e. has no pole on $z_r = 0$). Then as $d\alpha = 0$, $\frac{dz_r}{z_r} \wedge d\beta$ has no pole on $z_r = 0$, and as β does not depend on z_r, this implies that $d\beta = 0$. But β clearly has logarithmic singularities along $D' := \{z \mid z_1 \cdots z_{r-1} = 0\}$, and thus by induction on r we can write $\beta = \beta' + d\phi$, where β' has constant coefficients and ϕ has logarithmic singularities along D'. Thus, $\frac{dz_r}{z_r} \wedge \beta = \frac{dz_r}{z_r} \wedge \beta' - d\left(\frac{dz_r}{z_r} \wedge \phi\right)$ and $\frac{dz_r}{z_r} \wedge \phi$ has logarithmic singularities along D. Finally, as γ is holomorphic in z_r, γ is a section of $\Omega^k_X(\log D')$, and as γ is closed, by induction on r we may assume that γ is cohomologous in $\Gamma(\Omega^{\cdot}_X(\log D'))$ to a logarithmic form with constant coefficients. Thus, $\alpha = \frac{dz_r}{z_r} \wedge \beta + \gamma$ is cohomologous in $\Gamma(\Omega^{\cdot}_X(\log D))$ to a logarithmic form with constant coefficients. □

8.3 Filtrations and spectral sequences

8.3.1 Filtered complexes

Let $\mathcal{A}$ be an abelian category, and A an object of $\mathcal{A}$. A decreasing filtration on A is given by a family of subobjects

$$\cdots F^p A \hookrightarrow F^{p-1} A \hookrightarrow F^0 A = A,$$

for $p \geq 0$.

If $(A^\cdot, d)$ is a complex in the category $\mathcal{A}$, a decreasing filtration on $A^\cdot$ is given by a decreasing filtration $F^p A^k$ on each A^k such that d sends $F^p A^k$ to $F^p A^{k+1}$. In particular, for each p, we have a subcomplex $F^p A^\cdot$ of $A^\cdot$.

Such a filtration naturally induces a filtration F^p on the cohomology of $A^\cdot$. We set

$$F^p H^i(A^\cdot) = \operatorname{Im}(H^i(F^p A^\cdot) \to H^i(A^\cdot)).$$

Let F be a left-exact functor from $\mathcal{A}$ to $\mathcal{B}$. Assume that $\mathcal{A}$ has sufficiently many injective objects. We thus have derived functors $R^i F(A^\cdot)$ for every left-bounded complex $A^\cdot$. If $A^\cdot$ is filtered, the morphisms of complexes $F^p A^\cdot \to A^\cdot$ induce morphisms

$$\alpha_{p,i} : R^i F(F^p A^\cdot) \to R^i F(A^\cdot). \tag{8.4}$$

Definition 8.20 *We define the filtration $F^p R^i F(A^\cdot)$ on $R^i F(A^\cdot) \in \operatorname{Ob} \mathcal{B}$ by*

$$F^p R^i F(A^\cdot) = \operatorname{Im} \alpha_{p,i}.$$

$R^i F(A^\cdot)$ can be computed as the cohomology of the complex $F(I^\cdot)$, where $A^\cdot \to I^\cdot$ is a quasi-isomorphism and $I^\cdot$ is an injective complex. When $A^\cdot$ is filtered, we can construct an injective complex $I^\cdot$, quasi-isomorphic to $A^\cdot$, equipped with a filtration $F^p I^\cdot$ by injective complexes which are quasi-isomorphic to $F^p A^\cdot$. Then the cohomology of the subcomplex $F(F^p I^\cdot)$ computes $R^i F(F^p A^\cdot)$. Thus, the filtration $F^p R^i F(A^\cdot)$ is in fact the filtration induced in cohomology by the filtration

$$F^p(F(I^\cdot)) = F(F^p I^\cdot)$$

on the complex $F(I^\cdot)$.

We will encounter the following two types of filtration:

The "naive" filtration. Let $A^\cdot$ be a complex. Set $F^p A^\cdot = A^{\geq p}$. This is the complex which is zero in degree less than p, and equal to A in degree greater than or equal to p.

The filtration of a double complex. Let $(A^{p,q}, D_1, D_2)$ be a double complex supported in positive degree. Let $(A^{\cdot}, D)$ be the associated simple complex.

$$A^k = \bigoplus_{p+q=k} A^{p,q}, \quad D = D_1 + (-1)^p D_2.$$

Set $F^p A^k = \bigoplus_{r+s=k,\, r\geq p} A^{r,s}$. As $DA^{p,q} \subset A^{p+1,q} \oplus A^{p,q+1}$, F^p defines a decreasing filtration of the complex $A^{\cdot}$.

These two filtrations are closely related. Indeed, if we start from a complex $A^{\cdot}$, and equip it with a resolution by a double complex $(I^{\cdot\cdot}, D_1, D_2)$, i.e. if we have an injection

$$i^p : A^p \hookrightarrow I^{p,\cdot}$$

such that $i^p \circ d_A = D_1 \circ i^{p-1}$ and $\operatorname{Im} i^p = \operatorname{Ker} D_2$, the associated simple complex $(I^{\cdot}, D = D_1 + (-1)^p D_2)$ is quasi-isomorphic to $A^{\cdot}$ by lemma 8.5. For the same reason, the complex $F^p I^{\cdot}$ is quasi-isomorphic to $F^p A^{\cdot}$, where we put the filtration of the double complex on the simple complex, and the "naive" filtration on the second. Thus we have a filtered quasi-isomorphism between $(A^{\cdot}, F^p A^{\cdot})$ and $(I^{\cdot}, F^p I^{\cdot})$.

8.3.2 Spectral sequences

Consider a filtered complex $(A^{\cdot}, F^p A^{\cdot})$ in the category of abelian groups. We will assume that the filtration F^p satisfies the property:

$$\forall k,\ \exists l,\ F^l A^k = 0.$$

By successive approximations, we will write the filtration induced by F^p on $H^k(A^{\cdot})$, and more precisely, the successive quotients

$$\mathrm{Gr}_p^F H^i(A^{\cdot}) = F^p H^i(A^{\cdot}) / F^{p+1} H^i(A^{\cdot}).$$

Theorem 8.21 *There exist complexes*

$$\left(E_r^{p,q}, d_r\right), \quad d_r : E_r^{p,q} \to E_r^{p+r,q-r+1}$$

satisfying the conditions:

(i) $E_0^{p,q} = \mathrm{Gr}_p^F A^{p+q} := F^p A^{p+q} / F^{p+1} A^{p+q}$ *and* d_0 *is induced by* d.

(ii) $E_{r+1}^{p,q}$ *can be identified with the cohomology of* $(E_r^{p,q}, d_r)$ *i.e. with*

$$\operatorname{Ker}\left(d_r : E_r^{p,q} \to E_r^{p+r,q-r+1}\right) / \operatorname{Im}\left(d_r : E_r^{p-r,q+r-1} \to E_r^{p,q}\right).$$

(iii) For $p+q$ fixed and r sufficiently large, we have

$$E_r^{p,q} = \mathrm{Gr}_p^F H^{p+q}(A^\cdot).$$

The series of complexes $(E_r^{p,q}, d_r)$ is called the spectral sequence associated to the filtration F.

Remark 8.22 *The indices may appear bizarre. They are particularly well adapted to the case of a simple complex associated to a double complex. In general, one should recall that $p+q$ is the degree and p the level of the filtration.*

Proof Set

$$Z_r^{p,q} = \{x \in F^p A^{p+q} \mid dx \in F^{p+r} A^{p+q+1}\}.$$

$Z_r^{p,q}$ naturally contains $Z_{r-1}^{p+1,q-1}$ and $dZ_{r-1}^{p-r+1,q+r-2}$. Let

$$B_r^{p,q} := Z_{r-1}^{p+1,q-1} + dZ_{r-1}^{p-r+1,q+r-2} \subset Z_r^{p,q}$$

and $E_r^{p,q} = Z_r^{p,q}/B_r^{p,q}$. As d sends $Z_r^{p,q}$ to $Z_r^{p+r,q-r+1}$ and $B_r^{p,q}$ to $B_r^{p+r,q-r+1}$, we have a differential

$$d_r : E_r^{p,q} \to E_r^{p+r,q-r+1}$$

which clearly satisfies $d_r^2 = 0$ since it is induced by d.

It remains to check the various properties.

(i) We have $Z_0^{p,q} = F^p A^{p+q}$ and $B_0^{p,q} = F^{p+1} A^{p+q}$, so $E_0^{p,q} = \mathrm{Gr}_p^F A^{p+q}$. Moreover, the differential $d_0 : Z_0^{p,q} \to Z_0^{p,q+1}$ is simply d, and thus d_0 is simply the induced differential

$$\overline{d} : \mathrm{Gr}_p^F A^{p+q} \to \mathrm{Gr}_p^F A^{p+q+1}.$$

(iii) Fix n. For $k \geq n$ sufficiently large, we have $F^k A^n = 0$, $F^k A^{n-1} = 0$, $F^k A^{n+1} = 0$. We then have $E_{k+1}^{p,q} = \mathrm{Gr}_p^F H^n(A^\cdot)$ for $p+q=n$. Indeed, as $F^{p+k+1} A^{n+1} = 0$, we have $Z_{k+1}^{p,q} = \mathrm{Ker}\,(d : F^p A^n \to F^p A^{n+1})$. Furthermore, for the same reason, we have $Z_k^{p+1,q-1} = \mathrm{Ker}\,(d : F^{p+1} A^n \to F^{p+1} A^{n+1})$, and finally, as $p-k \leq 0$, $dZ_k^{p-k,q+k-1}$ is of course equal to $F^p A^n \cap \mathrm{Im}\, d$. These two properties exactly express the fact that

$$E_{k+1}^{p,q} \cong \frac{\mathrm{Im}\,(H^n(F^p A^\cdot) \to H^n(A^\cdot))}{\mathrm{Im}\,(H^n(F^{p+1} A^\cdot) \to H^n(A^\cdot))} = \mathrm{Gr}_p^F H^n(A^\cdot),$$

i.e. (iii).

It remains to see (ii), which follows almost immediately from the definitions: indeed, note first that if $z \in Z_{r+1}^{p,q}$, then $dz \in Z_{r-1}^{p+r+1,q-r} \subset B_r^{p+r,q-r+1}$. Thus, if $\overline{z}$ is the class of z in $E_r^{p,q}$, we have $d_r\overline{z} = 0$. Furthermore, if $z = z_1 + dz_2 \in B_{r+1}^{p,q} = Z_r^{p+1,q-1} + dZ_r^{p-r,q+r-1}$, as $Z_r^{p+1,q-1} \subset Z_{r-1}^{p+1,q-1} \subset B_r^{p,q}$, we have $\overline{z} = \overline{dz_2}$, and as $z_2 \in Z_r^{p-r,q+r-1}$, we have $\overline{dz_2} = d_r(\overline{z_2})$. Thus we have a map

$$Z_{r+1}^{p,q}/B_{r+1}^{p,q} \to \operatorname{Ker}(d_r : E_r^{p,q} \to E_r^{p+r,q-r+1})/\operatorname{Im}(d_r : E_r^{p-r,q+r-1} \to E_r^{p,q}). \tag{8.5}$$

Finally, if $z \in Z_r^{p,q}$ and $\overline{z} \in \operatorname{Ker} d_r$, then $dz \in B_r^{p+r,q-r+1} = dZ_{r-1}^{p+1,q-1} + Z_{r-1}^{p+r+1,q-r}$. So $d(z-w) \in Z_{r-1}^{p+r+1,q-r}$ with $w \in Z_{r-1}^{p+1,q-1}$. But then $d(z-w) \in F^{p+r+1}A^{p+q+1}$ and $z - w \in Z_{r+1}^{p,q}$. As $\overline{w} = 0$ since $Z_{r-1}^{p+1,q-1} \subset B_r^{p,q}$, we also have $\overline{z} = \overline{z-w}$. Thus, the map (8.5) is surjective. We show equally easily that it is injective. □

Remark 8.23 *If the initial complex is a complex of A-modules, where A is a commutative ring, and if the filtration is a filtration by A-submodules, then the same holds for the $E_r^{p,q}$.*

The following result is very useful.

Lemma 8.24 *We have $E_1^{p,q} = H^{p+q}(\operatorname{Gr}_F^p A^\cdot)$, and the differential*

$$d_1 : E_1^{p,q} \to E_1^{p+1,q}$$

can be identified with the connection map

$$\delta : H^{p+q}(\operatorname{Gr}_F^p A^\cdot) \to H^{p+q+1}(\operatorname{Gr}_F^{p+1} A^\cdot)$$

which appears in the long exact sequence associated to the short exact sequence

$$0 \to \operatorname{Gr}_F^{p+1} A^\cdot \to F^p A^\cdot / F^{p+2} A^\cdot \to \operatorname{Gr}_F^p A^\cdot \to 0.$$

Proof of theorem 8.21 The first statement is obvious, by the description of $E_0^{p,q} = \operatorname{Gr}_F^{p+q}$ with d_0 induced by d, and by property (ii) $E_1^{p,q} = H^{p,q}(E_0^{\cdot,\cdot}, d_0)$. Thus it suffices to understand d_1. But if $\overline{z} \in E_1^{p,q}$, then $d_1(\overline{z})$ consists in taking a representative z of $\overline{z}$ in $Z_1^{p,q}$, applying d to it and considering the result modulo $B_1^{p+1,q}$. But this is exactly the definition of the map δ, taking into account the fact that the lifting $z \in Z_1^{p,q}$ also gives a lifting z' of $\overline{z} \in H^{p+q}(\operatorname{Gr}_F^p A^\cdot)$ to $F^p A^{p+q}/F^{p+2}A^{p+q}$, and that we then have $\delta(\overline{z}) = dz \bmod dF^{p+1}A^{p+q}$. □

In the case of the filtration of the simple complex associated to a double complex, these results take the following particularly simple form.

Proposition 8.25 *Let $K^\cdot$ be the simple complex associated to a double complex $(K^{p,q}, D_1, D_2)$. Then if $F^pK^n = \bigoplus_{r\geq p, r+s=n} K^{r,s}$ is the filtration introduced in section 8.3.1, the spectral sequence associated to F has first terms given by*
- $E_0^{p,q} = K^{p,q}$, $d_0 = (-1)^p D_2$,
- $E_1^{p,q} = H^q(K^{p,\cdot})$,

and the differential $d_1 : H^q(K^{p,\cdot}) \to H^q(K^{p+1,\cdot})$ is induced by the morphism of complexes

$$D_1 : K^{p,\cdot} \to K^{p+1,\cdot}.$$

Proof The statement concerning (E_0, d_0) is immediate, since $E_0^{p,\cdot} = \mathrm{Gr}_p^F K^\cdot = K^{p,\cdot}$ with the differential induced by D. Now, if $x \in F^pK^\cdot$, then $Dx = (-1)^p D_2x \bmod F^{p+1}K^{\cdot+1}$.

The computation of E_1 follows immediately, since $E_1^{p,q} = H^{p,q}(E_0, d_0)$. The only point to check is the computation of d_1. But if $\overline{z} \in E_1^{p,q}$, then $d_1(\overline{z})$ is obtained by taking a representative of $\overline{z}$ in $Z_1^{p,q}$, applying D to it and considering the result modulo $B_1^{p+1,q}$. Now, such a representative is given by a D_2-closed element z of $K^{p,q}$, and we then have $Dz = D_1z$. □

Definition 8.26 *We say that the spectral sequence associated to a filtered complex $(K^\cdot, F)$ degenerates at E_r if $\forall k \geq r$, we have $d_k = 0$. We then have $E_r^{p,q} = E_\infty^{p,q} = \mathrm{Gr}_p^F H^{p+q}(K^\cdot)$.*

8.3.3 The Frölicher spectral sequence

Let X be a complex manifold, and let $(\Omega_X^\cdot, \partial)$ be its holomorphic de Rham complex. This complex is equipped with the "naive" filtration (cf. section 8.3.1)

$$F^p\Omega_X^\cdot = \Omega_X^{\geq p}.$$

We have the corresponding filtration on the de Rham complex of X:

$$F^p\mathcal{A}_X^k = \bigoplus_{r\geq p, r+s=k} \mathcal{A}_X^{r,s},$$

and $(F^p\mathcal{A}_X^\cdot, d)$ is the double complex associated to the fine resolutions (i.e. resolutions by fine sheaves, cf. definition 4.35) $(\mathcal{A}_X^{q,\cdot}, \overline{\partial})$ of Ω_X^q, $q \geq p$. Thus, $(F^p\mathcal{A}_X^\cdot, d)$ is quasi-isomorphic to $F^p\Omega_X^\cdot$ by lemma 8.5, and the cohomology of the complex of its global sections $(F^pA_X^\cdot, d)$ is equal to the hypercohomology of $F^p\Omega_X^\cdot$ by proposition 8.12. The spectral sequence associated to the filtration

F^p on the de Rham complex $(A^{\cdot}(X), d)$ is called the Frölicher spectral sequence. The term $E_1^{p,q}$ of this spectral sequence is easy to compute. Indeed, we apply proposition 8.25, which says that $E_1^{p,q} = H^q(A^{p,\cdot}(X), \overline{\partial})$. Now, by corollary 4.38, we have

$$H^q(A^{p,\cdot}(X), \overline{\partial}) = H^q(X, \Omega_X^p).$$

Finally, we know that the differential $d_1 : E_1^{p,q} \to E_1^{p+1,q}$ is induced by ∂, and thus is simply the map

$$\partial : H^q(X, \Omega_X^p) \to H^q(X, \Omega_X^{p+1}).$$

Assume now that X is a compact Kähler manifold. Then we have the Hodge decomposition

$$H^k(X, \mathbb{C}) = \bigoplus_{p+q=k} H^{p,q}(X),$$

and the Hodge filtration

$$F^p H^k(X, \mathbb{C}) = \bigoplus_{r+q=k, r\geq p} H^{r,q}(X).$$

By proposition 7.5, the Hodge filtration $F^p H^k(X, \mathbb{C})$ is exactly induced by the filtration F^p on the de Rham complex. Recall moreover (cf. lemma 6.18) that we have a natural isomorphism

$$\mathrm{Gr}_F^p H^{p+q}(X) = H^{p,q}(X) \cong H^q(X, \Omega_X^p). \tag{8.6}$$

Remark 8.27 *If X is not Kähler, there is not, in general, a map*

$$F^p H^k(X, \mathbb{C})/F^{p+1} H^k(X, \mathbb{C}) \to H^q(X, \Omega_X^p).$$

These two spaces are related via the theory of spectral sequences, which says that $F^p H^k(X, \mathbb{C})/F^{p+1} H^k(X, \mathbb{C}) = E_\infty^{p,q}$ can be identified with a quotient of a subspace of $E_1^{p,q} = H^q(X, \Omega_X^p)$.

We deduce the following result.

Theorem 8.28 *If X is a compact Kähler manifold, the Frölicher spectral sequence of X degenerates at E_1.*

Proof As noted above, the theory of spectral sequences says that for sufficiently large r,

$$F^p H^k(X, \mathbb{C})/F^{p+1} H^k(X, \mathbb{C}) = E_\infty^{p,q} = E_r^{p,q}$$

is computed starting from $E_1^{p,q}$, using $E_i^{p,q}$, $1 \leq i \leq r$, each $E_i^{p,q}$ being identified with the cohomology $\mathrm{Ker}\, d_i / \mathrm{Im}\, d_i$ in bidegree (p, q). Thus, we have $\dim E_i^{p,q} \leq \dim E_{i-1}^{p,q}$, with equality for every p, q if and only if $d_i = 0$. The equality $\dim E_\infty^{p,q} = \dim E_1^{p,q}$ given by (8.6) is thus equivalent to $d_i = 0$, $\forall i \geq 1$. □

Remark 8.29 *The degeneracy at E_1 of the Frölicher spectral sequence is, by the preceding argument, equivalent to the fact that*

$$F^p H^k(X, \mathbb{C}) / F^{p+1} H^k(X, \mathbb{C}) = H^q(X, \Omega_X^p),$$

as well as to the equality $b_k = \sum_{p+q=k} h^{p,q}$, where $b_k = \dim H^k(X, \mathbb{C})$ and $h^{p,q}(X) = \dim H^q(X, \Omega_X^p)$. Thus, it is a property which is not far from being equivalent to the Hodge decomposition theorem. Unfortunately, it does not imply the symmetry of the Hodge numbers $h^{p,q} = h^{q,p}$, nor the Hodge decomposition in the form

$$H^k(X, \mathbb{C}) = \bigoplus_{p+q=k} H^{p,q}(X), \quad H^{p,q}(X) = F^p H^k \cap \overline{F^q H^k}.$$

A remarkable fact is that this formulation of the Hodge theorem, the degeneracy at E_1 of the Frölicher spectral sequence, is, at least in the case of projective manifolds (or more generally, of complete algebraic varieties) a statement of algebraic geometry. Indeed, if X is an algebraic variety, equipped with the Zariski topology (see Hartshorne 1977), we can define the de Rham complex of X to be the complex consisting of the sheaves of algebraic differential forms, equipped with the exterior differential. These sheaves are coherent sheaves, i.e. sheaves of $\mathcal{O}_X^{\mathrm{alg}}$-modules which (for the Zariski topology) locally have finite presentations, and are even locally free. To such a coherent sheaf $\mathcal{E}_{\mathrm{alg}}$, there corresponds the sheaf $\mathcal{E}_{\mathrm{an}}$ (over X equipped with the usual topology) of free $\mathcal{O}_X^{\mathrm{an}}$-modules obtained by taking the pullback of $\mathcal{E}_{\mathrm{alg}}$ by the map Id_X (which is a continuous map from X equipped with the usual topology to X equipped with the Zariski topology) and tensoring the result with $\mathcal{O}_X^{\mathrm{an}}$. We now have the following essential result, known as "Serre's GAGA principle" (Serre 1956).

Theorem 8.30 *The map $\mathcal{E}_{\mathrm{alg}} \mapsto \mathcal{E}_{\mathrm{an}}$ between algebraic coherent sheaves and analytic coherent sheaves over X is an equivalence of categories if X is a complete complex algebraic variety. Moreover, for every sheaf $\mathcal{E}_{\mathrm{alg}}$ over such a variety X, and for every q, we have*

$$H^q(X_{\mathrm{zar}}, \mathcal{E}_{\mathrm{alg}}) = H^q(X_{\mathrm{an}}, \mathcal{E}_{\mathrm{an}}),$$

where X_{zar} (resp. X_{an}) is X equipped with the Zariski topology (resp. the usual topology).

This theorem implies that the degeneracy at E_1 of the Frölicher (or "Hodge to de Rham") spectral sequence of the holomorphic de Rham complex holds true if and only if the corresponding spectral sequence for the algebraic de Rham complex degenerates at E_1.

Deligne and Illusie (1987; Illusie 1996) gave an algebraic proof of the degeneracy at E_1 for manifolds in characteristic 0, or in characteristic p satisfying a lifting hypothesis.

Remark 8.31 *The GAGA principle applied to hypercohomology gives a canonical isomorphism for a projective complex smooth manifold*

$$H^k(X, \mathbb{C}) \cong \mathbb{H}^k(X_{\rm zar}, \Omega^{\cdot}_{X,\rm alg}),$$

namely the composition of the isomorphism

$$H^k(X, \mathbb{C}) \cong \mathbb{H}^k(X_{\rm an}, \Omega^{\cdot}_{X,\rm an})$$

coming from the fact that $\Omega^{\cdot}_{X,\rm an}$ is a resolution of the constant sheaf in the usual topology, with the "GAGA" isomorphism. One needs to be attentive to the fact that the algebraic de Rham complex is not at all locally exact in the Zariski topology (its cohomology is discussed by Bloch and Ogus (1974)), and that in the left-hand term, it is essential to take the cohomology relative to the usual topology, since the cohomology in positive degree (relative to the Zariski topology) of an algebraic variety with values in a constant sheaf is zero. Indeed, constant sheaves are flasque in the Zariski topology (see Hartshorne 1977).

8.4 Hodge theory of open manifolds

8.4.1 Filtrations on the logarithmic complex

When we have an open manifold $U = X - D$, where X is a compact complex manifold and D is a normal crossing divisor, we can of course consider the Frölicher spectral sequence of U, using the isomorphism

$$H^k(U, \mathbb{C}) \cong \mathbb{H}^k(U, \Omega^{\cdot}_U).$$

Unfortunately, in general this gives rather weak information: for example, if U is affine, the sheaves Ω^q_U are acyclic, and thus we have

$$H^k(U, \mathbb{C}) \cong H^k(H^0(U, \Omega^{\cdot}_U)).$$

All the classes are then represented by holomorphic forms, and for the induced filtration in cohomology, we have $F^k H^k(U, \mathbb{C}) = H^k(U, \mathbb{C})$.

The introduction of the logarithmic de Rham complex with singularities now allows us to define a Hodge filtration on the cohomology of U which is different from the preceding one, and which gives a very rich structure in the case where X is Kähler.

The Hodge filtration. Let (X, D) be as above. We have shown that

$$H^k(U, \mathbb{C}) = \mathbb{H}^k(X, \Omega^\cdot_X(\log D)).$$

Since the complex $\Omega^\cdot_X(\log D)$ is equipped with the "naive" filtration, we have an induced filtration on $H^k(U, \mathbb{C})$, which will be called the Hodge filtration of $H^k(U, \mathbb{C})$:

$$F^p H^k(U, \mathbb{C}) = \operatorname{Im}\left(\mathbb{H}^k\left(X, \Omega_X^{\geq p}(\log D)\right) \to \mathbb{H}^k(X, \Omega^\cdot_X(\log D))\right).$$

The weight filtration. There exists another filtration, called the weight filtration, on the cohomology of U. It is essentially the filtration given by the Leray spectral sequence of the inclusion $j : U \to X$. (The Leray spectral sequence associated to a continuous map $\phi : X \to Y$ is obtained as follows: by section 8.1.3, we can compute $H^k(X, \mathbb{Z})$ as $\mathbb{H}^k(Y, \phi_* I^\cdot)$, where $I^\cdot$ is a flasque resolution of $\mathbb{Z}$. The Leray spectral sequence of ϕ is then the spectral sequence of hypercohomology of the complex $\phi_* I^\cdot$. It will be defined and studied in detail in the second volume of this book.)

We will now describe an increasing filtration W on the complex $\Omega^\cdot_X(\log D)$, which up to a change of indices, induces the Leray filtration on $H^*(U, \mathbb{C})$. This last point shows that the induced filtration W on $H^*(U, \mathbb{C})$ is defined over $\mathbb{Z}$.

For $0 \leq l \leq r$, we define $W_l \Omega^\cdot_X \subset \Omega^\cdot_X$ to be the subcomplex equal to

$$\bigwedge^l \Omega^1_X(\log D) \wedge \Omega_X^{\cdot - l}.$$

In other words, in local coordinates z_i in which D is described by the equation $z_1 \ldots z_r = 0$, a form α with logarithmic singularities lies in W_l if it is a combination with holomorphic coefficients of monomials $\frac{dz_I}{z_I} \wedge dz_J$, $I \subset \{1, \ldots, r\}$ with $|I| \leq l$. It is immediate to check that this defines a subcomplex of $\Omega^\cdot_X(\log D)$.

We also introduce the notation $W^k = W_{-k}$, which enables us to define a decreasing filtration; this is better if we want to apply the theory of spectral sequences.

8.4.2 First terms of the spectral sequence

We will make the following hypothesis (which is useful but not essential) to describe the graded complex associated to the filtration W:

The divisor D is globally normal crossing, i.e. we have $D = \bigcup_{i\in I} D_i$ where each $D_i \subset X$ is a smooth hypersurface, and the intersection of l hypersurfaces $D_{i_1}, \ldots, D_{i_l}$ is transverse for every $(i_1, \ldots, i_l)$. We equip I with a total order.

We then introduce the following notation: $D^{(k)}$ is the disjoint union over the subsets $K \subset I$ of cardinal k of $D_K := \bigcap_{i\in K} D_i$. Note that since D has normal crossings, D_K is either empty or a smooth complex submanifold of codimension k of X. We set $D^{(0)} = X$. Let $j_k : D^{(k)} \to X$ denote the natural morphism, and j_M its restriction to D_M.

Proposition 8.32 *There exists a natural isomorphism*

$$W_k\Omega_X^{\cdot}(\log D)/W_{k-1}\Omega_X^{\cdot}(\log D) \cong j_{k*}\Omega_{D^{(k)}}^{\cdot-k}. \tag{8.7}$$

Proof The map (8.7) is given by the residue. In local coordinates on an open set $V \subset X$, let us write $\alpha = \sum_K \alpha_{K,L} dz_L \wedge \frac{dz_K}{z_K} \in \Gamma(V, W_k\Omega_X^{\cdot}(\log D))$, with $K \subset \{1, \ldots, r\} \subset I$ and $|K| \le k$. Here we equip K with the order induced by that of I via μ, and the inclusion $\mu : \{1, \ldots, r\} \hookrightarrow I$ is obtained by noting that a local branch of D corresponds to a unique global component of D. Then $\operatorname{Res}\alpha$ is the section of $j_{k*}\Omega_{D^{(k)}}$ on V defined by

$$(\operatorname{Res}\alpha)_M = (2i\pi)^k \sum_L \alpha_{M,L} dz_{L|D_M\cap V}.$$

Clearly, this annihilates the sections of $W_{k-1}\Omega_X^{\cdot}(\log D)$. If, moreover, we change the local equations defining the divisors D_i, letting $z_i' = f_i z_i$, $i \le s$ for an invertible holomorphic function f_i defined in the neighbourhood of a point of X, where D admits the equation $z_1 \ldots z_s = 0$, we find that

$$\frac{dz_i'}{z_i'} = \frac{dz_i}{z_i} + \frac{df_i}{f_i},$$

where $\frac{df_i}{f_i}$ is a holomorphic form. Thus, we have

$$\frac{dz_K'}{z_K'} = \frac{dz_K}{z_K} \mod W_{k-1}\Omega_X^{\cdot}(\log D),$$

so $\operatorname{Res}\alpha$ does not depend on the choice of the coordinates z_i, $i \le s$. We also see that $\operatorname{Res}\alpha$ does not depend on the choice of the coordinates z_i, $i > s$.

It remains to see that Res induces an isomorphism

$$W_k\Omega_X^{\cdot}(\log D)/W_{k-1}\Omega_X^{\cdot}(\log D) \cong j_{k*}\Omega_{D^{(k)}}^{\cdot-k}.$$

But this is obvious by the local definition: indeed, if we take a holomorphic form α_M on each $D_M \cap V$, $|M| = k$, α_M extends locally to a holomorphic form $\tilde{\alpha}_M$

in the neighbourhood of D_M, and we have $\alpha_M = \mathrm{Res}_M \sum_M (\frac{1}{2i\pi})^k \frac{dz_M}{z_M} \wedge \tilde{\alpha}_M$. Conversely, if $\mathrm{Res}_M \alpha = 0$, $\forall M, |M| = k$, we see that all the $\alpha_{M,L}$, for $|M| = k$, vanish on $D_M \cap V$ for every M, and this implies that $\alpha \in W_{k-1}\Omega^\cdot_X(\log D)$.

Finally, it is clear by the local definition that the residue map is a morphism of complexes. □

Corollary 8.33 *For the spectral sequence $_WE$ associated to the decreasing filtration $W^\cdot$, we have*

$$_WE_1^{p,q} \cong H^{2p+q}\big(D^{(-p)}, \mathbb{C}\big).$$

Proof Applying lemma 8.24, we obtain

$$_WE_1^{p,q} = \mathbb{H}^{p+q}\big(X, \mathrm{Gr}^p_W \Omega^\cdot_X(\log D)\big).$$

To conclude, we then use the isomorphism (8.7):

$$\mathrm{Gr}^p_W \Omega^\cdot_X(\log D) \cong j_{-p*}\big(\Omega^{\cdot+p}_{D^{(-p)}}\big),$$

and the fact that the map j_{-p} is proper with finite fibres. It follows that for every complex $\mathcal{F}^\cdot$ of sheaves over $D^{(-p)}$, we have

$$\mathbb{H}^i\big(D^{(-p)}, \mathcal{F}^\cdot\big) = \mathbb{H}^i(X, (j_{-p})_*\mathcal{F}^\cdot),$$

since $R^k(j_{-p})_*\mathcal{F}^i = 0,\ k > 0$. We obtain

$$\begin{aligned} H^{2p+q}\big(D^{(-p)}, \mathbb{C}\big) &= \mathbb{H}^{2p+q}\big(D^{(-p)}, \Omega^\cdot_{D^{(-p)}}\big) \\ &= \mathbb{H}^{2p+q}\big(X, (j_{-p})_*\Omega^\cdot_{D^{(-p)}}\big) = \mathbb{H}^{p+q}\big(X, \mathrm{Gr}^p_W \Omega^\cdot_X(\log D)\big). \end{aligned}$$

□

A crucial point in the proof of Deligne's theorem 8.35 is the computation of the differential $d_1 : {_WE_1^{p,q}} \to {_WE_1^{p+1,q}}$.

Proposition 8.34 *Taking corollary 8.33 into account, the differential*

$$\begin{array}{llll} d_1 : & H^{2p+q}\big(D^{(-p)}, \mathbb{C}\big) & \to & H^{2p+q+2}\big(D^{(-p-1)}, \mathbb{C}\big) \\ & \parallel & & \parallel \\ & \bigoplus_{|K|=-p} H^{2p+q}(D^K, \mathbb{C}) & \to & \bigoplus_{|L|=-p-1} H^{2p+q+2}(D^L, \mathbb{C}) \end{array}$$

has component $d_1{}^L_K$ equal to zero for $L \not\subset K$, and equal to $(-1)^{q+s} j^L_{K*}$ when $K = \{i_1 < \cdots < i_p\}$ and $L = K - \{i_s\}$, where j^L_K is the inclusion of D^K into D^L and j^L_{K*} is the corresponding Gysin morphism.

Proof We give the proof in the case where D is smooth, as the general case is not substantially different. We use lemma 8.24. It says that the differential

$$d_1 : E_1^{p,q} = H^{p+q}(W^pA^\cdot/W^{p+1}A^\cdot) \to E_1^{p+1,q} = H^{p+q+1}(W^{p+1}A^\cdot/W^{p+2}A^\cdot)$$

of the spectral sequence associated to a filtered complex $(A^\cdot, W^*A^\cdot)$ is the connection map δ of the long exact sequence of cohomology associated to the short exact sequence of complexes

$$0 \to W^{p+1}A^\cdot/W^{p+2}A^\cdot \to W^pA^\cdot/W^{p+2}A^\cdot \to W^pA^\cdot/W^{p+1}A^\cdot \to 0.$$

We will apply this to the complex of global sections of a filtered acyclic resolution of $\Omega_X^\cdot(\log D)$, so that

$$H^{p+q}(W^pA^\cdot/W^{p+1}A^\cdot) \cong \mathbb{H}^{p+q}(X, W^p\Omega_X^\cdot(\log D)/W^{p+1}\Omega_X^\cdot(\log D)).$$

In the case we are considering, the filtration W^* has three steps: $W^{-1},\ W^0,\ W^1$. The first is the whole complex, the second is $\Omega_X^\cdot$, and $W^1 = 0$. The exact sequence

$$\begin{aligned} 0 \to W^0\Omega_X^\cdot(\log D)/W^1\Omega_X^\cdot(\log D) &\to W^{-1}\Omega_X^\cdot(\log D)/W^1\Omega_X^\cdot(\log D) \\ &\to W^{-1}\Omega_X^\cdot(\log D)/W^0\Omega_X^\cdot(\log D) \to 0 \end{aligned}$$

can be identified with the following exact sequence via the isomorphism (8.7):

$$0 \to \Omega_X^\cdot \to \Omega_X^\cdot(\log D) \overset{\text{Res}}{\to} \Omega_D^{\cdot-1} \to 0, \tag{8.8}$$

and we must show that the induced morphism

$$\begin{array}{rccc} \delta : & \mathbb{H}^p(D, \Omega_D^\cdot) & \to & \mathbb{H}^{p+2}(X, \Omega_X^\cdot) \\ & \| & & \| \\ & H^p(D, \mathbb{C}) & \to & H^{p+2}(X, \mathbb{C}) \end{array}$$

can be identified with the Gysin morphism $(-1)^{p+1}j_*$, where j is the inclusion of D into X.

Let $\mathcal{A}_D^\cdot$ be the de Rham complex of D, and let $\mathcal{A}_X^\cdot(\log D)$ be the complex of sheaves over X generated locally by the C^∞ differential forms and by the $\alpha \wedge \frac{df}{f}$, where f is a local equation for D, and α is a C^∞ differential form. As f is defined up to multiplication by an invertible function, $\frac{df}{f}$ is defined up to a holomorphic form, so that the set of these forms does not depend on the choice of f, and we have thus defined a sheaf. We can clearly define the residue

$$\text{Res} : \mathcal{A}_X^\cdot(\log D) \to \mathcal{A}_D^\cdot$$

by $\text{Res}(\omega) = \frac{1}{2i\pi}\alpha_{|D}$ for $\omega = \beta + \alpha \wedge \frac{df}{f}$. As in the holomorphic case, we check that this does not depend on the choice of α and β. Now, let $\mathcal{A}_X'^{\cdot}$ be the kernel of the map Res. We have an obvious inclusion $\mathcal{A}^{\cdot}_X \subset \mathcal{A}'^{\cdot}_X$. Moreover, we have a commutative diagram

$$\begin{array}{ccccccc} 0 \to & \Omega^{\cdot}_X & \to & \Omega^{\cdot}_X(\log D) & \overset{\text{Res}}{\to} & \Omega^{\cdot -1}_D & \to 0 \\ & \downarrow & & \downarrow & & \downarrow & \\ 0 \to & \mathcal{A}'^{\cdot}_X & \to & \mathcal{A}^{\cdot}_X(\log D) & \overset{\text{Res}}{\to} & \mathcal{A}^{\cdot -1}_D & \to 0, \end{array} \tag{8.9}$$

where the last vertical map is a quasi-isomorphism. The lower exact sequence induces a map

$$\delta' : \mathbb{H}^k(D, \mathcal{A}^{\cdot}_D) \to \mathbb{H}^{k+2}(X, \mathcal{A}'^{\cdot}_X), \tag{8.10}$$

and as the complexes $\mathcal{A}^{\cdot}_D$ and $\mathcal{A}'^{\cdot}_X$ are complexes of sheaves of $\underline{\mathcal{C}}^\infty$-modules, their hypercohomology is equal to the cohomology of their complexes of global sections, which we denote by $A^{\cdot}$. The map (8.10) is thus a map

$$\delta' : H^k(A^{\cdot}_D) \to H^{k+2}(A'^{\cdot}_X).$$

We can now construct a form η of type $(1,0)$ on X which is singular along D, with compact support contained in a tubular neighbourhood $T \overset{\pi}{\to} D$ of D in X, and satisfying the condition that in the neighbourhood of D, η is equal to $\frac{1}{2i\pi}\frac{df}{f}$ modulo the $\mathcal{C}^\infty$ forms, where f is (any) local equation for D.

Then if $\alpha \in A^p_D$ is a closed form, $\pi^*\alpha \wedge \eta$ is a section of $A^{p+1}_X(\log D)$ on T with compact support in T, and thus it extends to a global section of $A^{p+1}_X(\log D)$. Moreover, clearly $\text{Res}(\pi^*\alpha \wedge \eta) = \alpha$. Thus, the closed form $d(\pi^*\alpha \wedge \eta) \in A'^{\,p+2}_X$ represents the class $\delta'([\alpha])$. Now, the form $d(\pi^*\alpha \wedge \eta)$ is in fact a $\mathcal{C}^\infty$ form, since $\pi^*\alpha$ is closed and $d\eta$ is $\mathcal{C}^\infty$. Thus, $d(\pi^*\alpha \wedge \eta)$ has a class in $H^{p+2}(A^{\cdot}_X)$. To see that this class is equal to $\delta([\alpha])$, it now suffices to note that the inclusion $\mathcal{A}^{\cdot}_X \subset \mathcal{A}'^{\cdot}_X$ is a quasi-isomorphism, which can be seen by using the diagram (8.9) and showing that the inclusion $\Omega^{\cdot}_X(\log D) \to \mathcal{A}^{\cdot}_X(\log D)$ is a quasi-isomorphism. The inclusion $A^{\cdot}_X \subset A'^{\cdot}_X$ is then also a quasi-isomorphism, since we are considering fine sheaves.

Thus, we have found the explicit representative $d(\pi^*\alpha \wedge \eta)$ for $\delta([\alpha])$, and it simply remains to show that $\mu := d(\pi^*\alpha \wedge \eta)$ also represents $(-1)^{p+1} j_*([\alpha])$. But this is easy: by definition of the Gysin morphism, we want to show that if $\beta \in H^{2n-p}(X, \mathbb{C})$, $n = \dim X$, then

$$\langle [\alpha], j^*\beta \rangle_D = (-1)^{p+1} \langle [\mu], \beta \rangle_X.$$

But β is represented by a closed form γ of degree $2n - p$ on X, and this is then equivalent to

$$\int_D \alpha \wedge j^*\gamma = (-1)^{p+1} \int_X d(\pi^*\alpha \wedge \eta) \wedge \gamma. \tag{8.11}$$

The right-hand term is equal to

$$(-1)^{p+1}\lim_{\epsilon\to 0}\int_{X_\epsilon} d(\pi^*\alpha\wedge\eta)\wedge\gamma,$$

where $X_\epsilon = X - T_\epsilon$ and T_ϵ is a tubular neighbourhood of D of radius ϵ.

As the form $\pi^*\alpha\wedge\eta$ is C^∞ in X_ϵ, Stokes' formula shows that

$$\int_{X_\epsilon} d(\pi^*\alpha\wedge\eta)\wedge\gamma = -\int_{\partial T_\epsilon}\pi^*\alpha\wedge\eta\wedge\gamma = (-1)^{p+1}\int_{\partial T_\epsilon}\pi^*\alpha\wedge\gamma\wedge\eta,$$

and thus the equality (8.11) follows from

$$\lim_{\epsilon\to 0}\int_{\partial T_\epsilon}\pi^*\alpha\wedge\gamma\wedge\eta = \int_D \alpha\wedge\gamma,$$

which follows from Fubini's theorem and the residue theorem, thanks to the local form of η. □

8.4.3 Deligne's theorem

As mentioned above, the complex $\Omega_X^\cdot(\log D)$ is equipped with its "naive" Hodge filtration, which induces a filtration on $H^k(U,\mathbb{C})\cong\mathbb{H}^k(X,\Omega_X^\cdot(\log D))$ also called the Hodge filtration, and denoted by $F^pH^k(U)$. As for the holomorphic de Rham complex, the spectral sequence associated to the Hodge filtration on $\Omega_X^\cdot(\log D)$ has first term equal to ${}_FE_1^{p,q} = H^q(X,\Omega_X^p(\log D))$, where the differential is induced by ∂.

Moreover, proposition 8.34 shows that the term ${}_WE_2^{p,q}$ of the spectral sequence associated to the filtration $W^\cdot$ on $\Omega_X^\cdot(\log D)$ is the cohomology of a complex whose terms are Hodge structures and whose differentials are morphisms of Hodge structures. By the results of section 7.3.1, it follows that ${}_WE_2^{p,q}$ is equipped with a Hodge structure (of weight $q+2p,\ p\le 0$). It is better to shift this Hodge structure by the bidegree $(-p,-p)$, i.e. to see it as a Hodge structure of weight q, since the morphism (8.7) causes the Hodge level to decrease by $-p$.

Theorem 8.35 *(Deligne 1971) Assume that X is a compact Kähler manifold and $D\subset X$ is a globally normal crossing divisor. Then*

(a) The spectral sequence associated to the weight filtration degenerates at E_2.

(b) The spectral sequence associated to the Hodge filtration degenerates at E_1.

(c) On each

$${}_WE_2^{p,q} = \mathrm{Gr}^W_{-p}H^k(U,\mathbb{C}),$$

the Hodge filtration on $\mathbb{H}^k(X, \Omega^{\cdot}_X(\log D))$ *induces the Hodge filtration described above.*

The proof of this theorem only uses the analysis of the two spectral sequences given above, and of course the Hodge theory of the Kähler manifolds $D^{(k)}$.

This result leads to the following definition.

Definition 8.36 *A mixed Hodge structure of weight* k *is a free abelian group of finite type* $V_{\mathbb{Z}}$, *together with an increasing filtration (by weight)* W_l *on* $V_{\mathbb{Z}}$, *and a decreasing filtration (the Hodge filtration)* $F^p V_{\mathbb{C}}$ *on* $V_{\mathbb{C}}$, *such that the filtration induced by* F *on each* $\mathrm{Gr}_r^W V_{\mathbb{C}}$ *defines a Hodge structure of weight* $k + r$ *on* $\mathrm{Gr}_r^W V_{\mathbb{Z}}$.

Deligne's theorem shows the existence of a mixed Hodge structure of weight k on the cohomology $H^k(U, \mathbb{Z})$, where $U \subset X$ is the complement of a normal crossing divisor. One can show that this mixed Hodge structure depends only on U and not on the compactification X.

Remark 8.37 *The mixed Hodge structures of smooth manifolds have the property that the filtration* W *is supported in positive degrees. By the description of the weight filtration, for such a manifold* U, *we have* $\mathrm{Gr}_0^W H^k(U) = \mathrm{Im}\,(H^k(X) \to H^k(U))$, *where* X *is a Kähler compactification such that* $X - U$ *is a normal crossing divisor. By studying the cohomology of singular manifolds, we can obtain more general mixed Hodge structures (Deligne 1975).*

Exercises

1. Let S be a compact complex surface.
   (a) Show that the Frölicher spectral sequence of S degenerates at E_3 for degree reasons. What are the possibly non-zero differentials d_2?
   (b) By studying the integrals

$$\int_S \alpha \wedge \overline{\alpha}$$

   for α a holomorphic form of degree 2, show that the holomorphic 1-forms on S are closed.
   (c) More generally, show that the map

$$H^0(S, \Omega_S^2) \to H^2(S, \mathbb{C})$$

which to a holomorphic 2-form associates its de Rham class, is injective. Deduce from this that the differential

$$d_2^{0,1} : E_2^{0,1} \to E_2^{2,0}$$

vanishes. (Notice that the last space is equal to $H^0(S, \Omega_S^2)$ by b).)

(d) Show that the differentials

$$d_1 : H^p\big(S, \Omega_S^q\big) \to H^p\big(S, \Omega_S^{q+1}\big)$$

and

$$d_1 : H^{2-p}\big(S, \Omega_S^{2-q-1}\big) \to H^{2-p}\big(S, \Omega_S^{2-q}\big)$$

are dual up to sign with respect to Serre's duality. Deduce from this and (b) that

$$d_1 = \partial : H^2(S, \mathcal{O}_S) \to H^2(S, \Omega_S)$$

is equal to 0.

(e) Show that the map

$$H^0(S, \Omega_S) \to H^1(S, \mathbb{C})$$

which to a holomorphic 1-form (note this is closed by (b)) associates its de Rham class, is injective.

(f) Let $\delta := \partial : H^1(S, \mathcal{O}_S) \to H^1(S, \Omega_S)$. Show that we have the relations

$$b_1(S) = h^{1,0}(S) + h^{0,1}(S) - rk\,\delta, \quad b_3(S) \leq h^{2,1}(S) + h^{1,2}(S) - rk\,\delta,$$

with equality if and only if $d_2^{0,2} = 0$. Deduce from this that the Frölicher spectral sequence of S degenerates at E_2 and that it degenerates at E_1 if and only if $\delta = 0$.

(g) Show that we have the relations

$$b_3(S) = b_1(S) = h^{1,0}(S) + h^{0,1}(S) - \mathrm{rank}\,\delta,$$
$$b_2(S) = h^{2,0}(S) + h^{0,2}(S) + h^{1,1}(S) - 2\mathrm{rank}\,\delta.$$

NB. One can show that δ is always 0, that is the Frölicher spectral sequence of a compact complex surface degenerates at E_1.

2. *Spherical spectral sequences.* Let $(M^\cdot, F)$ be a filtered complex in an abelian category. We assume that the decreasing filtration F satisfies the finiteness condition $F^pM^k = 0$ for p sufficiently large. We assume that there exist two integers $q < q'$ such that the complexes $\mathrm{Gr}_F^k M^\cdot$ are exact for $k \neq q, q'$.
(a) Show that the differentials d_i, $i \geq 1$ of the spectral sequence of $(M^\cdot, F)$ are 0 for $i \neq q' - q =: r$.

(b) Show that $E_{r+1}^{k,l} = E_\infty^{k,l}$ is equal to 0 for $k \neq q, q'$ and that one has an exact sequence

$$0 \to E_\infty^{q',l} \to H^{q'+l}(M^\cdot) \to E_\infty^{q,q'+l-q} \to 0.$$

(c) Show that

$$E_\infty^{q',l} = E_r^{q',l}/\operatorname{Im} d_r = E_1^{q',l}/\operatorname{Im} d_r,$$
$$E_\infty^{q,l} = \operatorname{Ker} d_r \subset E_r^{q,l} = E_1^{q,l}.$$

(d) Deduce from this that one has a long exact sequence

$$\cdots \to H^k(M^\cdot) \to E_1^{q,k-q} \overset{d_r}{\to} E_1^{q',k-q'+1} \to H^{k+1}(M^\cdot) \to \cdots$$

where $E_1^{q,k-q} = H^k(\operatorname{Gr}_F^q M^\cdot)$, $E_1^{q',k-q'} = H^k(\operatorname{Gr}_F^{q'} M^\cdot)$.

These spectral sequences appear as Lerays spectral sequences of the sphere bundles. The Leray spectral sequences will be introduced and studied in the second volume of this book.

Part III

Variations of Hodge Structure

9

Families and Deformations

This chapter and the next one are devoted to variations of the Hodge structure, which will be one of the main objects of study of the second volume. Here, we content ourselves with establishing their essential properties. The preceding chapters allowed us to show the existence of a Hodge structure on the cohomology of a Kähler manifold, depending only on its complex structure. Now, we wish to describe how this Hodge structure varies with the complex structure.

In this chapter, we will establish various results from the theory of deformations of a compact complex manifold, which will enable us in the following chapter to formalise the notion of a period map (or a variation of Hodge structure), and to study its infinitesimal properties. Starting from the notion of a family of compact complex manifolds, we show that by Ehresmann's theorem, such a family can be considered locally as a family of complex structures on a fixed differentiable manifold. In particular, the cohomology groups of the fibres X_t of this family can be considered locally as constant spaces by these trivialisations, and this will allow us to locally define the period map in the following chapter: indeed, the Hodge structure on the cohomology of the fibre X_t can be considered as a variable Hodge structure on a constant lattice.

The notion of a family of complex manifolds will give rise to the notion of a holomorphic deformation of the complex structure. We will concentrate here on the study of these families to first order, or on the functor of infinitesimal deformation of the complex structure. We show that the infinitesimal deformations of a complex compact manifold X are represented by $H^1(X, T_X)$, where T_X is the sheaf of holomorphic vector fields over X. We give different constructions of the Kodaira–Spencer map, with values in $H^1(X, T_X)$, the classifying space for infinitesimal deformations.

The second section of this chapter will be devoted to generalities on local systems and the Gauss–Manin connection. Its goal is to give a global and intrinsic meaning to the notion of a locally constant cohomology class for a family of manifolds. The local trivialisations of such a family $\mathcal{X} \to B$ show that the cohomology of the fibres is locally constant. Another way to formulate this is to say that the sheaves $R^k\pi_*\mathbb{Z}$ are locally constant. We introduce the bundle $\mathcal{H}^k$ over B, whose sections are the families of cohomology classes in the fibres X_b varying differentiably or holomorphically with the point b. This bundle is equipped with a flat connection ∇, the Gauss–Manin connection, and the flat sections of $\mathcal{H}^k$ are by definition the sections of the sheaf $R^k\pi_*\mathbb{C}$, i.e. the locally constant families of cohomology classes in the fibres.

Although it is equivalent to consider locally constant sheaves and vector bundles equipped with a flat connection, as we will see, it is essential to have both languages at our disposal. The point of view of locally constant sheaves enables us to define the period map, whereas the language of vector bundles equipped with a flat connection is much more algebraic, so that we can work within the framework of algebraic geometry.

In the final section, we will prove some continuity results for the Hodge filtration on the cohomology of the fibres X_t of a family of Kähler manifolds. We prove the semicontinuity theorem for the numbers $h^{p,q} = \dim H^q(X_t, \Omega^p_{X_t})$, which by a spectral sequence argument implies that in the neighbourhood of a Kähler fibre X_0, the numbers $h^{p,q}(X_t)$ are constant. We also deduce the following result.

Theorem 9.1 *Small deformations of a Kähler manifold X remain Kähler.*

A reference for this chapter is the book by Kodaira (1986).

9.1 Families of manifolds

Let $\mathcal{X}$ be a complex manifold, B a complex manifold and $\phi : \mathcal{X} \to B$ a holomorphic map. Let $X_t := \phi^{-1}(t)$ denote the fibre of ϕ above the point $t \in B$.

Definition 9.2 *We say that $\mathcal{X} \stackrel{\phi}{\to} B$ is a family of complex manifolds if ϕ is a proper holomorphic submersion.*

If B is connected and $0 \in B$ is a reference point, we say that $\mathcal{X}$ is a family of deformations of the fibre X_0. Each fibre X_t, $t \in B$, is called a deformation of X.

9.1.1 Trivialisations

The following result applies to families of complex manifolds as defined above.

Theorem 9.3 (Ehresmann) *Let $\phi : \mathcal{X} \to B$ be a proper submersion between two differentiable manifolds, where B is a contractible manifold equipped with* *a base point* 0. *Then there exists a diffeomorphism*

$$T : \mathcal{X} \cong X_0 \times B$$

over B, *i.e. such that* $\mathrm{pr}_2 \circ T = \phi$.

Remark 9.4 *Giving such a trivialisation is thus equivalent to giving its first component* $T_0 : \mathcal{X} \to X_0$, *which induces a diffeomorphism* $X_t \cong X_0$ *for each* t, *since the second component of* T *is equal to* ϕ. *Up to composing* T *with* $(T_{0|X_0})^{-1}$, *we may assume that* $T_{0|X_0} = \mathrm{Id}$, *i.e. that* T_0 *is a retraction of* $\mathcal{X}$ *onto* X_0.

In the complex case, we cannot in general choose the trivialisation T to be holomorphic, since the fibres are not holomorphically equivalent. However, via each diffeomorphism $T_0 : X_t \to X_0$, a C^∞ trivialisation enables us to consider the complex structure on X_t as a complex structure on X_0 varying with t.

Proof of theorem 9.3 Locally, this is a special case of the theorem of tubular neighbourhoods. This result says that a neighbourhood W of X_0 in $\mathcal{X}$ is diffeomorphic to a neighbourhood of X_0 in its normal bundle, and in particular that there exists a differentiable retraction $T_0 : W \to X_0$. But then, consider the map

$$(T_0, \phi) : W \to X_0 \times B.$$

This map has a differential (which is invertible along X_0. As X_0 is compact, there exists an open set $W' \subset W$ containing X_0 such that $(T_0, \phi)_{|W'}$ is an embedding. Finally, as ϕ is proper, W' contains an open set W'' of the form $\phi^{-1}(U)$, where $U \subset B$ is a neighbourhood of 0. Then clearly $(T_0, \phi)(W'') = X_0 \times U$ and we have shown that $T = (T_0, \phi)$ is a diffeomorphism from $\phi^{-1}(U)$ to $X_0 \times U$, which obviously satisfies $\mathrm{pr}_2 \circ T = \phi$.

For the global case (which we do not actually need here) as B is contractible, we can introduce a vector field χ on B whose associated flow Φ_t exists for every t and satisfies $\mathrm{Im}\, \Phi_t \subset U$ for sufficiently large t. But then χ lifts to a vector field χ' on $\mathcal{X}$, and the associated flow Ψ_t exists for every time t, since ϕ is proper. Moreover, we have $\phi \circ \Psi_t = \Phi_t \circ \phi$ since $\phi_*(\chi') = \chi$. Thus, Ψ_t gives a diffeomorphism compatible with ϕ between $\mathcal{X}$ and $\phi^{-1}(U')$, with $U' \subset U$, and it then suffices to apply the local result. □

In the complex case, we have a more precise trivialisation statement.

Proposition 9.5 *Let* $\mathcal{X} \xrightarrow{\phi} B$ *be a family of complex manifolds. Let* $0 \in B$ *be a point of* B. *Then up to replacing* B *by a neighbourhood of* 0, *there exists a* C^∞

trivialisation $T = (T_0, \phi) : \mathcal{X} \to X_0 \times B$ *having the following property: that the fibres of* T_0 *are complex submanifolds of* $\mathcal{X}$.

These fibres are submanifolds of $\mathcal{X}$ which are diffeomorphic to B. The fact that they are complex does not, of course, imply that T_0 is holomorphic, since these submanifolds do not vary holomorphically with the point $x \in X_0$, but it does say that the family of complex structures on X_0 parametrised by B, which to t associates the complex structure of

$$X_t \overset{T_{0|X_t}}{\cong} X_0,$$

varies holomorphically with $t \in B$.

Proof[1] We use the fact that since $\mathcal{X}$ and X_0 are in particular real analytic manifolds, there exists an isomorphism ψ between a neighbourhood W of the zero section of the normal bundle N of X_0 in $\mathcal{X}$ and a neighbourhood of X_0 in $\mathcal{X}$, such that:

(i) ψ induces the inclusion on X_0 (identified with the zero section).

(ii) The differential of ψ along X_0 induces the canonical isomorphism between $N_{X_0/N}$ and $N_{X_0/\mathcal{X}}$.

(iii) The map ψ is real analytic in the neighbourhood of 0 on each fibre (which is an open set of a vector space) of the natural projection $\pi : W \subset N \to X_0$.

This can be shown for example by putting a real analytic metric on $\mathcal{X}$ in the neighbourhood of X_0, and using the geodesic map, which is defined on a neighbourhood W of the zero section of the normal bundle N, and which to a pair (x, u) consisting of a point $x \in X_0$ and a normal vector u at x, which can be considered as a tangent vector to $\mathcal{X}$, orthogonal to $T_{X_0,x}$, associates the endpoint of the unique geodesic starting at x with tangent vector u.

Now, for each $x \in X_0$, since the map

$$\psi_x : W_x := \pi^{-1}(x) \to \mathcal{X}$$

is real analytic, it admits expansions of the local holomorphic coordinates on X, as power series of the $\mathbb{R}$-linear coordinates on W_x. Now, N_x naturally has the structure of a complex vector space. Let $\psi_x^h : W_x \to \mathcal{X}$ be the holomorphic map obtained by taking the holomorphic part of this power series expansion. One can check that ψ_x^h does not depend on the choice of local coordinates on $\mathcal{X}$ and is C^∞ in x. Thus, we have a C^∞ map

$$\psi^h : W \to \mathcal{X}$$

[1] This proof was pointed out to me by J.-P. Demailly.

which is holomorphic on the fibres of π. Moreover, by property (ii), the differential of $\phi \circ \psi$ along the section 0 is a $\mathbb{C}$-linear (surjective) morphism, and thus is equal to the differential of $\phi \circ \psi^h$ along X_0, which is thus also surjective. Thus, by property (i) and the local inversion theorem, ψ^h is a local diffeomorphism from W' to $\mathcal{X}$, where W' is a neighbourhood of X_0 in N. Then $T = (\pi \circ \psi^{h^{-1}}, \phi)$ gives the desired trivialisation. □

9.1.2 The Kodaira–Spencer map

Let $\phi : \mathcal{X} \to B$ be a family of complex manifolds. The differential ϕ_* is a morphism of holomorphic vector bundles $T_{\mathcal{X}} \to \phi^*(T_B)$. Its restriction to $X = \phi^{-1}(0)$ has kernel equal to the holomorphic tangent bundle T_X. Therefore, we have an exact sequence of holomorphic vector bundles over X:

$$0 \to T_X \to T_{\mathcal{X}|X} \to \phi^* T_{B|X} \to 0. \tag{9.1}$$

Now, $\phi^* T_{B|X}$ is the trivial holomorphic vector bundle of fibre $T_{B,0}$. The exact sequence (9.1) thus gives an extension of the holomorphic bundle T_X by the trivial bundle of fibre $T_{B,0}$. By theorem 4.50, this extension is characterised by the map

$$\rho : T_{B,0} = H^0(X, \phi^* T_{B|X}) \to H^1(X, T_X)$$

induced by the long exact sequence associated to (9.1).

Definition 9.6 *The map $\rho : T_{B,0} \to H^1(X, T_X)$ is called the Kodaira–Spencer map at 0 of the family $\mathcal{X} \to B$.*

We will now explain why the Kodaira–Spencer map can be seen as the differential of "the map $t \mapsto$ complex structure on $X_t \cong X$", or the classifying map for the first order deformation of X induced by the deformation $\mathcal{X}$.

The first point of view consists in studying the subscheme (see Hartshorne 1977) X_ϵ of $\mathcal{X}$ defined by $X_\epsilon = \phi^{-1}(B_\epsilon)$, where B_ϵ is the first order neighbourhood of 0 in B. Subschemes are analytic subsets defined as the locus of the zeros of a coherent sheaf of ideals $\mathcal{I}$ of $\mathcal{O}_{\mathcal{X}}$, equipped with the sheaf of functions $\mathcal{O}_{\mathcal{X}}/\mathcal{I}$.

In the case we are considering, the sheaf of ideals of B_ϵ is the square $\mathcal{M}_0^2$ of the maximal ideal $\mathcal{M}_0$ of $\mathcal{O}_{B,0}$ consisting of the holomorphic functions which vanish at 0, and is thus generated in the neighbourhood of 0 by the holomorphic functions which have the property that both they and their first derivatives vanish at 0. Thus, B_ϵ is a point, but it is equipped with an extended sheaf of functions which parametrises the vectors tangent to B at 0. Similarly, X_ϵ is equal to X

as a topological space, but its sheaf of holomorphic functions differs from $\mathcal{O}_X$, since it contains the functions $\phi^* z_i \bmod \phi^*\mathcal{M}_0^2$ which describe the directions in $\mathcal{X}$ which are normal to X.

If we want to give an abstract description of the first order deformation $\phi : X_\epsilon \to B_\epsilon$ with $\dim B = 1$ so that the ring of the functions on B_ϵ is equal to $\mathbb{C}[\epsilon]/(\epsilon^2)$, we simply note that by the local holomorphic inversion theorem, the family $\mathcal{X} \to B$ is locally isomorphic to a product along X, or that we can cover X by open sets U_i such that there exists $V_i \subset \mathcal{X}$, with $V_i \cap X = U_i$, and a holomorphic isomorphism compatible with ϕ, $V_i \cong U_i \times B_i$ equal to Id on U_i, where B_i is an open set of B containing 0. But then the restriction $\mathcal{O}_{X_\epsilon|U_i}$ is the quotient of the sheaf of holomorphic functions on $U_i \times B_i$ by the ideal generated by ϵ^2, where ϵ is a coordinate centred at 0. Thus, in fact we have an isomorphism

$$\theta_i : \mathcal{O}_{X_\epsilon|U_i} \cong \mathcal{O}_{U_i}[\epsilon]/(\epsilon^2) \tag{9.2}$$

which intrinsically describes the subscheme $V_i \cap X_\epsilon$.

On the open set $V_i \cap V_j$, the change of trivialisation then gives an automorphism of sheaves of rings

$$\theta_{ij} := \theta_i \circ \theta_j^{-1} : \mathcal{O}_{U_{ij}}[\epsilon]/(\epsilon^2) \cong \mathcal{O}_{U_{ij}}[\epsilon]/(\epsilon^2)$$

satisfying the property that θ_{ij} is compatible with the exact sequence

$$0 \to \mathcal{O}_{U_{ij}}.\epsilon \to \mathcal{O}_{U_{ij}}[\epsilon]/(\epsilon^2) \to \mathcal{O}_{U_{ij}} \to 0.$$

Indeed, this follows from the fact that the trivialisations are compatible with ϕ, i.e. preserve the function ϵ and are equal to the identity on U_i.

Now, such an automorphism θ_{ij} of sheaves of rings is determined by a derivation of $\mathcal{O}_{U_{ij}}$, which we denote by χ_{ij}, given by $\theta_{ij}(\phi) = \phi + \epsilon\chi_{ij}(\phi)$. It is immediate to check that θ_{ij} is a morphism of rings if and only if χ_{ij} is a derivation, i.e. by section 2.1.2 a holomorphic vector field on U_{ij}. Finally, as $\theta_{ij} = \theta_i \circ \theta_j^{-1}$ we have $\theta_{ij} \circ \theta_{jk} \circ \theta_{ki} = \mathrm{Id}$, i.e.

$$\chi_{ij} + \chi_{jk} + \chi_{ki} = 0$$

on U_{ijk}. In conclusion, a first order deformation trivialised in the open sets U_i is characterised by a Čech cocycle relative to the covering U_i and with values in the holomorphic tangent bundle T_X. Furthermore, if we change the trivialisations θ_i to θ_i', we have $\theta_i' = \theta_i \circ \mu_i$, where μ_i is an automorphism of $\mathcal{O}_{U_i}$ preserving ϵ and equal to Id modulo ϵ. Thus, μ_i is given by a derivation χ_i, and the cocycle χ_{ij} is modified by the coboundary $\chi_i - \chi_j$. Taking the limit over all coverings

U_i, we have thus shown that the first order deformations of X parametrised by $B_\epsilon = \operatorname{Spec} \mathbb{C}[\epsilon]/(\epsilon^2)$ are exactly parametrised by the group $H^1(X, T_X)$.

The data above describe the general first order deformation of X. It is important to note that, in general, there exist first order deformations which are not first order neighbourhoods of X in a deformation $\mathcal{X} \to B$ of X. Such deformations are said to be "obstructed". When there exists a universal family of deformations of X (cf. section 10.3.1) over a base which is a germ of analytic sets $(B, 0)$, the first order deformations of X are parametrised by $T_{B,0}$, and the existence of obstructions is equivalent to the fact that B is singular at 0 (see Kodaira 1986).

Let us now show that the map described above, which associates a class $\alpha \in H^1(X, T_X)$ to such a deformation, is exactly the Kodaira–Spencer map ρ (definition 9.6), applied to the tangent vector $\frac{\partial}{\partial \epsilon}$ to B_ϵ at 0. For this, we first note that $\rho : T_{B_\epsilon,0} \to H^1(X, T_X)$ is determined by the scheme X_ϵ. Indeed, given X_ϵ, we define $T_{X_\epsilon|X}$ as the sheaf of derivations of $\mathcal{O}_{X_\epsilon}$ with values in $\mathcal{O}_X$. If we also define $T_{B_\epsilon,0}$ in this way, we see that $T_{B_\epsilon,0}$ is simply generated by $\frac{\partial}{\partial \epsilon}$. It is also easy to check that $T_{X_\epsilon|X}$ is a sheaf of free $\mathcal{O}_X$-modules freely generated in the local trivialisations (9.2) by T_X and $\frac{\partial}{\partial \epsilon}$. We then have the exact sequence

$$0 \to T_X \to T_{X_\epsilon|X} \stackrel{\phi_*}{\to} T_{B_\epsilon,0} \otimes \mathcal{O}_X \to 0,$$

where ϕ_* consists in looking at the effect of a derivation on the function ϵ. In the case where X_ϵ is the first order neighbourhood of X in $\mathcal{X}$, this exact sequence can be identified with (9.1).

The trivialisations (9.2) provide splittings of this exact sequence in the open sets U_i, given by the sections

$$\chi_i = \theta_i^*\left(\frac{\partial}{\partial \epsilon}\right) := \frac{\partial}{\partial \epsilon} \circ \theta_i$$

of $T_{X_\epsilon|U_i}$. Now, by definition of the χ_{ij}, we have

$$\chi_{ij} = \theta_j^*\left(\theta_{ij}^*\left(\frac{\partial}{\partial \epsilon}\right) - \frac{\partial}{\partial \epsilon}\right) = \theta_i^*\left(\frac{\partial}{\partial \epsilon}\right) - \theta_j^*\left(\frac{\partial}{\partial \epsilon}\right) = \chi_i - \chi_j$$

as vector fields on U_{ij}. Thus, $\rho(\frac{\partial}{\partial \epsilon})$ is also represented by the Čech cocycle χ_{ij} with values in T_X.

There exists a somewhat different view of deformations, which is more analytic, but very useful, and which consists in using a C^∞ trivialisation of the family $\mathcal{X} \to B$ to consider this family as a family of complex structures on X parametrised by B.

Consider a $\mathcal{C}^\infty$ trivialisation $T = (\pi, \phi) : \mathcal{X} \cong X_0 \times B$ of the family $\phi : \mathcal{X} \to B$, which satisfies the conditions of proposition 9.5. For $t \in B$, consider the diffeomorphism $\pi_t : X_t \cong X_0 = X$. For $x \in X$, we have a family of complex structures $t \mapsto I_t$ on $T_{X,x}$, where I_t is deduced via π_{t*} from the complex structure on T_{X_t,x_t}, with $x_t = \pi_t^{-1}(x)$. The complex structure I_0 at x is equivalent to giving the complex subspace $T^{1,0}_{X,x} \subset T_{X,x,\mathbb{C}}$, or the direct sum decomposition

$$T_{X,x,\mathbb{C}} = T^{1,0}_{X,x} \oplus T^{0,1}_{X,x}. \tag{9.3}$$

For t near 0, the complex subspace

$$\left(T^{1,0}_{X,x}\right)_t \subset T_{X,x,\mathbb{C}}$$

corresponding to the complex structure I_t is then parametrised by the form $\alpha_t \in \Omega^{0,1}_{X,x} \otimes T^{1,0}_{X,x}$, which is identified, up to sign, with the composition

$$T^{0,1}_{X,x} \cong \left(T^{0,1}_{X,x}\right)_t \to T^{1,0}_{X,x},$$

where the first map is the inverse of the projection $(T^{0,1}_{X,x})_t \to T^{0,1}_{X,x}$ given by the decomposition (9.3), and the second map is the projection to $T^{1,0}_{X,x}$ restricted to $(T^{0,1}_{X,x})_t$.

Conversely, given α_t, the vectors of type (0, 1) for $I_{t,x}$ are the vectors of the form $u - \alpha_t(u)$, where u is of type (0, 1) for $I_{0,x}$. Clearly the form $\alpha_t \in \mathcal{A}^{0,1}(T_X)$ thus constructed is $\mathcal{C}^\infty$. Moreover, it vanishes for $t = 0$, and under the hypotheses on T, it is holomorphic in t.

The following result shows even more concretely (if possible) that the Kodaira–Spencer map classifies the first order deformations of the complex structure.

Proposition 9.7 *The map $T_{B,0} \to A^{0,1}(T_X)$ given by $u \mapsto d_u(\alpha_t)$ has values in the set of the $\overline{\partial}$-closed sections of $\mathcal{A}^{0,1}(T_X)$, and for $u \in T_{B,0}$, the Dolbeault cohomology class of $d_u(\alpha_t)$ in $H^1(T_X)$ is equal to $\rho(u)$.*

Proof Consider the trivialisation $T^{-1} : X_0 \times B \cong \mathcal{X}$. As each submanifold $T^{-1}(x \times B)$ of $\mathcal{X}$ is a holomorphic submanifold by the hypothesis on the trivialisation, we obtain a $\mathcal{C}^\infty$ complex subbundle of $T^{1,0}_{\mathcal{X}}$, which is isomorphic via ϕ_* to $\phi^* T_B$, by considering $T_*^{-1}(T_B)$. This subbundle thus gives a $\mathcal{C}^\infty$ map

$$\sigma : \phi^* T_B \to T_{\mathcal{X}},$$

which gives a $\mathcal{C}^\infty$ splitting of the exact sequence

$$0 \to T_{\mathcal{X}/B} \to T_{\mathcal{X}} \to \phi^* T_B \to 0,$$

and thus, restricted to X, gives a $\mathcal{C}^\infty$ splitting of the exact sequence (9.1).

By definition 9.6 and the representation in Dolbeault cohomology of the connection morphism, the Kodaira–Spencer map $\rho : T_{B,0} \to H^1(T_X)$ is then described in Dolbeault cohomology by

$$\rho(u) = \overline{\partial}\sigma(u)$$

for $u \in T_{B,0}$. Proposition 9.7 then follows from the equality

$$\overline{\partial}\sigma(u) = d_u(\alpha_t) \in A^{0,1}(T_X), \quad \forall u \in T_{B,0}. \tag{9.4}$$

This equality is local. We can thus assume that we have local holomorphic coordinates t_i centred at 0 on B, and functions $z_1, \dots, z_n$ on $\mathcal{X}$, such that $z_1, \dots, z_n, \phi^* t_1, \dots, \phi^* t_r$ give a coordinate system on $\mathcal{X}$. The map ϕ is then given in these coordinates by $(z_1, \dots, z_n, t_1, \dots, t_r) \mapsto (t_1, \dots, t_r)$. The map $\pi : \mathcal{X} \to X$ is given in these coordinates by an n-tuple of differentiable functions

$$(\pi_1(z_1, \dots, z_n, t_1, \dots, t_r), \dots, \pi_n(z_1, \dots, z_n, t_1, \dots, t_r)),$$

holomorphic in the t_i. Then the vector fields of type (0, 1) for I_t are generated at the point $\pi(z_1, \dots, z_n, t)$ by the $\pi_* \left(\frac{\partial}{\partial \overline{z_i}}\right) = \sum_j \frac{\partial \pi_j}{\partial \overline{z_i}} \frac{\partial}{\partial z_j} + \sum_j \frac{\partial \overline{\pi}_j}{\partial \overline{z_i}} \frac{\partial}{\partial \overline{z_j}}$. Thus, we have

$$\alpha_t \left(\sum_j \frac{\partial \overline{\pi}_j}{\partial \overline{z_i}} \frac{\partial}{\partial \overline{z_j}} \right) = -\sum_j \frac{\partial \pi_j}{\partial \overline{z_i}} \frac{\partial}{\partial z_j} \tag{9.5}$$

at the point $\pi((z_1, \dots, z_n, t))$. But clearly $\alpha_0 = 0$, and $\pi_{|X} = \mathrm{Id}$, so that to the first order in t, (9.5) gives

$$\alpha_t \left(\frac{\partial}{\partial \overline{z_i}} \right) = -\sum_j \frac{\partial \pi_j}{\partial \overline{z_i}} \frac{\partial}{\partial z_j}$$

at the point $(z_1, \dots, z_n, 0)$. Differentiating this identity with respect to t_k, we find

$$\frac{\partial}{\partial t_k}(\alpha_t)\Big|_{t=0} \left(\frac{\partial}{\partial \overline{z_i}} \right) = -\frac{\partial}{\partial \overline{z_i}} \left(\sum_j \frac{\partial \pi_j}{\partial t_k} \frac{\partial}{\partial z_j} \right). \tag{9.6}$$

But moreover, the vector field $\sigma\left(\frac{\partial}{\partial t_k}\right)$ is the unique vector field of type $(1, 0)$ which is annihilated by π_* and which is sent by ϕ_* to the vector field $\frac{\partial}{\partial t_k}$ on B.

As $\pi_*(\frac{\partial}{\partial t_k}) = \sum_j \frac{\partial \pi_j}{\partial t_k}\frac{\partial}{\partial z_j}$, and $\pi_* = \mathrm{Id}$ along X, we thus find that on T_X, in the coordinates (z_i, t_j), $\sigma\left(\frac{\partial}{\partial t_k}\right)$ can be written

$$\sigma\left(\frac{\partial}{\partial t_k}\right) = \frac{\partial}{\partial t_k} - \sum_j \frac{\partial \pi_j}{\partial t_k}\frac{\partial}{\partial z_j}. \tag{9.7}$$

Comparing (9.6) and (9.7), we thus see that

$$\frac{\partial}{\partial t_k}(\alpha_t)\Big|_{t=0}\left(\frac{\partial}{\partial \overline{z_i}}\right) = \frac{\partial}{\partial \overline{z_i}}\left(\sigma\left(\frac{\partial}{\partial t_k}\right)\right).$$

This proves the equality (9.4). □

9.2 The Gauss–Manin connection

9.2.1 Local systems and flat connections

Let B be a topological space.

Definition 9.8 *A local system over B is a sheaf of abelian groups locally isomorphic to a constant sheaf of stalk G, where G is a fixed abelian group.*

This local system can thus be trivialised in the open sets U_i of an open cover of B, and gives rise to transition isomorphisms $M_{ij} \in \mathrm{Aut}(G)$. When G is a vector space, a local system of vector spaces (of rank equal to rank (G)) is a local system of stalk G, whose transition isomorphisms are vector space automorphisms.

Given a local system H of abelian groups over B, we can consider the associated sheaf of free $\underline{\mathcal{C}}^0(B)$-modules $\mathcal{H}$ defined by

$$\mathcal{H} = H \otimes_{\mathbb{Z}} \underline{\mathcal{C}}^0(B).$$

If H is a local system of $\mathbb{R}$-vector spaces, we can also define

$$\mathcal{H} = H \otimes_{\mathbb{R}} \underline{\mathcal{C}}^0(B).$$

When B is a differentiable manifold, or a complex manifold, we can define the associated sheaves of free $\underline{\mathcal{C}}^\infty(B)$-modules or of free $\mathcal{O}_B$-modules. The $\mathcal{C}^\infty$ holomorphic vector bundles obtained in this manner are equipped with an additional structure: a flat connection.

Indeed, let us define the following connection $\nabla : \mathcal{H} \to \mathcal{H} \otimes \Omega_B$ on $\mathcal{H}$. For $\sigma \in \mathcal{H}$, $\sigma = \sum_i \alpha_i \sigma_i$ in a basis σ_i of a local trivialisation of H, we set

$$\nabla\sigma = \sum_i \sigma_i \otimes d\alpha_i \in \mathcal{H} \otimes \Omega_B.$$

This expression does not depend on the choice of trivialisation, since another local trivialisation of H is deduced from the first one by a transition matrix with constant coefficients, which commutes with the derivations.

In the C^∞ case, this construction gives a C^∞ vector bundle equipped with a C^∞ connection. In the holomorphic case, we obtain a holomorphic vector bundle equipped with a holomorphic connection (i.e. $\nabla\sigma \in \mathcal{H} \otimes_{\mathcal{O}_B} \Omega_B$ for $\sigma \in \mathcal{H}$, where Ω_B is the sheaf of holomorphic differential forms).

Given a connection ∇, we define its curvature

$$\Theta : \mathcal{H} \to \mathcal{H} \otimes \bigwedge^2 \Omega_B$$

as follows: ∇ gives a map

$$\nabla : \mathcal{H} \otimes \Omega_B \to \mathcal{H} \otimes \bigwedge^2 \Omega_B$$

defined by $\nabla(\sigma \otimes \alpha) = \nabla\sigma \wedge \alpha + \sigma \otimes d\alpha$.

Definition 9.9 *The curvature of* ∇ *is then defined by* $\Theta = \nabla \circ \nabla$.

One checks easily that Θ is a $\mathcal{O}_B$-linear map, i.e. that $\Theta(f\sigma) = f\Theta(\sigma)$, so that Θ is in fact a section of $\operatorname{End} \mathcal{H} \otimes \bigwedge^2 \Omega_B$. Of course, all of this can also be done in the differentiable framework.

Definition 9.10 *We say that a connection is flat if it is of curvature zero.*

The curvature of the connection associated to a local system is zero. Indeed, this follows from the fact that ∇ can be identified in the local trivialisations of H with the usual differentiation, and the fact that $d \circ d = 0$ on forms. The vector bundle associated to a local system of vector spaces is thus equipped with a canonical flat connection.

Proposition 9.11 *The correspondence constructed in this way is a bijective correspondence between isomorphism classes of* C^∞ *(or holomorphic in the case where* B *is a complex manifold) vector bundles equipped with a flat connection and isomorphism classes of local systems of vector spaces.*

(In the second case, the vector spaces are complex; in the first case they can be real if we consider real bundles equipped with a real connection.) All the isomorphisms are natural: an isomorphism of vector bundles equipped with connections is an isomorphism of bundles which preserves the connection. Isomorphisms of local systems are isomorphisms of the corresponding sheaves.

Proof The inverse map associates to $(\mathcal{H}, \nabla)$ the local system H of the flat sections of $\mathcal{H}$, i.e. those annihilated by ∇. We need to see that H is a local system, and that we have $\mathcal{H} = H \otimes \underline{\mathcal{C}}^\infty(X)$ (or $\mathcal{H} = H \otimes \mathcal{O}_X$ in the holomorphic case).

This follows from the following fact.

Lemma 9.12 *If $(\mathcal{H}, \nabla)$ is a flat connection, the flat sections σ of $\mathcal{H}$ in the neighbourhood of each point x of B can be identified by restriction to x with the fibre at x of the vector bundle associated to $\mathcal{H}$.*

Recall that this fibre can also be identified, as a complex vector space, with $\mathcal{H}_x \otimes_{\mathcal{O}_{B,x}} \mathcal{O}_{B,x}/\mathcal{M}_x$, where $\mathcal{M}_x \subset \mathcal{O}_{B,x}$ is the ideal of holomorphic functions vanishing at x. The map sending the flat sections to the fibre at x is then the composition

$$H \hookrightarrow \mathcal{H} \to \mathcal{H}_x \otimes_{\mathcal{O}_{B,x}} \mathcal{O}_{B,x}/\mathcal{M}_x.$$

Proof of lemma 9.12 This is an application of Frobenius' theorem 2.20. Indeed, a connection ∇ on $\mathcal{H}$ gives a distribution D on the vector bundle $\pi : \tilde{H} \to B$ corresponding to $\mathcal{H}$ as follows. Let $x \in B$ and $\sigma_0 \in \tilde{H}_x$, and let σ be a section of $\mathcal{H}$ in the neighbourhood of x, such that $\sigma_0 = \sigma_{|x}$. Set

$$D_{(x,\sigma_0)} = \mathrm{Im}\,(\sigma_* - \nabla(\sigma)) : T_{B,x} \to T_{\tilde{H},(x,\sigma_0)},$$

where $\nabla\sigma$ at the point x is considered as an element of $\mathrm{Hom}\,(T_{B,x}, T_{\tilde{H},(x,\sigma_0)})$ via the natural inclusion $\tilde{H}_x \subset T_{\tilde{H},(x,\sigma_0)}$, whose image is exactly the tangent space to the fibre of π. Leibniz' formula shows immediately that the subspace $D_{(x,\sigma_0)}$ of $T_{\tilde{H},(x,\sigma_0)}$ constructed in this way is independent of the choice of σ.

Now, by a local computation, one can check that the curvature of ∇ vanishes if and only if the distribution D satisfies the integrability condition of the Frobenius theorem. This theorem then says that $\tilde{H}$ is locally (in the neighbourhood of the point 0 above x) fibred by integral submanifolds of D, which are then locally isomorphic to B via π, as D is transverse to the fibres of π. Moreover, these integral manifolds are, at least locally, in bijective correspondence with a neighbourhood of 0 in the fibre $\tilde{H}_x$ of π at x. Now, by the definition of D, the integral manifolds of D can be locally identified with the ∇-flat sections of $\mathcal{H}$ (as the distribution D is transverse to the fibres of π, the integral manifolds of D project locally isomorphically onto B, and can be seen locally as sections of π.) The group generated by these sections of course generates the fibre $\tilde{H}_x$ by restriction, since it contains an open set in the fibre of π, and as the sum of two flat sections is flat, we have shown that locally the flat sections generate the fibre of $\tilde{H}$ at each point. Furthermore, a flat section which is zero at a point is zero everywhere, by uniqueness of the solution of the differential equation $\nabla\sigma = 0$ with fixed value σ_x. Lemma 9.12 and proposition 9.11 are thus proved. □

Let $\pi : \mathcal{X} \to B$ be a proper submersive map from one manifold to another. By Ehresmann's theorem 9.3, $\mathcal{X}$ is isomorphic in the neighbourhood of $X_0 = \pi^{-1}(0)$ to $X_0 \times B_0$, where B_0 is a neighbourhood of 0 in B. Consider the sheaves $H^k_A := R^k\pi_*A$, where A is a ring of coefficients (usually $\mathbb{Z}$, $\mathbb{Q}$, $\mathbb{R}$ or $\mathbb{C}$), considered as the constant sheaf of stalk A, and $R^k\pi_*$ is the kth derived functor of the functor π_* from the category of sheaves over $\mathcal{X}$ to the category of sheaves over B. In general, it is not difficult to show that $R^k\pi_*\mathcal{F}$ is the sheaf associated to the presheaf $U \mapsto H^k(\pi^{-1}(U), \mathcal{F}_{|\pi^{-1}(U)})$. In our case, as B is locally contractible, we have $H^k(X_0 \times B_0, A) \cong H^k(X_0, A)$ for a fundamental system of neighbourhoods B_0 of 0, and we deduce that $R^k\pi_*A$ is a local system, isomorphic in the neighbourhood of 0 to the constant sheaf of stalk $H^k(X_0, A)$. Note that the stalk of this local system at a point $t \in B$ is canonically isomorphic to $H^k(X_t, A)$ by restriction.

Definition 9.13 *The flat connection*

$$\nabla : \mathcal{H}^k \to \mathcal{H}^k \otimes \Omega_B$$

on the vector bundle associated to the local system H^k_A is called the Gauss–Manin connection.

9.2.2 The Cartan–Lie formula

Let $\pi : \mathcal{X} \to B$ be a proper submersion, and let $\mathcal{H}^k$ be the (C^∞ or holomorphic according to the context) bundle associated to the local system $H^k_{\mathbb{Z}}$. Let ∇ be the Gauss–Manin connection. Let Ω be a complex differential form of degree k on $\mathcal{X}$ such that $\forall b \in B$, $\Omega_b = \Omega_{|X_b}$ is closed. Then we have a section

$$\omega : b \mapsto [\Omega_b] \in H^k(X_b, \mathbb{C}) \cong H^k(\mathcal{X}_U, \mathbb{C}),$$

where the last equality holds in a contractible neighbourhood U of b. It is easily seen that this map, which has values in $H^k(\mathcal{X}_U, \mathbb{C})$ for $b \in U$, is C^∞ if Ω is. Thus, ω is a section of $\mathcal{H}^k$. We want to compute $\nabla\omega$. For this, we may assume we have a trivialisation $T : \mathcal{X}_{|B} \cong X_0 \times B$, shrinking B is necessary. The fibres X_b become diffeomorphic to X_0 via $T_{|X_b}$, and we can consider $(\Omega_b)_{b\in B}$ as a family of differential forms ϕ_b on X_0 which vary in a C^∞ way with $b \in B$. By pullback, the diffeomorphisms $T_{|X_b} : X_b \cong X_0$ also induce isomorphisms $H^k(X_0, \mathbb{C}) \cong H^k(X_b, \mathbb{C})$, which renders the following diagram of isomorphisms compatible:

$$\begin{array}{ccc} H^k(\mathcal{X}, \mathbb{C}) & \to & H^k(X_0, \mathbb{C}) \\ \downarrow & & \downarrow \\ H^k(\mathcal{X}, \mathbb{C}) & \to & H^k(X_b, \mathbb{C}) \end{array}$$

It follows immediately that $\nabla\omega_{|0} \in H^k(X_0, \mathbb{C}) \otimes \Omega_{B,0}$ is simply the map $u \mapsto \text{class}(d_u(\phi_b)_{|b=0}) \in H^k(X_0, \mathbb{C})$, defined on $T_{B,0}$.

In the trivialisation $T : \mathcal{X} \cong X_0 \times B$, and for coordinates t_i on B, let us write $\Omega = \Phi + \sum_i dt_i \wedge \psi_i + \Omega'$, where $\Phi_{|X_0 \times b} = \phi_b$, the dt_i do not occur in the forms Φ and ψ_i, and Ω' lies in $\text{pr}_2^* \bigwedge^2 \Omega_B \wedge \Omega_{\mathcal{X}}^{k-2}$. Then, since the forms ϕ_b are closed, we have

$$d\Omega = \sum_i dt_i \wedge \frac{\partial \phi_b}{\partial t_i} - \sum_i dt_i \wedge d\psi_i + d\Omega'.$$

We thus obtain

$$\text{int}\left(\frac{\partial}{\partial t_i}\right)(d\Omega)_{|X_0} = \frac{\partial \phi_b}{\partial t_i}_{|b=0} - d\psi_{i|X_0}.$$

As $d\psi_{i|X_0}$ is exact, and the vector field $\frac{\partial}{\partial t_i}$ along X_0 can be replaced here by any field $v_i \in T_{\mathcal{X}|X_0}$ satisfying $\phi_* v_i = \frac{\partial}{\partial t_i}$ since $d\Omega_{|X_0} = 0$, we have shown the following result, which can be considered as a version of the Cartan–Lie formula.

Proposition 9.14 *If $u \in T_{B,0}$ and $v \in \Gamma(T_{\mathcal{X}|X_0})$ is such that $\phi_*(v) = u$, we have*

$$\nabla(\omega)_{|0}(u) = \text{class}(\text{int}(v)(d\Omega)_{|X_0}). \tag{9.8}$$

9.3 The Kähler case

9.3.1 Semicontinuity theorems

Let $\phi : \mathcal{X} \to B$ be a family of complex compact manifolds, with fibre X_b, $b \in B$. Let $\mathcal{F}$ be a holomorphic vector bundle over $\mathcal{X}$. We have the following result.

Theorem 9.15 *The function $b \mapsto \dim H^q(X_b, \mathcal{F}_{|X_b})$ is upper semicontinuous. In other words, we have $\dim H^q(X_b, \mathcal{F}_{|X_b}) \leq \dim H^q(X_0, \mathcal{F}_{|X_0})$ for b in a neighbourhood of $0 \in B$.*

Proof If the fibration is projective algebraic, we can argue as follows. Let $\mathcal{O}_{\mathcal{X}}(1)$ be a line bundle, ample over the fibres of ϕ, i.e. there exists (at least locally) a holomorphic embedding of $\mathcal{X}$ in $\mathbb{P}^K \times B$ over B such that the pullback of

$\mathcal{O}_{\mathbb{P}^K}(1)$ for some k, by this embedding is a multiple of $\mathcal{O}_{\mathcal{X}}(1)$. Up to replacing $\mathcal{O}_{\mathcal{X}}(1)$ by $\mathcal{O}_{\mathcal{X}}(N)$ for sufficiently large N, we may assume by applying Kodaira's vanishing theorem that $H^q(X_b, \mathcal{O}_{X_b}(l)) = 0$ for $l \geq 1$, $q > 0$ and b near 0. Moreover, there exists a resolution

$$0 \to \mathcal{F} \to \mathcal{O}_{\mathcal{X}}(l_0)^{N_0} \to \mathcal{O}_{\mathcal{X}}(l_1)^{N_1} \to \cdots \to \mathcal{O}_{\mathcal{X}}(l_{n+1})^{N_{n+1}} \to \mathcal{F}' \to 0,$$

where $n = \dim X_b$, $l_i > 0$.

To see this, we will say that a holomorphic vector bundle $\mathcal{K}$ over a complex manifold $\mathcal{X}$ is generated by its global sections if the evaluation map

$$H^0(\mathcal{X}, \mathcal{K}) \to \mathcal{K}_x \otimes_{\mathcal{O}_{\mathcal{X},x}} \mathcal{O}_{\mathcal{X},x}/\mathcal{M}_x$$

is surjective at every point $x \in \mathcal{X}$.

As in the proof of theorem 7.11, Kodaira's vanishing theorem applied to the manifold $\mathbb{P}(\mathcal{F}_{|X_0})$ implies that for sufficiently large l_0, the restriction to X_0, $\mathcal{F}_0^*(l_0) := \mathcal{F}^* \otimes \mathcal{O}_{X_0}(l_0)$ is generated by its global sections and satisfies $H^q(X_0, \mathcal{F}_0^*(l_0)) = 0$, $\forall q > 0$. It follows (Hartshorne 1977, III.12) that $\mathcal{F}^*(l_0)$ is also generated by its global sections in the neighbourhood of X_0, and even by a finite number N_0 of global sections.

We then have a surjection

$$\mathcal{O}_{\mathcal{X}}^{N_0} \to \mathcal{F}^*(l_0)$$

in a neighbourhood of X_0, and thus an injection $\mathcal{F} \to \mathcal{O}_{\mathcal{X}}(l_0)^{N_0}$ whose cokernel is locally free. It suffices to iterate this reasoning $n+1$ times to obtain the desired resolution.

This resolution gives an isomorphism

$$H^q(\mathcal{F}_{|X_b}) \cong \mathbb{H}^q\left(\mathcal{O}_{X_b}(l_0)^{N_0} \to \cdots \to \mathcal{O}_{X_b}(l_{n+1})^{N_{n+1}} \to \mathcal{F}'_{|X_b} \to 0\right).$$

As $\mathcal{O}_{X_b}(l_i)$ is acyclic, we then obtain an isomorphism (for $q \leq n$, which is the only interesting case, since the cohomology in degree $> n$ is zero by corollary 4.39)

$$H^q(X_b, \mathcal{F}_{|X_b}) = \frac{\operatorname{Ker}(H^0(X_b, \mathcal{O}_{X_b}(l_q)^{N_q}) \to H^0(X_b, \mathcal{O}_{X_b}(l_{q+1})^{N_{q+1}}))}{\operatorname{Im}(H^0(X_b, \mathcal{O}_{X_b}(l_{q-1})^{N_{q-1}}) \to H^0(X_b, \mathcal{O}_{X_b}(l_q)^{N_q}))}. \quad (9.9)$$

The semicontinuity theorem is then implied by the following lemma.

Lemma 9.16 *For a sufficiently large multiple $\mathcal{O}_{\mathcal{X}}(l)$ of an invertible ample bundle, $R^0\phi_*(\mathcal{O}_{\mathcal{X}}(l))$ is a holomorphic vector bundle, whose fibre at the point b is isomorphic to $H^0(X_b, \mathcal{O}_{X_b}(l))$ by restriction.*

Indeed, this lemma together with equality (9.9) shows that there exists a complex of holomorphic vector bundles $\mathcal{E}^\cdot$ over B satisfying the property

$$\forall q \leq n, \ H^q(X_b, \mathcal{F}_{|X_b}) \cong H^q(\mathcal{E}^\cdot_{|b}).$$

Now, it is elementary to show that the function $b \mapsto \dim H^q(\mathcal{E}^\cdot_{|b})$ is upper semicontinuous on B, where $\mathcal{E}^\cdot$ is a complex of holomorphic vector bundles over B. Indeed, via a local trivialisation of the bundles $\mathcal{E}^i$, the differentials of the complex $\mathcal{E}^\cdot$ are represented by matrices with holomorphic coefficients, and this statement thus follows from the lower semicontinuity of the rank of a matrix with variable coefficients. □

In the general (non-projective) case, we can use Hodge theory to prove the semicontinuity theorem 9.15 in the following way. Let us put Hermitian metrics on $\mathcal{X}$ and on $\mathcal{F}$. Then for every $b \in B$, we have induced Hermitian metrics on X_b and $\mathcal{F}_{|X_b}$. Let $\Delta_{F,b}$ be the Laplacian associated to the operator $\overline{\partial}$ on the sections of $\mathcal{A}^{0,q}_{X_b}(F_b)$, where F_b is the holomorphic vector bundle associated to $\mathcal{F}_{|X_b}$. Then by theorem 5.24, we have $H^q(X_b, \mathcal{F}_{|X_b}) \cong \mathcal{H}^{0,q}(F_b)$, the space of $\Delta_{F,b}$-harmonic forms of type $(0, q)$ with values in F_b.

Now, the bundles $\Omega^{0,q}_{X_b} \otimes F_b$ are the restrictions to the fibres X_b of the $\mathcal{C}^\infty$ bundle $\Omega^{0,q}_{\mathcal{X}/B} \otimes F$ on $\mathcal{X}$, where

$$\Omega^{0,q}_{\mathcal{X}/B} = \bigwedge^q \Omega^{0,1}_{\mathcal{X}/B}, \quad \Omega^{0,1}_{\mathcal{X}/B} = \Omega^{0,1}_{\mathcal{X}}/\phi^*\Omega^{0,1}_B,$$

and it is clear that the family of elliptic operators $\Delta_{F,b}$ gives a $\mathcal{C}^\infty$ differential operator on $\mathcal{X}$. Thus, we can apply the following proposition (see Kodaira 1986).

Proposition 9.17 *Let $\phi : \mathcal{X} \to B$ be a family of manifolds (i.e. ϕ is proper and submersive). Let $G \to \mathcal{X}$ be a vector bundle, and let $\Delta = (\Delta_b)_{b\in B}$ be a relative differential operator (i.e. in local coordinates x_i, t_j such that $\phi(x,t) = t$, Δ is a combination with $\mathcal{C}^\infty$ coefficients of the $\frac{\partial^k}{\partial x_I}$, $|I| = k$) acting on G. Then if each operator Δ_b is elliptic of fixed order independent of b, the function $b \mapsto$* $\dim \operatorname{Ker} \Delta_b$ *is upper semicontinuous.*

Remark 9.18 *By locally trivialising the family $\mathcal{X}$ and the bundle G, we can consider Δ as a family of operators acting on a fixed bundle over the fibre X_0.*

This concludes the proof of theorem 9.15 in the general case. □

Applying this theorem to the holomorphic vector bundle of relative differential forms $\Omega^p_{\mathcal{X}/B}$ defined by

$$\Omega^p_{\mathcal{X}/B} = \bigwedge^p \Omega_{\mathcal{X}/B}, \quad \Omega_{\mathcal{X}/B} = \Omega_{\mathcal{X}}/\phi^*\Omega_B,$$

which clearly satisfies $(\Omega^p_{\mathcal{X}/B})_{|X_b} \cong \Omega^p_{X_b}$ (where the isomorphism is given by restriction of differentials), we obtain the following.

Corollary 9.19 *The function* $b \mapsto h^{p,q}(X_b) := \dim H^q(X_b, \Omega^p_{X_b})$ *is upper semicontinuous.*

9.3.2 The Hodge numbers are constant

Let $\phi : \mathcal{X} \to B$ be a family of complex manifolds. Assume that $X = X_0,\ 0 \in B$ is a Kähler manifold. We do not assume, a priori, that this holds for the neighbouring fibres, although this is actually the case, as we will see later.

Proposition 9.20 *For b near 0, we have $h^{p,q}(X_b) = h^{p,q}(X_0)$. Moreover, the Frölicher spectral sequence of X_b degenerates at E_1.*

Proof For b near 0, we have $h^{p,q}(X_b) \leq h^{p,q}(X_0)$ by corollary 9.19. Now, we have $H^q(X_b, \Omega^p_{X_b}) = E_1^{p,q}(X_b)$, where $E_r^{p,q}(X_b)$ is the Frölicher spectral sequence of X_b. Moreover, as observed in section 8.3.3, we have

$$\dim E_\infty^{p,q} \leq \dim E_1^{p,q} \quad \text{with} \quad E_\infty^{p,q} = F^p H^{p+q}(X_b)/F^{p+1}H^{p+q}(X_b).$$

This last identity gives

$$\dim H^k(X_b, \mathbb{C}) = \sum_{p+q=k} \dim E_\infty^{p,q}(X_b).$$

Now, as X_b is diffeomorphic to X, we have $\dim H^k(X_b, \mathbb{C}) = \dim H^k(X, \mathbb{C}) := b_k$, and thus the chain of inequalities

$$b_k = \sum_{p+q=k} \dim E_\infty^{p,q}(X_b) \leq \sum_{p+q=k} \dim E_1^{p,q}(X_b) \tag{9.10}$$

$$= \sum_{p+q=k} h^{p,q}(X_b) \leq \sum_{p+q=k} h^{p,q}(X) = b_k, \tag{9.11}$$

where the last equality holds by theorem 8.28, since X is Kähler. The equality of the two extreme terms thus implies that all the inequalities are equalities, so we have $h^{p,q}(X_b) = h^{p,q}(X)$ and $E_1^{p,q}(X_b) = E_\infty^{p,q}(X_b)$. □

In fact, we can also see that the existence of the Hodge decomposition still holds for X_b, if b is sufficiently near 0.

Proposition 9.21 *For b near* 0, *we have*

$$H^k(X_b, \mathbb{C}) = \bigoplus_{p+q=k} H^{p,q}(X_b),$$

with $H^{p,q}(X_b) = \overline{H^{q,p}(X_b)}$ *and* $H^{p,q}(X_b) \cong H^q\big(X_b, \Omega^p_{X_b}\big)$.

Proof The subspace $F^p H^k(X_b, \mathbb{C}) \subset H^k(X_b, \mathbb{C}) \cong H^k(X, \mathbb{C})$, which is of dimension independent of b by proposition 9.20, varies in a C^∞ way, by an application of proposition 9.22 below. For $b = 0$, we have

$$H^k(X, \mathbb{C}) = F^p H^k(X, \mathbb{C}) \oplus \overline{F^{q+1} H^k(X, \mathbb{C})}, \tag{9.12}$$

when $p + q = k$. By continuity, this also holds for b near 0. Set $H^{p,q}(X_b) = F^p H^k(X_b, \mathbb{C}) \cap \overline{F^q H^k(X_b, \mathbb{C})}$. As these two spaces generate $H^k(X_b, \mathbb{C})$, the dimension of $H^{p,q}(X_b)$ is equal to that of $H^{p,q}(X)$. Now, property (9.12) for X_b shows that the inclusion $H^{p,q}(X_b) \subset F^p H^k(X_b, \mathbb{C})$ followed by the projection $F^p H^k(X_b, \mathbb{C}) \to F^p H^k(X_b, \mathbb{C})/F^{p+1} H^k(X_b, \mathbb{C}) \cong H^q(X_b, \Omega^p_{X_b})$ is an isomorphism.

Finally, we clearly have $H^k(X_b, \mathbb{C}) = \bigoplus_{p+q=k} H^{p,q}(X_b)$ by property (9.12) for X_b, and $H^{p,q}(X_b) = \overline{H^{q,p}(X_b)}$. □

Proposition 9.22 (See Kodaira 1986) *Let* $\Delta = (\Delta_b)_{b\in B}$ *be a relative differential operator acting on a vector bundle* $F \to \mathcal{X}$, *such that each induced operator* Δ_b *on* F_b *is elliptic of fixed order. Then if* dim Ker Δ_b *is independent of b, the subspace* Ker $\Delta_b \subset C^\infty(F_b)$ *varies in a* C^∞ *way with b.*

This means that up to shrinking B near b, there exist C^∞ sections $(\eta^i_b)_{b\in B}$ of F over $\mathcal{X}$ whose restrictions to X_b for fixed b form a basis of Ker Δ_b.

9.3.3 Stability of Kähler manifolds

We will use proposition 9.21 to show the following result.

Theorem 9.23 *Let* $\phi : \mathcal{X} \to B$ *be a family of complex manifolds,* $0 \in B$. *If the fibre* X_0 *is Kähler, then so is* X_b *for all b sufficiently near* 0.

Proof By proposition 9.20, the function $b \mapsto \dim H^1(X_b, \Omega_{X_b})$ is constant in the neighbourhood of 0. If we put a Hermitian metric on $\mathcal{X}$, we have an induced Hermitian metric on each X_b, and $H^1(X_b, \Omega_{X_b})$ can be identified with the forms of type (0, 1) with values in Ω_{X_b} which are harmonic for the Laplacian associated to the operator $\overline{\partial}$. We can thus apply proposition 9.22 to the family of elliptic operators $(\Delta_{\overline{\partial}})_{b\in B}$ acting on $\Omega^{0,1}_{X_b} \otimes \Omega^{1,0}_{X_b}$, noting that $\Omega^{0,1}_{X_b} \otimes \Omega^{1,0}_{X_b}$ is the restriction to X_b of the bundle $\Omega^{0,1}_{\mathcal{X}/B} \otimes \Omega^{1,0}_{\mathcal{X}/B}$. This shows the following result.

Corollary 9.24 *If* $\dim H^1(X_b, \Omega_{X_b}) = \dim H^1(X_0, \Omega_{X_0})$ *for b near 0, then for every $\Delta_{\overline{\partial}}$-harmonic form ω of type* $(1, 1)$ *on X_0, there exists a C^∞ section* $(\omega_b)_{b\in B}$, $\omega_0 = \omega$ *of* $\Omega_{\mathcal{X}/B} \otimes \Omega^{0,1}_{\mathcal{X}/B}$ *such that ω_b is $\overline{\partial}$-closed for every b.*

Here, we may assume that the induced Hermitian metric on X_0 is Kähler, so that the form ω is in fact d-closed.

Proposition 9.20 now says that the Frölicher spectral sequence of X_b degenerates at E_1, and thus that $\partial\omega_b = \overline{\partial}(\eta_b)$ for a form η_b of type $(2, 0)$ on X_b. In fact, applying a somewhat more precise version of proposition 9.22 (Kodaira 1986), one can even assume that η_b and its derivatives tend uniformly to 0 with b, since $\partial\omega_0 = 0$.

The form $\partial\eta_b$ is a form of type $(3, 0)$ which is $\overline{\partial}$-closed, and thus holomorphic. Now, its complex conjugate is $\overline{\partial}$-exact, and as we know by the degeneracy at E_1 of the Frölicher spectral sequence of X_b and by proposition 9.21 that $H^{3,0}(X_b)$ can be identified with the holomorphic forms of type $(3, 0)$ and that complex conjugation sends $H^{3,0}(X_b)$ to $H^{0,3}(X_b)$ isomorphically, it follows that $\partial\eta_b = 0$.

It then follows that $\overline{\partial}(\overline{\eta_b}) = 0$ and thus that $\overline{\eta_b}$ has a class in $H^{0,2}(X_b)$. Now, we know by proposition 9.21 that $H^{0,2}(X_b) = \overline{H^{2,0}(X_b)}$, where $H^{2,0}$ is the set of holomorphic forms of type $(2, 0)$ on X_b. Thus we can write $\overline{\eta_b} = \overline{\alpha_b} + \overline{\partial}\gamma_b$ with α_b holomorphic and γ_b of type $(0, 1)$ on X_b. Once again, we may assume that γ_b as well as all its derivatives converge uniformly to 0 with b.

We then have the equality

$$\partial\omega_b = \overline{\partial}(\alpha_b + \partial\overline{\gamma_b}) = \overline{\partial}\partial(\overline{\gamma_b}) = -\partial\overline{\partial}\overline{\gamma_b},$$

where the form $\overline{\partial}\overline{\gamma_b}$ is of type $(1, 1)$. Thus, $\omega'_b = \omega_b + \overline{\partial}\overline{\gamma_b}$ is both ∂ and $\overline{\partial}$-closed, and is of type $(1, 1)$ on X_b. Moreover, when b tends to 0, ω'_b tends uniformly to ω.

Assume now that ω is the Kähler form of a Kähler metric on X_0. We have a closed form ω'_b of type $(1, 1)$ on X_b, which converges uniformly with b to ω. As ω is real, $\Re\,\omega'_b$ also converges uniformly with b to ω. Moreover, it is a real closed form of type $(1, 1)$ on X_b. As ϕ is proper and ω is positive on X_0, $\Re\,\omega'_b$ is positive on X_b for b near 0, and thus X_b is Kähler. □

Remark 9.25 *The fact that the neighbouring fibres X_b are Kähler also implies that they satisfy the property of degeneracy at E_1 of the Frölicher spectral sequence, but this was the first step in the proof of theorem 9.23, so that it was essential to give a proof of proposition 9.20 which was independent of theorem 9.23.*

10

Variations of Hodge Structure

In this chapter, we introduce the period domain and the period map for a family of Kähler manifolds. The period domain parametrises the Hodge filtrations with fixed Hodge numbers on a fixed vector space V. It is an open set in a flag space, which can also be considered as a submanifold of a product of Grassmannians, via the map which associates the sequences of spaces $F^i V$ to a filtration $F^{\cdot} V$ on V. We thus devote a section to the construction of Grassmannians as complex manifolds, and the description of their tangent space.

Proposition 10.1 *The tangent space of the Grassmannian of the subspaces of V at the point $W \subset V$ is canonically isomorphic to* $\mathrm{Hom}\,(W, V/W)$.

We then proceed to the study of the local period map $\mathcal{P}^k : B \to \mathcal{D}$, defined for a family of Kähler manifolds parametrised by a simply connected base B. This map associates to $t \in B$ the Hodge filtration on $H^k(X_t, \mathbb{C})$, considered as the constant space $H^k(X_0, \mathbb{C})$.

Theorem 10.2 *The period map is holomorphic and satisfies the transversality condition.*

This last property is essential. Writing $\mathcal{P}^{p,k} : B \to G = \mathrm{Grass}(h_p, V)$ for the map which to $t \in B$ associates the subspace $F^p H^k(X_t)$ of $H^k(X_t, \mathbb{C})$ viewed as a constant space V, consider the differential

$$d\mathcal{P}^{p,k} : T_{B,0} \to T_{G,F^p H^k(X_0)}.$$

By proposition 10.1, this space can be identified with

$$\mathrm{Hom}\,(F^p H^k(X_0), H^k(X_0, \mathbb{C})/F^p H^k(X_0)).$$

The transversality property then says that $\operatorname{Im} d\mathcal{P}^{p,k}$ is contained in

$$\operatorname{Hom}(F^pH^k(X_0), F^{p-1}H^k(X_0)/F^pH^k(X_0)).$$

We also translate this result in terms of Hodge bundles and the Gauss–Manin connection, which will be the form used in the second volume, where concepts will be formulated in terms of variations of Hodge structure rather than period maps.

Theorem 10.3 *The Hodge filtration on the cohomology of the fibres gives a filtration of $\mathcal{H}^k$ by holomorphic subbundles, called Hodge subbundles, and written $F^p\mathcal{H}^k \subset \mathcal{H}^k$. These bundles satisfy the transversality property*

$$\nabla F^p\mathcal{H}^k \subset F^{p-1}\mathcal{H}^k \otimes \Omega_B.$$

We also give the explicit computation of the differential of the period map at the point $0 \in B$, which by theorem 10.2 is given by a family of maps indexed by p, from $T_{B,0}$ to

$$\operatorname{Hom}(F^pH^k(X_0)/F^{p+1}H^k(X_0), F^{p-1}H^k(X_0)/F^pH^k(X_0)).$$

Noting that we have the canonical isomorphism

$$F^pH^k(X_0)/F^{p+1}H^k(X_0) \cong H^q\left(X_0, \Omega^p_{X_0}\right),$$

where $p + q = k$, we have the following.

Theorem 10.4 *$d\mathcal{P}^{k,p}$ is the composition of the Kodaira–Spencer map*

$$\rho : T_{B,0} \to H^1(X_0, T_{X_0})$$

with the map given by the cup-product and the interior product

$$H^1(X_0, T_{X_0}) \to \operatorname{Hom}\left(H^q\left(X_0, \Omega^p_{X_0}\right), H^{q+1}\left(X_0, \Omega^{p-1}_{X_0}\right)\right).$$

As an application of this result, we obtain the generic Torelli theorem for curves of genus at least 5.

All the results presented here are due to Griffiths (1968).

10.1 Period domain and period map

10.1.1 Grassmannians

Let W be a complex vector space. Let Grass(k, W) denote the set of complex vector subspaces of dimension k of W. For example, if $k = 1$, then Grass(k, W) is the complex projective space $\mathbb{P}(W)$.

Proposition 10.5 *The Grassmannian* $G = \mathrm{Grass}(k, W)$ *naturally has the structure of a compact complex (and even projective) manifold of dimension* $k(w-k)$*, where* $w = \dim W$.

Proof Let $W = V \oplus K$ be a decomposition into a direct sum of complex subspaces, where $\dim K = k$. Let $\pi_V : W \to V$ and $\pi_K : W \to K$ be the projections onto each factor. Let G_V be the subset of G consisting of the vector subspaces Z of W of dimension k such that $Z \cap V = \{0\}$. Such a subspace $Z \subset W$ is then isomorphic to K via the projection π_K, and can be identified with the graph of the $\mathbb{C}$-linear map

$$h_Z := \pi_V \circ \pi_{K|Z}^{-1} : K \to V.$$

Thus, if K is a given supplementary subspace of V, then G is covered by the subsets G_V which admit a natural bijection $\phi_{V,K} : G_V \to \mathrm{Hom}_{\mathbb{C}}(K, V)$ where the right-hand spaces are $\mathbb{C}$-vector spaces of dimension $k(w-k)$. If G_V is equipped with the vector space topology induced by any bijection $\phi_{V,K}$, then obviously $G_V \cap G_{V'}$ is open in G_V. Thus, G has a topology for which the G_V are open sets and which induces the vector space topology on each G_V.

The complex structure on G will be the complex structure which induces on each G_V its complex structure as a complex vector space. To justify this definition, we must show that the change of chart morphisms

$$\phi_{V',K'} \circ \phi_{V,K}^{-1} : \mathrm{Hom}\,(K, V)_{V'} \to \mathrm{Hom}\,(K', V')_V$$

are holomorphic, where $\mathrm{Hom}\,(K, V)_{V'} := \phi_{V,K}(G_V \cap G_{V'})$. But

$$\mathrm{Hom}\,(K, V)_{V'} = \{\psi \in \mathrm{Hom}\,(K, V) \mid \phi := \pi_{K'} \circ (1_K + \psi) : K \to K'$$

is an isomorphism}, and we have

$$\phi_{V',K'} \circ \phi_{V,K}^{-1}(\psi) = \pi_{V'} \circ (1_K + \psi) \circ \phi^{-1}$$

for $\psi \in \mathrm{Hom}\,(K, V)_{V'}$. Thus, $\phi_{V',K'} \circ \phi_{V,K}^{-1}$ is holomorphic.

The compactness of the Grassmannian can be shown by induction on k, using the compactness of the projective space and introducing the incidence variety $P = \{(x, V) \in \mathbb{P}(V) \times G \mid x \in V\}$.

One sees easily that P is a projective bundle over G, and a bundle of Grassmannians $\mathrm{Grass}(k-1, W')$, with $w' = w - 1$, over $\mathbb{P}(W)$. The induction hypothesis then shows that P is compact and thus G is also compact. □

Remark 10.6 *The Grassmannian is in fact a projective manifold. The simplest embedding into projective space is the Plücker embedding*

$$G(k, W) \to \mathbb{P}\left(\bigwedge^k W\right),$$

which to $Z \subset W$ associates $\alpha_Z = \langle e_1 \wedge \cdots \wedge e_k \rangle$, where the e_i form a basis of Z. In other words, α_Z is the line $\wedge^k Z$ in $\wedge^k W$. The fact that this map is injective follows from the fact that Z is determined by α_Z, by the formula

$$Z = \left\{ u \in W \mid \alpha_Z \wedge u = 0 \text{ in } \bigwedge^{k+1} W \right\}.$$

The fact that this map is a holomorphic immersion is obvious in the charts $G(k, W)_V \cong \mathrm{Hom}\,(K, V)$.

Let us now describe the tangent bundle of the Grassmannian. Let $K \in G(k, W)$ and let $V \subset W$ be a supplementary subspace of K in W. Then we have the open set $G_V \cong \mathrm{Hom}\,(K, V)$ of G, and the tangent space of G at K can be identified (as a complex space) with that of $\mathrm{Hom}_{\mathbb{C}}(K, V)$ at 0, i.e. with $\mathrm{Hom}_{\mathbb{C}}(K, V)$.

In fact, V is isomorphic to the quotient W/K, which is a complex vector space, and we have the following result.

Lemma 10.7 *The identification $T_{G,K} \cong \mathrm{Hom}\,(K, W/K)$ obtained as the composition*

$$T_{G,K} \cong T_{\mathrm{Hom}\,(K,V),0} \cong \mathrm{Hom}_{\mathbb{C}}(K, W/K) \tag{10.1}$$

is canonical, i.e. independent of the choice of V.

Proof We will give a more canonical description of this identification. Over the Grassmannian G, we have a tautological vector subbundle $\mathcal{S}$ of the trivial bundle of fibre W, whose fibre at the point $K \in G$ is the subspace $K \subset W$. Using the local charts $G_V \cong \mathrm{Hom}_{\mathbb{C}}(K, V)$, we easily check that this is a holomorphic vector subbundle. Indeed, on G_V, we have the tautological holomorphic map

$$\Psi : G_V \cong \mathrm{Hom}\,(K, V) \to \mathrm{Hom}\,(K, V), \quad \Psi(\psi) = \psi.$$

Thus, we have the injective morphism of trivial holomorphic vector bundles over G_V:

$$1 + \Psi : K \otimes \mathcal{O}_{G_V} \to W \otimes \mathcal{O}_{G_V},$$

and clearly $\mathcal{S}_{|G_V} = \mathrm{Im}\,(1 + \Psi)$.

Now let $K \in G$, and let u be a tangent vector to G at K. Let $(\sigma_1, \ldots, \sigma_k)$ be a basis of K and let $\tilde{\sigma}_i$ be holomorphic sections of $\mathcal{S}$ in the neighbourhood of

K such that $\tilde{\sigma}_i(K) = \sigma_i$. Consider the $\mathbb{C}$-linear map $h_u : K \to W/K$ defined by

$$h_u(\sigma_i) = u(\tilde{\sigma}_i) \bmod K,$$

where $u(\tilde{\sigma}_i)$ is the derivative with respect to u of the section $\tilde{\sigma}_i$ considered as a function on G with values in W.

We check that h_u depends neither on the choice of the basis σ_i, nor on the choice of $\tilde{\sigma}_i$; this last point follows from Leibniz' rule, since if σ is a section of $\mathcal{S}$ which vanishes at a point K of G, we can locally write $\sigma = \sum_i f_i \tilde{\sigma}_i$, where the f_i are functions which are holomorphic in the neighbourhood of $K \in G$ and zero at K. But then

$$u(\sigma) = \sum_i u(f_i)\tilde{\sigma}_i(K),$$

and this is in $K \subset W$. Thus, $u(\sigma) = 0$ in W/K if σ vanishes at the point K.

This shows that the map $u \mapsto h_u$ is canonically defined. It remains to see that it is given in the charts G_V by the composition (10.1). Now, as the subbundle $\mathcal{S}$ is naturally identified with the trivial bundle with fibre K over $G_V \cong \mathrm{Hom}\,(K, V)$ via the morphism

$$(1 + \Psi) : K \otimes \mathcal{O}_{G_V} \to W \otimes \mathcal{O}_{G_V},$$

if the $\sigma_i \in K$ are as above, we can take the sections $\tilde{\sigma}_i(\psi) = \sigma_i + \psi(\sigma_i)$ as extensions on G_V. Then if u is a vector tangent to $\mathrm{Hom}\,(K, V)$ at $\psi = 0$, identified with an element $\tilde{u}$ of $\mathrm{Hom}\,(K, V)$, we have

$$h_u(\sigma_i) = u(\tilde{\sigma}_i) = u(\sigma_i + \psi(\sigma_i))_{\psi=0} = \tilde{u}(\sigma_i) \bmod K,$$

and thus $u \mapsto h_u$ can be identified with the composition of the maps $u \mapsto \tilde{u}$ and $\tilde{u} \mapsto \pi \circ \tilde{u}$, where π is the natural isomorphism between V and W/K. □

10.1.2 The period map

Let X be a Kähler manifold and $\phi : \mathcal{X} \to B$ a family of deformations of X. By proposition 9.20, we may assume, up to restricting B, that the fibres X_b satisfy the property of degeneracy at E_1 of the Frölicher spectral sequence, and also satisfy $\dim F^p H^k(X_b, \mathbb{C}) = \dim F^p H^k(X_0, \mathbb{C}) =: b^{p,k}$. (We may even assume, by theorem 9.23, that X_b is Kähler, but this will not play any further role.) Furthermore, also up to restricting B, we may assume that B is contractible, which by Ehresmann's theorem 9.3 gives a canonical identification

$H^k(X_b, \mathbb{C}) \cong H^k(X_0, \mathbb{C})$, $b \in B$, coming from the restrictions

$$H^k(\mathcal{X}, \mathbb{C}) \cong H^k(X_0, \mathbb{C}), \quad H^k(\mathcal{X}, \mathbb{C}) \cong H^k(X_b, \mathbb{C}).$$

Definition 10.8 *The period map*

$$\mathcal{P}^{p,k} : B \to \mathrm{Grass}(b^{p,k}, H^k(X, \mathbb{C}))$$

is the map which to $b \in B$ associates the subspace

$$F^p H^k(X_b, \mathbb{C}) \subset H^k(X_b, \mathbb{C}) \cong H^k(X, \mathbb{C}).$$

The following result is due to Griffiths (1968).

Theorem 10.9 *The period map $\mathcal{P}^{p,k}$ is holomorphic for all p, k, $p \leq k$.*

Proof The map $\mathcal{P}^{p,k}$ is at least C^∞, by an application of proposition 9.22. To see that it is holomorphic, it now suffices to show that its differential is $\mathbb{C}$-linear, which is equivalent to saying that the $\mathbb{C}$-linear extension of its differential to $T_{B,b} \otimes \mathbb{C}$ vanishes on the vectors of type (0, 1).

Now, by the results of the preceding section, the differential

$$d\mathcal{P}^{p,k} : T_{B,b} \to \mathrm{Hom}\,(F^p H^k(X_b), H^k(X, \mathbb{C})/F^p H^k(X_b, \mathbb{C}))$$

is obtained by choosing for $\sigma \in F^p H^k(X_b, \mathbb{C})$ a differentiable function $\tilde{\sigma}$ on B with values in $H^k(X_b, \mathbb{C})$, satisfying $\tilde{\sigma}(b) = \sigma$ and $\tilde{\sigma}(b') \in F^p H^k(X_{b'}, \mathbb{C})$ for all $b' \in B$. We then have

$$d\mathcal{P}^{p,k}(u)(\sigma) = u(\tilde{\sigma}) \bmod F^p H^k(X_b, \mathbb{C}). \tag{10.2}$$

(More intrinsically, $\tilde{\sigma}$ must be viewed as a section of the bundle $\mathcal{H}^k$, and $u(\tilde{\sigma})$ as $\nabla_u \tilde{\sigma}$.)

Now, applying proposition 9.22 and using the fact that the cohomology of the complex $F^p A^k(X_b)$ is of constant rank, we see that there exists a differential form Ω on $\mathcal{X}$ in the neighbourhood of X_b satisfying the conditions

(i) $\Omega \in F^p A^k(\mathcal{X})$.

(ii) $\Omega_{|X_{b'}}$ is closed and its class in $F^p H^k(X_{b'}, \mathbb{C})$ is equal to $\tilde{\sigma}(b')$ for b' near b.

More precisely, proposition 9.22 gives the existence of such a relative form (i.e. a form in $F^p A^k_{\mathcal{X}/B}$), and it suffices to take a lifting in $F^p A^k(\mathcal{X})$.

We then apply formula (9.8). Let us take a C^∞ decomposition $T_{\mathcal{X}|X_b} \cong T_{X_b} \oplus M$ along X_b, where M is a complex subbundle isomorphic to $\phi^* T_{B,b}$. We can also see this decomposition as a decomposition of the real tangent bundle, compatible with the complex structure. Let $u \in T_{B,b}$, and let v be the section of M such that $\phi_*(v) = u$. Formula (9.8) then gives

$$\begin{aligned} d\mathcal{P}^{p,k}(u)(\sigma) &= \nabla_u(\tilde{\sigma}) \bmod F^p H^k(X_b) \\ &= \mathrm{class}\big(\mathrm{int}(v)(d\Omega)_{|X_b}\big) \bmod F^p H^k(X_b). \end{aligned} \tag{10.3}$$

This formula obviously remains valid when u is a complexified tangent vector, since the subspace $M \subset T_{\mathcal{X}}$ is a complex subspace at every point.

Now, if u is a vector of type $(0, 1)$, then the vector field v is of type $(0, 1)$ along X_b. As $d\Omega \in F^p A^{k+1}(\mathcal{X})$, we have $\mathrm{int}(v)(d\Omega) \in F^p A^k(\mathcal{X})$. Thus, $\mathrm{int}(v)(d\Omega)_{|X_b} \in F^p A^k(X_b)$, and as it is a closed form, the class of $\mathrm{int}(v)(d\Omega)_{|X_b}$ is in $F^p H^k(X_b, \mathbb{C})$. So for such a u, we have

$$d\mathcal{P}^{p,k}(u)(\sigma) = \nabla_u(\tilde{\sigma}) \bmod F^p H^k(X_b, \mathbb{C}) = 0,$$

and theorem 10.9 is proved. □

We can give a shorter and more conceptual proof of this theorem by using the following theorem, known as the base change theorem, which is proved by the same arguments as theorem 9.15.

Theorem 10.10 *Let $\phi : \mathcal{X} \to B$ be a family of compact complex manifolds, and let $\mathcal{E}^\cdot$ be a complex of vector bundles over $\mathcal{X}$ satisfying the following condition: the hypercohomology vector space $\mathbb{H}^k(X_b, \mathcal{E}^\cdot_{|X_b})$ is of constant rank over B. Then $R^k\phi_*\mathcal{E}^\cdot$ is a sheaf of free $\mathcal{O}_B$-modules of finite rank with fiber at b isomorphic to $\mathbb{H}^k(X_b, \mathcal{E}^\cdot_{|X_b})$. If $\mathcal{F}^\cdot$ is a subcomplex (consisting of vector subbundles) of $\mathcal{E}^\cdot$ satisfying the same condition and such that for every b, the arrow*

$$\mathbb{H}^k\big(X_b, \mathcal{F}^\cdot_{|X_b}\big) \to \mathbb{H}^k\big(X_b, \mathcal{E}^\cdot_{|X_b}\big) \tag{10.4}$$

is injective, then the natural map $R^k\phi_*\mathcal{F}^\cdot \to R^k\phi_*\mathcal{E}^\cdot$ is the inclusion of a subsheaf of free $\mathcal{O}_B$-modules, and is injective at every point. Moreover, its value at every point b can be identified with the inclusion (10.4).

Remark 10.11 *Let us underline here the property of injectivity at every point, which is not a simple consequence of the injectivity of the morphism of sheaves; this is actually the real content of the base change theorem.*

Indeed, an injective morphism of sheaves of free $\mathcal{O}_X$-modules may very well not be injective at each point x, i.e. after tensoring with $\mathcal{O}_X/\mathcal{M}_x$, as shown by the example of multiplication by a function $f \in \mathcal{O}_X : \mathcal{O}_X \to \mathcal{O}_X$.

To deduce theorem 10.9 from this statement, we take for $\mathcal{E}^\cdot$ the relative holomorphic de Rham complex $\Omega^\cdot_{\mathcal{X}/B}$, and for $\mathcal{F}^\cdot$ the truncated de Rham complex $F^p\Omega^\cdot_{\mathcal{X}/B}$. It remains only to note that $R^k\phi_*\mathcal{E}^\cdot$ is isomorphic to $R^k\phi_*\mathbb{C}\otimes\mathcal{O}_B$, which follows from the fact that $\Omega^\cdot_{\mathcal{X}/B}$ is a resolution of $\phi^{-1}\mathcal{O}_B$. □

Formula (10.3) implies much more than the fact that $\mathcal{P}^{p,k}$ is holomorphic; indeed, it implies the following "Griffiths transversality" property.

Proposition 10.12 *The map*

$$d\mathcal{P}^{p,k} : T_{B,b} \to \mathrm{Hom}\,(F^pH^k(X_b), H^k(X_b,\mathbb{C})/F^pH^k(X_b))$$

has values in $\mathrm{Hom}\,(F^pH^k(X_b), F^{p-1}H^k(X_b)/F^pH^k(X_b))$.

Proof With the notation of the preceding proof, we have $d\mathcal{P}^{p,k}(u)(\sigma) = \nabla_u(\tilde{\sigma}) \bmod F^pH^k(X_b)$ and formula (9.8):

$$\nabla_u(\tilde{\sigma}) = \mathrm{class}(\mathrm{int}(v)(d\Omega)_{|X_b}).$$

Now, Ω lies in $F^pA^k(\mathcal{X})$, so $d\Omega$ is in $F^pA^{k+1}(\mathcal{X})$ and $\mathrm{int}(v)(d\Omega)$ is in $F^{p-1}A^k(\mathcal{X})$. Thus, $\nabla_u(\tilde{\sigma})$ is representable by a closed form in $F^{p-1}A^k(X_b)$, and thus it lies in $F^{p-1}H^k(X_b,\mathbb{C})$. So we have

$$d\mathcal{P}^{p,k}(u)(\sigma) \in F^{p-1}H^k(X_b,\mathbb{C})/F^pH^k(X_b,\mathbb{C}). \tag{10.5}$$

□

Remark 10.13 *The above result holds in an equivalent way for the tangent space of type* $(1,0)$ *or the real tangent space of* B *at* b*, since we know that* $\mathcal{P}^{p,k}$ *is holomorphic. In what follows, since we consider holomorphic maps, the notation* T_B *will indicate the holomorphic tangent bundle.*

10.1.3 The period domain

Let X be a Kähler manifold, and k a positive integer. Let

$$b^{p,k} := \dim F^pH^k(X,\mathbb{C}).$$

We have the Hodge filtration

$$0 = F^{k+1}H^k(X) \subset \cdots \subset F^pH^k(X) \subset F^{p-1}H^k(X) \cdots \subset F^0H^k(X) = H^k(X,\mathbb{C})$$

by complex subspaces of dimension $b^{p,k}$. Such filtrations are parametrised by the flag space (or space of filtrations) $F_{b^\cdot,k}(H^k(X,\mathbb{C}))$ determined by the numbers $b^{p,k}$. This space parametrises the decreasing filtrations $F^{\cdot}$ on $H^k(X,\mathbb{C})$ such that $\dim F^p H^k(X,\mathbb{C}) = b^{p,k}$.

In general, if W is a complex space and b^p, $1 \leq p \leq k$ is a decreasing sequence of numbers less than $\dim W$, we can realise the flag space $F_{b^{\cdot}}(W)$ as the subset of $\prod_{0<p\leq k} \mathrm{Grass}(b^p, W)$ consisting of the k-tuples of subspaces $(W^1,\ldots,W^k)$ of W satisfying the condition $W^i \subset W^{i-1}$, via the map which to a filtration $F^k W \subset \cdots \subset F^1 W$ associates the k-tuple $(F^k W,\ldots,F^1 W)$. It is easy to check that this defines a complex submanifold of $\prod_{0<p\leq k} \mathrm{Grass}(b^p, W)$.

The tangent space of $F_{b^{\cdot}}(W)$ at a point $F = (F^k W \subset \cdots \subset F^1 W)$ of $F_{b^{\cdot}}(W)$ is described by the following lemma.

Lemma 10.14 *$T_{F_{b^{\cdot}}(W),F} \subset \bigoplus_i T_{G(b^i,W),F^{\cdot}W}$ is equal to*

$$\left\{(h_1,\ldots,h_k) \in \bigoplus_i \mathrm{Hom}\,(F^i W, W/F^i W) \,\middle|\, h_{i|F^{i+1}W} = h_{i+1} \bmod F^i W\right\}.$$

Proof Let (σ_i) be a basis of $F^1 W$ adapted to the filtration $F^{\cdot}$ (i.e. for each p, we can extract a basis of $F^p W$ from this basis). We can extend σ_i to a basis $\tilde{\sigma}_i$ of $\mathcal{F}^1 W$ adapted to the filtration $\mathcal{F}^{\cdot}$, where $\mathcal{F}^p W$ is the pullback of the tautological subbundle over Grass (b^p, W). If u is a vector tangent to $F_{b^{\cdot}}(W)$ at F, its image $(h_1,\ldots,h_k) \in \bigoplus_i \mathrm{Hom}\,(F^i W, W/F^i W)$ is given by the results of the preceding section by

$$h_i(\sigma_l) = u(\tilde{\sigma}_l) \bmod F^i W,$$

for $\sigma_l \in F^i W$. Thus, clearly

$$h_{i|F^{i+1}W} = h_{i+1} \bmod F^i W.$$

This shows the inclusion $\subset$. The desired equality then follows from the equality of the dimensions. The dimension of the flag space can easily be computed by induction on k, starting from the formula $\dim G(k, W) = k(w-k)$, $w = \dim W$ and noting that $F_{b^1,\ldots,b^k}$ is fibred above $F_{b^1,\ldots,b^{k-1}}$ in Grassmannians $G(b^k, W_{k-1})$ with $\dim W_{k-1} = b^{k-1}$. □

If $\mathcal{X} \to B$ is a family of deformations of X, then (at least locally) we have the period map

$$\mathcal{P}^k : B \to F_{b^{\cdot}}(H^k(X,\mathbb{C}))$$

defined by $\mathcal{P}^k(b) = (\mathcal{P}^{1,k}(b),\ldots,\mathcal{P}^{k,k}(b))$. The preceding results show that $\mathcal{P}^k$ is holomorphic.

If we restrict ourselves to the points $b \in B$ such that X_b is Kähler, the Hodge filtration must also satisfy the condition

$$F^p H^k(X_b, \mathbb{C}) \oplus \overline{F^{k-p+1} H^k(X_b, \mathbb{C})} = H^k(X_b, \mathbb{C}).$$

This condition clearly defines an open set $\mathcal{D}$ of $F_{b^\cdot}(H^k(X, \mathbb{C}))$.

Definition 10.15 *$\mathcal{D}$ is called a (non-polarised) period domain.*

The polarised period domain, which has much more interesting properties as a complex manifold, is introduced naturally in the study of families $\phi : \mathcal{X} \to B$ of polarised manifolds. One assumes that there exists a class $\omega \in H^2(\mathcal{X}, \mathbb{Z})$ such that the restriction $\omega_{|X_b}$ is a Kähler class for every b. The cup-product by the class ω then gives a morphism of local systems

$$L : R^i \phi_* \mathbb{C} \to R^{i+2} \phi_* \mathbb{C},$$

and the Lefschetz decomposition

$$R^k \phi_* \mathbb{C} = \bigoplus_{k \geq 2r \geq 2k-2n} L^r R^{k-2r} \phi_* \mathbb{C}_{\mathrm{prim}},$$

where $R^i \phi_* \mathbb{C}_{\mathrm{prim}} := \operatorname{Ker} L^{n-i+1}$, $n = \dim X_b$, $i \leq n$. We know that for each b, this decomposition of $H^k(X_b, \mathbb{C})$ is compatible with the Hodge decomposition, i.e. this direct sum is a direct sum of Hodge structures. Furthermore, we have the intersection form $Q(\alpha, \beta) = \langle L^{n-k} \alpha, \beta \rangle$ on $H^k(X_b, \mathbb{Z})$, which is compatible with the local identifications $H^k(X_b, \mathbb{Z}) \cong H^k(X_0, \mathbb{Z})$. Indeed, on the one hand, the class $\omega_{|X_b}$ is locally constant, i.e. compatible with these identifications, and on the other hand, these identifications preserve the intersection form. Finally, by the results of section 6.3.2, the Hodge filtration satisfies the following conditions relative to Q:

(i) $F^p H^k(X_b, \mathbb{C}) = F^{k-p+1} H^k(X_b, \mathbb{C})^\perp$.
(ii) $H^k(X_b, \mathbb{C}) = F^p H^k(X_b, \mathbb{C}) \oplus \overline{F^{k-p+1} H^k(X_b, \mathbb{C})}$.
(iii) On $H^{p,q}(X_b)_{\mathrm{prim}} = F^p H^k(X_b)_{\mathrm{prim}} \cap \overline{F^q H^k(X_b)_{\mathrm{prim}}}$, $p + q = k$, we have

$$(-1)^{\frac{k(k-1)}{2}} i^{p-q} Q(\alpha, \overline{\alpha}) > 0.$$

Setting $W = H^k(X, \mathbb{C})_{\mathrm{prim}}$, equipped with its intersection form Q, we are thus led to define the polarised period domain $\mathcal{D}$ as the set of Hodge filtrations on W satisfying conditions (i), (ii) and (iii) above. The first condition is a condition described by holomorphic equations on the set of flags on W. The last two conditions are open conditions on the set of filtrations satisfying the first condition.

The (local) polarised period map defined on B (which we assume contractible), with values in $\mathcal{D}$, then associates to $b \in B$ the Hodge filtration on $H^k(X_b, \mathbb{C})_{\text{prim}} \cong H^k(X_0, \mathbb{C})_{\text{prim}}$. It is obviously holomorphic as a component of $\mathcal{P}^k$.

Example 10.16 *The simplest period domain is the one which parametrises the Hodge structures of weight* 1*, or complex tori (cf. section 7.2.2). In this case, the Hodge filtration is simply described by* $F^1H \subset H$ *with* $H = F^1H \oplus \overline{F^1H}$. *If* $2g = \dim H$*, then* $\mathcal{D}$ *is an open set in the Grassmannian* $\text{Grass}(g, H)$.

In the polarised case, the space H *is equipped with an alternating form* Q*, and the subspace* $F^1H \subset H$ *must satisfy the conditions:*
(i) F^1H *is totally isotropic for* Q.
(ii) The Hermitian form $iQ(\alpha, \overline{\alpha})$ *is positive definite on* F^1H.

Remark 10.17 *The integral structure on the cohomology plays an essential role in the notion of Hodge structure, whereas it disappears on the level of the local period map. The point is that the integral structure is mainly used to rigidify the structure, allowing one to determine the position of the* F^pH^k *with respect to the integral lattice. When we study the period map, we want to understand how the subspaces* F^pH^k *vary with the point b, and the essential point is the canonical identification of the cohomologies of the fibres* X_b*, which rigidifies the situation sufficiently. The integral structure is not actually lost, since it is flat, i.e. compatible with these identifications (cf. section 9.2.1).*

10.2 Variations of Hodge structure

10.2.1 Hodge bundles

Let $\phi : \mathcal{X} \to B$ be a family of compact complex Kähler manifolds. Let k be a positive integer, and let $\mathcal{H}^k = R^k\phi_*\mathbb{C} \otimes \mathcal{O}_B$ be the holomorphic vector bundle (or sheaf of free $\mathcal{O}_B$-modules) constructed in section 9.2.1. The bundle $\mathcal{H}^k$ is equipped with the Gauss–Manin connection

$$\nabla : \mathcal{H}^k \to \mathcal{H}^k \otimes_{\mathcal{O}_B} \Omega_B$$

which is a holomorphic flat connection. The bundle $\mathcal{H}^k$ admits natural local ∇-flat trivialisations

$$\mathcal{H}^k_{|B_0} \cong H^k(X_0, \mathbb{C}) \otimes_{\mathbb{C}} \mathcal{O}_B,$$

and we have shown that the period map

$$\mathcal{P}^{p,k} : B_0 \to G := \text{Grass}(b^{p,k}, H^k(X_0, \mathbb{C}))$$

is holomorphic. Recall that the period map is defined on an open neighbourhood B_0 of 0 on which the local system $R^k\phi_*\mathbb{C}$ is trivial, and to $b \in B_0$, it associates

$$F^pH^k(X_b, \mathbb{C}) \subset H^k(X_b, \mathbb{C}) \cong H^k(X_0, \mathbb{C}).$$

This implies that there exists a holomorphic vector subbundle

$$F^p\mathcal{H}^k \subset \mathcal{H}^k$$

defined by the condition:

(*) *$F^p\mathcal{H}^k_b \subset \mathcal{H}^k_b$ can be identified with $F^pH^k(X_b, \mathbb{C}) \subset H^k(X_b, \mathbb{C})$ for every $b \in B$.*

(Here, we use the identification

$$\mathcal{H}^k_b = (R^k\phi_*\mathbb{C})_b \otimes (\mathcal{O}_B/\mathcal{M}_b\mathcal{O}_B) \cong H^k(X_b, \mathbb{C}).)$$

If we admit theorem 10.10, $F^p\mathcal{H}^k$ is simply equal to

$$R^k\phi_*(\Omega^{\geq p}_{\mathcal{X}/B}) \subset R^k\phi_*(\Omega^{\cdot}_{\mathcal{X}/B}).$$

More concretely, knowing that the map $\mathcal{P}^{p,k}$ is holomorphic, we locally define $F^p\mathcal{H}^k \subset H^k(X_0, \mathbb{C}) \otimes \mathcal{O}_B$ as the pullback $(\mathcal{P}^{p,k})^*(\mathcal{S})$, where $\mathcal{S} \subset H^k(X_0, \mathbb{C}) \otimes \mathcal{O}_G$ is the tautological subbundle over the Grassmannian already introduced in the preceding section. We must check that this definition does not depend on the choice of the trivialisation of the local system, but this is obvious from the fact that by the definition of the tautological subbundle, the bundle thus defined satisfies condition (*), which determines it uniquely.

The bundles $F^p\mathcal{H}^k$ are called the Hodge subbundles. Their successive quotients $\mathcal{H}^{p,q} := F^p\mathcal{H}^k/F^{p+1}\mathcal{H}^k$ satisfy

$$\begin{aligned}\mathcal{H}^{p,q}_b &= (F^p\mathcal{H}^k)_b/(F^{p+1}\mathcal{H}^k)_b = F^pH^k(X_b)/F^{p+1}H^k(X_b)\\ &= H^q(X_b, \Omega^p_{X_b}), \quad p+q=k.\end{aligned}$$

10.2.2 Transversality

Let $\phi : \mathcal{X} \to B$ be a family of complex compact Kähler manifolds, and k a positive integer. Thus, on B, we have the flat holomorphic vector bundle $(\mathcal{H}^k, \nabla)$ and its (decreasing) Hodge filtration by the holomorphic subbundles $F^p\mathcal{H}^k$. Formula (10.5) then implies the following result.

Proposition 10.18 *The subbundles $F^p\mathcal{H}^k$ satisfy the property*

$$\nabla F^p\mathcal{H}^k \subset F^{p-1}\mathcal{H}^k \otimes \Omega_B.$$

Proof A holomorphic section $\sigma \in F^p\mathcal{H}^k$ is, in particular, a C^∞ section of $\mathcal{H}^k$ satisfying the property that $\sigma(b) \in F^pH^k(X_b, \mathbb{C})$, $\forall b \in B$. Thus, proposition 10.18 is an immediate consequence of (10.5), since

$$d\mathcal{P}^k(u)(\sigma(b)) = \nabla_u(\sigma) \bmod F^pH^k(X_b), \quad \forall b \in B, \quad u \in T_{B,b}.$$

□

Consider the following commutative diagram defining $\overline{\nabla}$:

$$\begin{array}{cccc}
\nabla: & F^{p+1}\mathcal{H}^k & \rightarrow & F^p\mathcal{H}^k \otimes \Omega_B \\
& \downarrow & & \downarrow \\
\nabla: & F^p\mathcal{H}^k & \rightarrow & F^{p-1}\mathcal{H}^k \otimes \Omega_B \\
& \downarrow & & \downarrow \\
\overline{\nabla}^{p,q}: & \mathcal{H}^{p,q} & \rightarrow & \mathcal{H}^{p-1,q+1} \otimes \Omega_B \\
& \downarrow & & \downarrow \\
& 0 & & 0.
\end{array}$$

As ∇ satisfies Leibniz' rule $\nabla(f\sigma) = f\nabla(\sigma) + \sigma \otimes df$ for $f \in \mathcal{O}_B$, and $\sigma \in \mathcal{H}^k$, we have

$$\nabla(f\sigma) = f\nabla(\sigma) \bmod F^p\mathcal{H}^k \otimes \Omega_B \tag{10.6}$$

for $\sigma \in F^p\mathcal{H}^k$, and thus $\overline{\nabla}^{p,q}(f\sigma) = f\overline{\nabla}^{p,q}(\sigma)$ for a section σ of $\mathcal{H}^{p,q}$. In other words, $\overline{\nabla}^{p,q}$ is a morphism of $\mathcal{O}_B$-modules, so that in particular we can consider its value $\overline{\nabla}^{p,q}_b$ at each point $b \in B$. The collection of maps

$$\begin{array}{cccc}
\overline{\nabla}^{p,q}_b: & \mathcal{H}^{p,q}_b & \rightarrow & \mathcal{H}^{p-1,q+1}_b \otimes \Omega_{B,b} \\
& \| & & \| \\
& H^q\left(X_b, \Omega^p_{X_b}\right) & \rightarrow & H^{q+1}\left(X_b, \Omega^{p-1}_{X_b}\right) \otimes \Omega_{B,b}
\end{array}$$

is called the infinitesimal variation of Hodge structure at the point b.

10.2.3 Computation of the differential

In the proof of lemma 10.7, we constructed a natural identification of $T_{G,K}$ with $\mathrm{Hom}\,(K, W/K)$, where G is the Grassmannian Grass(k, W) and $K \in G$. This identification is obtained by differentiation of the holomorphic sections of the tautological subbundle in the neighbourhood of K.

If we apply this to the period map, noting that differentiating the sections of the trivial bundle $H^k(X_0, \mathbb{C}) \otimes \mathcal{O}_B$ is equivalent to applying the Gauss–Manin connection to the sections of $\mathcal{H}^k$, we obtain the following result.

Lemma 10.19 *The differential*

$$d\mathcal{P}_b^{p,k} : T_{B,b} \to T_{G,F^pH^k(X_b)} = \mathrm{Hom}\,(F^pH^k(X_b), H^k(X_b,\mathbb{C})/F^pH^k(X_b))$$

is the map constructed by adjunction starting from the map

$$\overline{\nabla}_b^p : F^pH^k(X_b) \to H^k(X_b,\mathbb{C})/F^pH^k(X_b) \otimes \Omega_{B,b},$$

which is the value at the point b of the morphism $\overline{\nabla}^p$ (which is $\mathcal{O}_B$-linear by (10.6)) defined as the composition

$$F^p\mathcal{H}^k \xrightarrow{\nabla} \mathcal{H}^k \otimes \Omega_B \to (\mathcal{H}^k/F^p\mathcal{H}^k)\otimes\Omega_B.$$

Finally, by proposition 10.12 applied to F^p and to F^{p+1}, we know that $\mathrm{Im}\, d\mathcal{P}^{p,k}$ is contained in the subspace

$$\mathrm{Hom}\,(F^pH^k(X_b)/F^{p+1}H^k(X_b), F^{p-1}H^k(X_b,\mathbb{C})/F^pH^k(X_b)).$$

Remark 10.20 *By lemma 10.14,*

$$\bigoplus_p \mathrm{Hom}\,(F^pH^k(X_b)/F^{p+1}H^k(X_b), F^{p-1}H^k(X_b,\mathbb{C})/F^pH^k(X_b))$$

is contained in the tangent space of the flag manifold at the point $(F^pH^k(X_b))$. This subspace is called the horizontal tangent space of the flag manifold, or of its open subset $\mathcal{D}$. It is different from $T_{\mathcal{D}}$, except in the case of Hodge structures of weight $2p-1$ having $h^{p,p-1}$ and $h^{p-1,p}$ as the only non-trivial Hodge numbers. In the polarised case, the intersection of $T_{\mathcal{D}}^{\mathrm{hor}}$ with the tangent space $T_{\mathcal{D}_{\mathrm{pol}}}$ is also in general a strict subspace of $T_{\mathcal{D}_{\mathrm{pol}}}$, except in the preceding case or in the case of Hodge structures of weight $2p$ having $h^{p+1,p-1} = 1 = h^{p-1,p+1}$ and $h^{p,p}$ as only non-zero Hodge numbers . This implies that outside of the cases described above, the period map, even polarised, is not surjective, so that a general Hodge structure is not the Hodge structure of a Kähler manifold.

Finally, it follows from lemma 10.19 and from the definition of $\overline{\nabla}^{p,q}$ that

$$d\mathcal{P}^{p,k} : T_{B,b} \to \mathrm{Hom}\left(\frac{F^pH^k(X_b)}{F^{p+1}H^k(X_b)}, \frac{F^{p-1}H^k(X_b)}{F^pH^k(X_b)}\right)$$

is constructed by adjunction starting from the map

$$\overline{\nabla}_b^{p,q} : \frac{F^pH^k(X_b)}{F^{p+1}H^k(X_b)} \to \frac{F^{p-1}H^k(X_b)}{F^pH^k(X_b)} \otimes \Omega_{B,b}.$$

We have now the following cohomological description of

$$d\mathcal{P}^{p,k}(u) \in \mathrm{Hom}\left(H^q(X_b, \Omega_{X_b}^p), H^{q+1}(X_b, \Omega_{X_b}^{p-1})\right).$$

Theorem 10.21 *(Griffiths) The map*

$$d\mathcal{P}^{p,k}(u) : H^q\big(X_b, \Omega^p_{X_b}\big) \to H^{q+1}\big(X_b, \Omega^{p-1}_{X_b}\big)$$

is equal to the cup-product with the class $\rho(u) \in H^1(X_b, T_{X_b})$, where ρ is the Kodaira–Spencer map, composed with the map induced on cohomology by the interior product $T_{X_b} \otimes \Omega^p_{X_b} \to \Omega^{p-1}_{X_b}$.

Proof Recall first (cf. section 5.3.2) that if $\mathcal{E}$, $\mathcal{F}$ are sheaves of free $\mathcal{O}_X$-modules, the cup-product

$$H^r(X, \mathcal{E}) \otimes H^s(X, \mathcal{F}) \to H^{r+s}(X, \mathcal{E} \otimes \mathcal{F})$$

is represented by the exterior product of the forms in Dolbeault cohomology

$$A^{0,r}(E) \otimes A^{0,s}(F) \to A^{0,r+s}(E \otimes F).$$

Moreover, as shown in section 9.1.2, the class $\eta = \rho(u) \in H^1(X_b, T_{X_b})$ is represented by the form $\alpha = \overline{\partial} v_{|X_b}$, where $v \in T_{\mathcal{X}}$ is a C^∞ vector field of type $(1,0)$ such that $\phi_*(v) = u \in T^{1,0}_{B,b}$.

Now, let $\sigma \in H^q(X_b, \Omega^p_{X_b})$, and let Ω be a section of $F^p\Omega^k_{\mathcal{X}}$, where $k = p+q$, such that $\Omega_{|X_t}$ is closed for every $t \in B$ near b, and such that the Dolbeault cohomology class of the component $\Omega^{p,q}$ of $\Omega_{|X_b}$ is equal to σ. The preceding results together with formula (9.8) show that

$$d\mathcal{P}^{p,k}(u)(\sigma) = \overline{\nabla}^{p,q}_u(\sigma) = \big[\mathrm{int}(v)(d\Omega)^{p-1,q+1}_{|X_b}\big], \tag{10.7}$$

where in the last term, the form $\mathrm{int}(v)(d\Omega)_{|X_b}$ is closed and lies in $F^{p-1}A^k(X_b)$, and $[\mathrm{int}(v)(d\Omega)^{p-1,q+1}_{|X_b}]$ denotes the class of its component $\mathrm{int}(v)(d\Omega)^{p-1,q+1}_{|X_b}$ of type $(p-1, q+1)$ in the Dolbeault cohomology group $H^{q+1}(X_b, \Omega^{p-1}_{X_b})$. By the decomposition into types and because v is of type $(1,0)$, we clearly have

$$\mathrm{int}(v)(d\Omega)^{p-1,q+1}_{|X_b} = (\mathrm{int}(v)\overline{\partial}\Omega^{p,q})_{|X_b}. \tag{10.8}$$

Furthermore, $\Omega^{p,q}_{|X_b}$ is $\overline{\partial}$-closed, and we have $\sigma = [\Omega^{p,q}_{|X_b}]$ in $H^q(X_b, \Omega^p_{X_b})$. Finally, we see that

$$\overline{\partial}(\mathrm{int}(v)(\Omega^{p,q})) = -\mathrm{int}(v)(\overline{\partial}\Omega^{p,q}) + \mathrm{int}(\overline{\partial} v)(\Omega^{p,q}),$$

where the last interior product with $\overline{\partial} v \in A^{0,1}(T_{\mathcal{X}})$ combines the interior product $T_{\mathcal{X}} \otimes \Omega^p_{\mathcal{X}} \to \Omega^{p-1}_{\mathcal{X}}$ and the exterior product on the forms. Restricting this equality to X_b, we obtain the equality of the Dolbeault cohomology classes of

$$[(\mathrm{int}(v)\overline{\partial}\Omega^{p,q})_{|X_b}] = \big[\mathrm{int}\big(\overline{\partial} v\big)\big(\Omega^{p,q}_{|X_b}\big)\big],$$

which together with the equalities (10.7) and (10.8) proves theorem 10.21. □

10.3 Applications

10.3.1 Curves

By theorem 10.21, the differential of the period map associated to the deformations of a compact Kähler manifold X is computed using the map given by the cup-product and the interior product

$$H^1(X, T_X) \to \mathrm{Hom}\left(H^q(X, \Omega_X^p), H^{q+1}(X, \Omega_X^{p-1})\right).$$

Consider the case where X is a complete curve of genus $g := h^{1,0}(X)$ (or a compact Riemann surface), and consider the variation of the Hodge structure on $H^1(X)$. We can show that X has a universal family of deformations $\phi : \mathcal{X} \to B$, where $(B, 0)$ is a germ of smooth complex manifolds with tangent space equal (via the Kodaira–Spencer map) to $H^1(T_X)$ and $X_0 \cong X$. This essentially means that the theory of the deformations of X is not obstructed. The base B of such a family represents the deformation functor of X which to a germ $(S, 0)$ of analytic spaces associates the set of isomorphism classes of proper flat families $\phi_S : \mathcal{X}_S \to S$ equipped with an isomorphism $X_0 \cong X$. The family $\phi : \mathcal{X} \to B$ is universal in the sense that every family $\phi_S : \mathcal{X}_S \to S$ as above is a Cartesian product $\mathcal{X}_S = \mathcal{X} \times_B S$ for a uniquely determined morphism of germs $(S, 0) \to (B, 0)$. Concretely, the points of B are in bijection with the isomorphism classes of small deformations of the complex structure of X.

The existence of this universal family is obvious in the case where the genus g is equal to 1, since then X is a 1-dimensional torus (or elliptic curve), and in the case $g \geq 2$, it follows from the fact that X has no infinitesimal automorphisms, i.e. no non-zero section on T_X. The smoothness of B follows from the fact that $H^2(X, T_X) = 0$, since X is a curve. The obstructions to extending the formal deformations of order k to order $k + 1$ actually "live" in $H^2(X, T_X)$ (Kodaira 1986).

The local period map is then defined by

$$\begin{aligned} \mathcal{P}^1 : B &\to \mathrm{Grass}(g, H^1(X, \mathbb{C})) \\ b &\mapsto H^{1,0}(X_b) \subset H^1(X_b, \mathbb{C}) \cong H^1(X, \mathbb{C}), \end{aligned}$$

where the last isomorphism is canonical whenever B is contractible. Its differential

$$d\mathcal{P}_b^1 : T_{B,b} \to \mathrm{Hom}\,(H^{1,0}(X_b), H^{0,1}(X_b))$$

at the point $b \in B$ can be identified by theorem 10.21 with the map

$$H^1(X_b, T_{X_b}) \to \mathrm{Hom}\,(H^0(\Omega_{X_b}), H^1(\mathcal{O}_{X_b})) \tag{10.9}$$

given by the cup-product and the contraction $T_{X_b} \otimes \Omega_{X_b} \to \mathcal{O}_{X_b}$. Note that in the 1-dimensional case, T_{X_b} is a holomorphic bundle of rank 1, dual to the bundle $\Omega_{X_b} = K_{X_b}$. Serre duality (theorem 5.32) also gives the isomorphisms

$$H^1(\mathcal{O}_{X_b}) \cong H^0(K_{X_b})^*, \quad H^1(X_b, T_{X_b}) \cong H^0(K_{X_b}^{\otimes 2})^*.$$

Lemma 10.22 *The map*

$$H^0(K_{X_b}) \otimes H^0(K_{X_b}) \to H^0(K_{X_b}^{\otimes 2})$$

obtained by dualising (10.9) and applying Serre duality is simply the product μ on the sections.

Proof Let $\eta = \alpha \otimes \beta \in H^0(K_{X_b})^{\otimes 2}$, and let $[u] \in H^1(T_{X_b})$, where u is a form of type (0, 1) with values in T_{X_b}. We have

$$\langle \mu(\eta), [u] \rangle = [u\alpha\beta] \in H^1(K_{X_b}) \cong \mathbb{C},$$

where $u\alpha\beta$ is the form of type (0, 1) obtained by contracting $u \in A^{0,1}(T_{X_b}) = A^{0,1}(K_{X_b}^*)$ and $\alpha\beta \in H^0(K_{X_b}^{\otimes 2})$. But clearly, this is also equal to $[(u\alpha)\cdot\beta]$, where $(u\alpha) \in A^{0,1}(X_b)$ is obtained by contracting u and α.

Now, $[(u\alpha)\cdot\beta] \in H^1(X_b, K_{X_b}) \cong \mathbb{C}$ is equal to the pairing given by Serre duality between $[u\alpha] \in H^1(X_b, \mathcal{O}_{X_b})$ and $\beta \in H^0(X_b, K_{X_b})$. Thus we have shown that

$$\langle \mu(\alpha \otimes \beta), [u] \rangle = \langle [u\alpha], \beta \rangle = \langle [u] \cdot \alpha, \beta \rangle,$$

where $[u] \cdot \alpha \in H^1(X_b, \mathcal{O}_{X_b})$ is obtained by cup-product and contraction of the classes $[u] \in H^1(X_b, T_{X_b})$, and $\alpha \in H^0(X_b, \Omega_{X_b})$. Lemma 10.22 is thus proved. □

We next have theorem 10.24, which concerns the canonical embedding of a curve of genus g, $\phi_{K_X} : X \to \mathbb{P}^{g-1}$ given by the holomorphic sections of the canonical bundle. Let us first introduce the following definitions.

Definition 10.23 *A complete curve X is called hyperelliptic if there exists a rational map $X \to \mathbb{P}^1$ of degree 2. It is called trigonal if there exists a rational map $X \to \mathbb{P}^1$ of degree 3.*

One can show that a generic curve of genus ≥ 3 is not hyperelliptic, and that a generic curve of genus ≥ 5 is not trigonal, nor, in genus 6, isomorphic to a smooth curve defined by an equation of degree 5 in $\mathbb{P}^2$. (Such a curve is called a planar quintic.) From the local point of view which we have adopted here, this means that for every curve X and every universal family of deformations $\mathcal{X} \to B$ of X, the set of points $b \in B$ such that X_b is hyperelliptic is a proper analytic subset of B if $g > 2$, and the set of points $b \in B$ such that X_b is trigonal or isomorphic to a smooth plane curve of degree 5 is a proper analytic subset of B if $g \geq 5$.

Theorem 10.24 (See Arbarello et al. 1985)
(a) (Noether) *Let X be a non-hyperelliptic curve. Then the map given by the product*

$$\mu : H^0(X, K_X)^{\otimes 2} \to H^0(X, K_X^{\otimes 2})$$

is surjective.
(b) (Petri) *Let X be a curve which is non-hyperelliptic, non-trigonal and not isomorphic to a planar quintic. Then X is determined by*

$$\operatorname{Ker} \mu : H^0(X, K_X)^{\otimes 2} \to H^0(X, K_X^{\otimes 2})$$

in the following way: the canonical map $\phi_{K_X} : X \to \mathbb{P}^{g-1}$ is an embedding since X is not hyperelliptic, and the symmetric elements of Ker μ *are exactly the homogeneous polynomials of degree 2 over $\mathbb{P}^{g-1}$ which vanish on X; when X is neither trigonal nor isomorphic to a planar quintic, $X \subset \mathbb{P}^{g-1}$ is isomorphic to the algebraic subscheme or complex submanifold defined by these equations.*

These algebraic statements now give us the following results on the period map for curves.

Corollary 10.25 (Infinitesimal Torelli theorem for curves) *Let X be a non-hyperelliptic curve. Then the local period map*

$$\mathcal{P} : B \to \mathrm{Grass}(g, H^1(X, \mathbb{C}))$$

is an embedding at the point $0 \in B$ corresponding to X.

Proof By theorem 10.24 and lemma 10.22, $d\mathcal{P}^1$ is injective at 0 when X is not hyperelliptic. □

Corollary 10.26 *If $g \geq 5$, a generic curve X is determined by its infinitesimal variation of Hodge structure and by the isomorphism*

$$H^{1,0}(X) \cong (H^1(X, \mathbb{C})/H^{1,0}(X))^*. \tag{10.10}$$

In particular, assume that we have two curves X and X', an isomorphism compatible with the intersection forms $i : H^1(X, \mathbb{C}) \cong H^1(X', \mathbb{C})$, and a germ of isomorphisms $j : (B, 0) \cong (B', 0')$ between the bases of the local universal deformations of X and X' respectively, giving an identification of the variations of Hodge structure in the neighbourhood of X and of X':

$$\begin{array}{ccccc} \mathcal{P}^1 : & B & \rightarrow & \mathrm{Grass}(g, H^1(X, \mathbb{C})) \\ & \downarrow^j & & \downarrow^i \\ \mathcal{P}^1 : & B' & \rightarrow & \mathrm{Grass}(g, H^1(X', \mathbb{C})). \end{array}$$

Then X and X' are isomorphic.

Proof The first statement follows from the fact that the differential $d\mathcal{P}^1$ at the point $0 \in B$ corresponding to X gives a symmetric map (relative to the Serre duality $H^{1,0}(X) \cong (H^1(X, \mathbb{C})/H^{1,0}(X))^*$):

$$W = T_{B,0} \rightarrow \mathrm{Hom}(H^{1,0}(X), H^1(X, \mathbb{C})/H^{1,0}(X)).$$

The isomorphism $H^{1,0}(X) \cong (H^1(X, \mathbb{C})/H^{1,0}(X))^*$ induced by the intersection form on $H^1(X, \mathbb{C})$ makes it possible to dualise this map to a symmetric map $\mu : H^{1,0}(X) \otimes H^{1,0}(X) \rightarrow W^*$. Lemma 10.22 and theorem 10.24 then show that if X is not hyperelliptic, trigonal or a planar quintic, X can be identified with the subscheme of $\mathbb{P}(H^{1,0}(X)^*)$ defined by the symmetric elements of $\mathrm{Ker}\,\mu$.

The second result follows immediately from this, since by differentiation, the commutative diagram above gives an identification of the infinitesimal variations of Hodge structures at the points b and $j(b)$, $\forall b \in B$, compatible with the duality isomorphisms (10.10) since i preserves the intersection form, i.e. a commutative diagram where the vertical arrows are isomorphisms:

$$\begin{array}{cccc} \mu_b^* & : & T_{B,b} & \rightarrow & \mathrm{Hom}(H^{1,0}(X_b), H^1(X_b, \mathbb{C})/H^{1,0}(X_b)) \\ & & {}^{j_*}\downarrow & & {}^{i_*}\downarrow \\ \mu_{j(b)}^* & : & T_{B',j(b)} & \rightarrow & \mathrm{Hom}\big(H^{1,0}(X_{j(b)}), H^1(X_{j(b)}, \mathbb{C})/H^{1,0}(X_{j(b)})\big). \end{array}$$

But then for generic b, we must have $X_b \cong X_{j(b)}$, since X_b and $X_{j(b)}$ are determined by $\mathrm{Ker}\,\mu_b$ and $\mathrm{Ker}\,\mu_{j(b)}$ respectively. It then follows that $X_b \cong X_{j(b)}$ for every $b \in B$. (The moduli space of the curves is separated (Deligne & Mumford 1967).) □

The result shown above is the generic Torelli theorem for curves of genus ≥ 5. It says essentially that the *global* period map, which to a curve X associates the polarised Hodge structure on $H^1(X, \mathbb{Z})$, is of degree 1 on its image. In the second volume, we will prove a similar statement due to Donagi for most of the families of smooth hypersurfaces in projective space. In the case of curves, the actual Torelli theorem, whose proof (Andreotti & Mayer 1967) uses delicate analysis of the Θ divisor of the Jacobian, says that the global polarised period map is injective, i.e. that two curves with isomorphic polarised Hodge structures are isomorphic. This statement is finer than the one proved above. Note, however, that Torelli's theorem necessitates considering *integral* Hodge structures, while the generic statement we prove is also valid for rational Hodge structures.

10.3.2 Calabi–Yau manifolds

Another application of theorem 10.21 concerns the infinitesimal Torelli problem, i.e. the question of whether the local period map is an immersion, for compact Kähler manifolds with trivial canonical bundle, also known as Calabi–Yau manifolds. We know (see Tian 1987; Friedman 1991) that such a manifold X admits a local universal deformation $\phi : \mathcal{X} \to B$ with B smooth, $T_{B,0} \cong H^1(X, T_X)$ (Bogomolov–Tian–Todorov theorem). If $n = \dim X$, then on B we have the period map $\mathcal{P}^n$, and its component

$$\mathcal{P}^{n,n} : B \to \mathrm{Grass}(h^{n,0}, H^n(X, \mathbb{C}))$$

which to $b \in B$ associates $H^{n,0}(X_b) \subset H^n(X_b, \mathbb{C}) \cong H^n(X, \mathbb{C})$. As K_X is trivial, we have

$$H^{n,0}(X) = H^0(X, K_X) = H^0(X, \mathcal{O}_X) \cong \mathbb{C}.$$

It follows easily that K_{X_b} remains trivial in a neighbourhood of 0, since by the fact that the Hodge numbers are locally constant, we must have $H^{n,0}(K_{X_b}) = \mathbb{C}$ for b near 0, and as the unique section of K_X has no zeros on X, this also holds for X_b with b near 0.

In this case, the Grassmannian is thus the projective space $\mathbb{P}(H^n(X, \mathbb{C}))$. When the dimension n is even, the symmetric intersection form Q on $H^n(X, \mathbb{C})$ imposes the condition $Q(\omega, \omega) = 0, \forall b \in B, \omega \in H^{n,0}(X_b)$. $\mathcal{P}^{n,n}$ then has values in the quadric defined by Q. (We saw the special case of dimension 2 in section 7.2.3.) We now have the following result.

Theorem 10.27 *Let X be an n-dimensional Calabi–Yau manifold. Then the local period map (defined on a simply connected local moduli space for X)*

$$\mathcal{P}^{n,n} : B \to \mathbb{P}(H^n(X, \mathbb{C}))$$

is an immersion.

Proof We must show that the differential

$$d\mathcal{P}^{n,n} : T_{B,b} \to \mathrm{Hom}(H^{n,0}(X_b), H^n(X_b, \mathbb{C})/H^{n,0}(X_b)) \quad (10.11)$$

is injective. The Kodaira–Spencer map $\rho : T_{B,b} \to H^1(X_b, T_{X_b})$ is an isomorphism, and thus by theorem 10.21, this differential can be identified with the map

$$\mu : H^1(X_b, T_{X_b}) \to \mathrm{Hom}\big(H^0(X_b, K_{X_b}), H^1(X_b, \Omega_{X_b}^{n-1})\big) \quad (10.12)$$

given by the cup-product and contraction, followed by the inclusion

$$\mathrm{Hom}\big(H^0(X_b, K_{X_b}), H^1(X_b, \Omega_{X_b}^{n-1})\big) \hookrightarrow$$
$$\mathrm{Hom}(H^{n,0}(X_b), H^n(X_b, \mathbb{C})/H^{n,0}(X_b)).$$

Now, the map μ of (10.12) is an isomorphism. Indeed, let $\Omega \in H^0(X_b, K_{X_b})$ be a generator. Ω is an everywhere non-zero holomorphic n-form, so by the interior product, it gives an isomorphism of holomorphic vector bundles

$$\Omega : T_{X_b} \cong \Omega_{X_b}^{n-1}.$$

Thus, we have an induced isomorphism

$$\Omega : H^1(X_b, T_{X_b}) \to H^1(X_b, \Omega_{X_b}^{n-1}),$$

and obviously $\Omega(u) = \mu(u)(\Omega)$. □

Remark 10.28 *In the case of surfaces with trivial canonical bundle, the period map $\mathcal{P}^{2,2}$ is a local isomorphism on the quadric defined by Q, as shown by the proof of theorem 10.27. In higher dimensions, this never holds, because of the transversality condition 10.12.*

Exercises

1. *Contact structure on $\mathbb{P}(V)$.* Let X be a complex manifold of dimension $2n-1$. A contact structure on X is determined by the local datum of a holomorphic 1-form α which is well-defined up to multiplication by an invertible holomorphic function and which satisfies the condition that the $(2n-1)$-form

$$\alpha \wedge (d\alpha)^{n-1} \in K_X$$

does not vanish at any point.

(a) Let $\Sigma \subset X$ be an integral submanifold of a contact structure. This means that locally the restriction of the 1-form α vanishes on Σ. Show that

$\dim \Sigma \leq n-1$. (Use the fact that the 2-form $(d\alpha)_x$ has a non-degenerate restriction to the hyperplane $\alpha_x = 0 \subset T_{X,x}$ for any $x \in X$.)

Let V be a complex vector space endowed with a non-degenerate 2-form ω. Recall that $T_{\mathbb{P}(V),v}$ is isomorphic to $V/\langle v \rangle$.

(b) Show that the form α defined (up to a multiplicative coefficient) by

$$\alpha_v(\cdot) = \omega(v, \cdot)$$

provides a contact structure on $\mathbb{P}(V)$.

2. *Periods of Calabi–Yau threefolds and contact structure.* Let X be a Calabi–Yau threefold, and $V := H^3(X, \mathbb{C})$. V is endowed with the non-degenerate 2-form given by the intersection form.

   Deduce from the transversality property (proposition 10.12) that the image of the period map

$$\mathcal{M} \to \mathbb{P}(V)$$

$$t \mapsto H^{3,0}(X_t) \subset H^3(X_t, \mathbb{C}) \cong V$$

   is an integral submanifold of $\mathbb{P}(V)$ for the contact structure. Here $\mathcal{M}$ is the basis of a universal local deformation of X.

Part IV

Cycles and Cycle Classes

11
Hodge Classes

The last two chapters of this volume form an introduction to a subject which is one of the major themes of the second volume: the interaction between algebraic cycles and the Hodge theory of a projective smooth complex variety. Here, we remain in the framework of Kähler geometry; thus we consider analytic cycles, which are combinations with integral coefficients of irreducible closed analytic subsets.

The first object associated to an analytic cycle in a compact complex manifold is its homology class. More generally, without any compactness hypothesis, we can define the cohomology class $[Z] \in H^{2k}(X, \mathbb{Z})$ of an analytic subset of codimension k of a complex manifold. When the components Z_i of the cycle $Z = \sum_i n_i Z_i$ are smooth, this class $[Z] = \sum_i n_i [Z_i]$ is easy to define, using a tubular neighbourhood of Z_i and Thom's theorem. In the singular case, we reduce to the preceding case by showing that the singular locus of Z_i is stratified by real submanifolds of codimension $\geq 2k + 2$ in X, so that we have $H^{2k}(X, \mathbb{Z}) = H^{2k}(X - \text{Sing}\, Z_i, \mathbb{Z})$.

An easy but important point is the fact that if X is a compact Kähler manifold, the class of an analytic cycle of codimension k is a Hodge class, i.e. an integral class which is of type (k, k) in the Hodge decomposition. In the second section, we give other examples of Hodge classes on a Kähler manifold. These are the Chern classes of holomorphic vector bundles. Thus, a Kähler manifold, which does not necessarily contain non-trivial analytic subsets, nevertheless has Hodge classes given by Chern classes of its tangent bundle. (Only complex tori have a tangent bundle all of whose Chern classes are zero.) In the case of projective manifolds, we can show (by a generalised Lelong formula), that the group generated by the Chern classes of holomorphic vector bundles is contained in the group generated by the classes of algebraic cycles. We have the Hodge conjecture:

Conjecture 11.1 *If X is a smooth complex projective variety, the rational Hodge classes on X are exactly the classes of algebraic cycles with rational coefficients.*

We also study the relation between the Hodge classes on a product and the morphisms of Hodge structures given by the Künneth decomposition.

11.1 Cycle class

11.1.1 Analytic subsets

Let X be a complex manifold, and let $Z \subset X$ be a closed subset.

Definition 11.2 *Z is called an analytic subset of X if X admits a covering by open sets $U \subset X$ such that $U \cap Z$ is the zero locus of holomorphic functions $f_1, \ldots, f_N \in \Gamma(U, \mathcal{O}_U)$.*

In particular, a closed complex submanifold is an analytic subset. If $Z' \subset Z \subset X$ are analytic subsets, we will say that Z' is an analytic subset of Z. An analytic subset is not smooth in general, but we have the following result.

Proposition 11.3 *Let $Z \subset X$ be an analytic subset. Then there exists a nowhere dense analytic subset $Z' \subset Z$ outside of which Z is a complex submanifold of X.*

Proof Clearly the set Z_{smooth} of the smooth points of Z, i.e. those in whose neighbourhood there exist holomorphic equations $f_1, \ldots, f_r$ defining Z with independent differentials, is an open set of Z. To see that it is dense, let U be an open set of Z, $U = V \cap X$, where V is an open set of X in which there exist holomorphic equations $f_1, \ldots, f_N$ defining Z. We will admit the fact (Narasimhan 1966) that up to shrinking V, there exists a finite number of holomorphic equations which generate the ideal $\mathcal{I}_Z$ of holomorphic functions vanishing on $Z \cap V$ as an $\mathcal{O}_V$-module. We can thus assume that the f_i generate $\mathcal{I}_Z$. Now, there exists a non-empty open set $U' \subset U \subset Z$ on which the df_i generate a subspace of constant rank, say equal to k, of Ω_V. Moreover, up to restricting U' and permuting the f_i, we may assume that $df_1, \ldots, df_k$ are independent along U'. Let $V' \subset V$ be an open set such that $V' \cap Z = U'$ and the differentials df_i, $i \leq k$ are independent in V'. Let $Z_1 \subset V'$ be the complex submanifold defined by $f_1, \ldots, f_k$. We have $U' \subset Z_1$, and we propose to show the following.

Lemma 11.4 *The analytic subset $U' \subset Z_1$ is a connected component of Z_1.*

Proof Consider the equations $g_i := f_{i|Z_1}$, $i > k$ defining U' in Z_1. By definition of the f_i, these equations generate the ideal of U' in Z_1. Let $x \in U'$. Up to replacing the g_i by linear combinations, we may assume that the multiplicities (i.e. the smallest degree of any homogeneous term of the series expansion at the point x) of the g_i in x are increasing. Now, the partial derivatives of g_{k+1} lie in the ideal of $U' \subset Z_1$, since along U', the vector subspace of Ω_X generated by the df_i is equal to the vector subspace generated by the df_i, $i \leq k$, which shows by restriction to Z_1 that the g_i have differential equal to zero along U'. These partial derivatives are thus a combination of $g_{k+1}, \ldots, g_N$ with holomorphic coefficients. As their multiplicity is strictly less than that of g_{k+1}, we conclude that we have a contradiction unless the g_i are identically zero in the neighbourhood of x. This proves lemma 11.4. □

We have thus shown that the open dense subset of Z consisting of the points in whose neighbourhood the rank of the subspace of Ω_X generated by the df_i is constant is contained in Z_{smooth}. As the inverse inclusion is obvious, we obtain the equality. Now, the complement of this open set is obviously defined locally in Z by holomorphic equations given by minors of the Jacobian matrices of $\{f_1, \ldots, f_N\}$. Thus, $Z - Z_{\text{smooth}}$ is indeed a nowhere dense analytic subset of Z, which we denote from now on by Z_{sing}, and proposition 11.3 is proved. □

Definition 11.5 *We say that Z is irreducible if Z_{smooth} is connected.*

Let us admit the fact (see Narasimhan 1966) that an analytic subset Z can be written locally as a finite union of irreducible analytic sets. This holds globally if Z is compact.

Definition 11.6 *The dimension of an irreducible analytic subset $Z \subset X$ is the dimension of the connected complex manifold Z_{smooth}.*

We will admit the following theorem, known as the Weierstrass preparation lemma (Narasimhan 1966).

Theorem 11.7 *Let $f(x_1, \ldots, x_N)$ be a holomorphic function defined in a neighbourhood U of 0 in $\mathbb{C}^N$. Suppose that $f(x_1, 0, \ldots, 0)$ is not identically zero. Let l be the vanishing order of the function $f(x_1, 0, \ldots, 0)$ of x_1 at 0. Then, up* *to restricting U, there exists a holomorphic invertible function $\phi(x_1, \ldots, x_N)$ such that*

$$\phi \cdot f = x_1^l + \sum_{0 \le i < l} x_1^i f_i,$$

where the f_i are holomorphic functions of $x_2, \ldots, x_N$, zero at 0.

We will use this theorem to prove the following very useful result.

Proposition 11.8 *Let $Z \subset X$ be an analytic subset, where X is an open set of $\mathbb{C}^N$. Then, in the neighbourhood of each point z of Z, there exists a linear map from $\mathbb{C}^N$ to $\mathbb{C}^n$, where $n = \dim_z Z$, whose restriction $\phi : Z \to \mathbb{C}^n$ to Z satisfies: there exists an open set U of $\mathbb{C}^n$ such that $\phi(z) \in U$ and the restriction $\phi : \phi^{-1}(U) \to U$ of ϕ to $\phi^{-1}(U)$ is proper with finite fibres. Moreover, for a generic choice of linear coordinates on $\mathbb{C}^N$, this property is satisfied by all the linear projections to $\mathbb{C}^n$ given by the coordinates.*

Here, the dimension of Z at z is defined as the maximal dimension of its components passing through z.

Proof We prove this by induction on N. Let f be a holomorphic equation which vanishes on Z. Up to an affine change of coordinates, we may assume that $z = 0$ and $f(x_1, 0, \ldots, 0)$ is not identically zero. We can then apply the Weierstrass preparation lemma and assume (at least locally in the neighbourhood of 0) that f is of the form

$$x_1^l + \sum_{0 \le i < l} x_1^i f_i(x_2, \ldots, x_N). \tag{11.1}$$

Then let ψ be the linear map

$$(x_1, \ldots, x_N) \mapsto (x_2, \ldots x_N),$$

and let χ be its restriction to the hypersurface Y defined by f. Clearly, if $U \subset \mathbb{C}^{N-1}$ is a neighbourhood of 0 on which the f_i are holomorphic, then $\chi : \chi^{-1}(U) \to U$ is proper and finite. Moreover, it is easy to see that $\chi^{-1}(U) \subset X$ if U is sufficiently small. As Z is closed in X, $Z' := Z \cap \chi^{-1}(U)$ is also closed in $\chi^{-1}(U)$, so the restriction of χ to Z' remains proper with finite fibres.

Lemma 11.9 *$\chi(Z')$ is an analytic subset of U.*

Proof Let g be a holomorphic function on X. We can define its norm $\operatorname{Nm} g$ relative to the algebraic equation (11.1) defined by $f = 0$, which is an equation with coefficients in $\Gamma(U, \mathcal{O}_U)$. Given an algebraic extension of rings $A \subset B$ determined by an equation of the form (11.1), the ring B is a free A-module of rank l. If $\phi \in B$, then multiplication by ϕ is an A-endomorphism of B. The norm $\mathrm{Nm}\,\phi$ of ϕ is the determinant of this endomorphism.

The function $\mathrm{Nm}\, g$ is a holomorphic function of $x_2, \ldots, x_N$, and it is obvious that $\mathrm{Nm}\, g$ vanishes on $z \in U$ if and only if there exists $z \in \chi^{-1}(U)$ such that $g(z) = 0$. Thus, $\chi(Z')$ is the analytic closed set described by the equations $\mathrm{Nm}\, g,\ g \in \mathcal{I}_{Z'}$. □

Lemma 11.9 allows us to conclude, by induction, that there exists a linear map ψ, proper with finite fibres, from a neighbourhood of 0 in $\chi(Z')$ to a neighbourhood of 0 in $\mathbb{C}^m$, $m = \dim \chi(Z')$.

To conclude, it remains to see that $m = \dim_Z Z$. Let us apply lemma 11.9 to the analytic subset Z'_{sing} of Z'. Thus, $\chi(Z'_{\mathrm{sing}})$ is an analytic subset of $\chi(Z')$. In fact, arguing as in the proof of corollary 11.10 below, we show that $\chi(Z'_{\mathrm{sing}})$ has empty interior in $\chi(Z')$. Then let $V \subset \chi(Z')$ be the dense open set $\chi(Z') - \chi(Z'_{\mathrm{sing}}) - \chi(Z')_{\mathrm{sing}}$. The map χ gives a surjective map with finite fibres between the complex manifolds $\chi^{-1}(V)$ and V. We thus have $\dim \chi^{-1}(V) = \dim V$ and

$$\dim_Z Z \geq \dim \chi^{-1}(V) = \dim V = \dim \chi(Z') = m.$$

Also, we have $m \geq \dim_Z Z$, since otherwise $\psi \circ \chi$ could not be finite on a component of dimension $> m$ of Z_{smooth}. Thus, $m = \dim_Z Z$, and the statement also holds for N. To conclude, it suffices to note that the induction hypothesis is certainly true when $n = N$, since then Z is an open set of $\mathbb{C}^N$.

Furthermore, the above proof shows that in fact, in order for a linear projection ψ to satisfy the desired conclusion in the neighbourhood of Z, it suffices for z to be an isolated point of $\psi^{-1}(\psi(z)) \cap Z$. The second statement can thus be proved by showing that this condition is satisfied for projections onto $\mathbb{C}^n$ given by choosing n coordinates, as long as the coordinates are chosen generically, which we admit. □

Corollary 11.10 *Let $Z \subset X$ be an analytic subset, and let $Z' \subset Z$ be an analytic subset. Then* $\dim Z' \leq \dim Z$ *and the inequality is strict if Z' has empty interior in Z.*

Proof Let $n = \dim Z$ and take $z \in Z$. Then there exists a finite and proper projection of a neighbourhood of z in Z onto an open set of $\mathbb{C}^n$. The restriction of this projection to Z' remains finite (i.e. with finite fibres) and proper, and thus remains finite on the open subset of smooth points of Z', which is a union of smooth manifolds of dimensions k_i with $\mathrm{Sup}_i k_i = n' := \dim Z'$. Considering the dense open subset of Z'_{smooth} on which this projection has constant rank, we immediately conclude that $\dim Z' \leq n = \dim Z$.

As for the second point, if $Z' \subset Z$ has empty interior, we may assume that

(*) *$Z' \subset Z$ is defined by a single holomorphic equation g which is not identically zero in the neighbourhood of any point of Z.*

Let $\phi : Z \to \mathbb{C}^n$ be a finite projection, and let $x_1, \ldots, x_n$ be linear coordinates on $\mathbb{C}^n$. We denote similarly the functions $x_i \circ \phi$ on Z. Iterating applications of the Weierstrass preparation lemma, we see that we may assume (at least locally in the neighbourhood of $z \in Z'$) that the holomorphic function g on Z satisfies an algebraic equation

$$g^l + \sum_{i<l} g^i f_i(x_1, \ldots, x_n) = 0, \tag{11.2}$$

where the f_i are holomorphic functions defined in a neighbourhood U of $\phi(z)$. Suppose now that

(**) $\dim Z' \geq n$.

Then, as Z' contains an open set which is a complex manifold of dimension n, and $\phi_{|Z'} : Z' \to \mathbb{C}^n$ is finite, the preceding rank argument shows that $\phi(Z')$ must contain an open set of $\mathbb{C}^n$. It follows immediately, by equation (11.2), and by the fact that g vanishes on Z', that the function $f_0(x_1, \ldots, x_n)$ is zero on this open set, and thus must be zero on U. Thus, the equation (11.2) can be written

$$g\left(g^{l-1} + \sum_{0<i<l} f_i g^{i-1}\right) = 0. \tag{11.3}$$

If $f := g^{l-1} + \sum_{0<i<l} f_i g^{i-1}$ does not vanish at the point z, g is zero on Z in the neighbourhood of z, which contradicts the condition (*). Otherwise, Z is the union of Z' and $Z'' \subset Z$, which is defined by the equation $f = 0$. As Z' has empty interior, Z'' must be everywhere dense, and thus we must in fact have $f = 0$ on Z in the neighbourhood of z. The function g then satisfies the algebraic equation of order $l - 1$

$$g^{l-1} + \sum_{0<i<l} f_i g^{i-1} = 0. \tag{11.4}$$

As we could assume that l was the minimal order of an algebraic equation satisfied by g, we can exclude this last possibility, which shows that the hypothesis (**) was absurd, and that we must have $\dim Z' < n$. □

Combining the preceding results, we obtain the following.

Theorem 11.11 *Let $Z \subset X$ be an analytic subset of dimension n. Then Z admits a filtration $\emptyset = Z_{n+1} \subset \cdots \subset Z_0 = Z$ by closed analytic subsets Z_k of dimension $n_k > n_{k+1}$ such that $Z_k - Z_{k+1}$ is a closed complex submanifold of dimension n_k of $X - Z_{k+1}$.*

Proof It suffices to apply proposition 11.3 and corollary 11.10, and to use induction on the dimension. Indeed, set $Z_1 = Z_{\text{sing}} \cup Z'$, where Z' is the union of the irreducible components of Z of dimension $< n$. □

Remark 11.12 *If we study the algebraic subsets Z of an affine algebraic variety, i.e. the subsets defined by polynomial equations in $\mathbb{C}^N$, these results are much easier to prove. Indeed, if A is the ring of polynomials and $I \subset A$ is the ideal of those which vanish on Z, we can translate these results into statements on the ring A/I of algebraic functions on Z, which is a $\mathbb{C}$-algebra of finite type. The dimension of Z at the point z is then the dimension of the local ring $(A/I)_z$, which is the localisation of A/I at the maximal ideal M_z of polynomials vanishing at z, and the analogue of proposition 11.8 is Noether's normalisation theorem (see Hartshorne 1977; Mumford 1988).*

11.1.2 Cohomology class

Let $Z \subset X$ be a closed n-dimensional analytic subset. Let $r = \dim X - n$ be the codimension of Z. We propose to construct a cohomology class $[Z] \in H^{2r}(X, \mathbb{Z})$. We will first use theorem 11.11 to reduce to the case where Z is smooth. For this, let us first prove the following lemma.

Lemma 11.13 *Let $Y \subset X$ be a closed complex submanifold of codimension $k > r$. Then the restriction map*

$$H^l(X, \mathbb{Z}) \to H^l(X - Y, \mathbb{Z})$$

is an isomorphism for $l \leq 2r$.

Proof We have the exact sequence of relative cohomology

$$\cdots H^l(X, X - Y, \mathbb{Z}) \to H^l(X, \mathbb{Z}) \to H^l(X - Y, \mathbb{Z}) \to H^{l+1}(X, X - Y, \mathbb{Z}) \cdots$$

and Thom's isomorphism theorem (7.4)

$$H^j(X, X - Y, \mathbb{Z}) \cong H^{j-2k}(Y, \mathbb{Z}).$$

It follows that $H^j(X, X - Y, \mathbb{Z}) = 0$ for $j < 2k$, and thus $H^l(X, X - Y, \mathbb{Z}) = 0 = H^{l+1}(X, X - Y, \mathbb{Z})$ for $l \leq 2r \leq 2k - 2$. □

Now let $Z \subset X$ be a closed analytic subset of codimension r, and let $\emptyset = Z_{n+1} \subset \cdots \subset Z_k \subset \cdots \subset Z_0 = Z$ be the filtration introduced in theorem 11.11. We can apply the preceding lemma to the inclusion

$$X - Z_k \subset X - Z_{k+1}, \quad k \geq 1,$$

since $Z_k - Z_{k+1}$ is a smooth closed complex submanifold of $X - Z_{k+1}$ of codimension strictly greater than r, and we conclude that we have an isomorphism given by the restriction

$$H^{2r}(X, \mathbb{Z}) \cong H^{2r}(X - Z_1, \mathbb{Z}).$$

The class $[Z] \in H^{2r}(X, \mathbb{Z})$ will then be defined as the image under the inverse isomorphism of the class of the smooth complex submanifold $Z - Z_1 \subset X - Z_1$ which we are about to define.

Let $Z \subset X$ be a closed smooth submanifold of X of real codimension k. (Here X and Z are not necessarily complex; the only necessary condition to define $[Z]$ is that the normal bundle of Z in X is oriented.) We define the class $[Z] \in H^k(X, \mathbb{Z})$ using the natural map

$$j_Z : H^k(X, X - Z, \mathbb{Z}) \to H^k(X, \mathbb{Z})$$

and the Thom isomorphism

$$T : H^k(X, X - Z, \mathbb{Z}) \cong H^0(Z, \mathbb{Z}).$$

Set $[Z] = j_Z(T^{-1}(1))$.

We obtain an explicit de Rham representative of $[Z]$ as follows: let $U \subset X$ be a neighbourhood of Z isomorphic to a neighbourhood V of the section $0_N \cong Z$ of the normal bundle $N_{Z/X}$. Let ω be a closed form of degree k with support in V proper over Z, satisfying the condition

$$\int_{V_z} \omega = 1, \quad \forall z \in Z.$$

Here $V_z = \pi^{-1}(z)$, where $\pi : V \to Z$ is the restriction of the structural map $N_{Z/X} \to Z$. The integral is well-defined since the normal bundle is oriented, and the support of ω is proper above Z, so that $\omega_{|V_z}$ has compact support in V_z.

The form ω on U then extends to a differentiable closed form on X, and we have the following.

Lemma 11.14 *The form ω thus constructed on X is a representative in de Rham cohomology of the class $[Z]$, or more precisely, of the image of $[Z]$ in $H^k(X, \mathbb{R})$.*

Proof The form ω, viewed as a form on V, exactly gives the generator 1 of

$$H^k(N_{Z/X}, N_{Z/X} - 0_N, \mathbb{R}) \cong H^0(Z, \mathbb{R}).$$

Here, ω must be seen as a closed form which vanishes on $N_{Z/X} - V'$, where V' is a neighbourhood which is a retraction of 0_N, in order to attribute a class to it in the group

$$H^k(N_{Z/X}, N_{Z/X} - 0_N, \mathbb{R}) \cong H^k(N_{Z/X}, N_{Z/X} - V', \mathbb{R}).$$

This fact follows from the Leray–Hirsch theorem for the relative cohomology $H^k(N_{Z/X}, N_{Z/X} - V', \mathbb{R})$, and the fact that each $\omega_{|N_{Z/X,z}}$ gives the generator

$$1 \in H^k(N_{Z/X,z}, V'_z, \mathbb{R}) \cong H^{k-1}(S^{k-1}, \mathbb{R}) \cong \mathbb{R},$$

where the first isomorphism is given by the exact sequence of relative cohomology and the last isomorphism is given by integration.

Finally, considering ω as a form on X which is zero outside of a tubular neighbourhood of Z comes down to using the excision isomorphism

$$H^k(X, X - Z) \cong H^k(U, U - Z) = H^k(V, V - Z)$$

which is implicit in the preceding statement of the Thom isomorphism. □

Corollary 11.15 *If Z and X are compact oriented manifolds, with orientations compatible with the orientation of the normal bundle of Z, and* $\dim Z = n$ *(so that* $\dim X = k + n$*), then for $\alpha \in H^n(X, \mathbb{R})$ represented by a closed form $\tilde{\alpha}$ of degree n on X, we have*

$$\langle [Z], \alpha \rangle_X = \int_Z \tilde{\alpha}_{|Z}. \tag{11.5}$$

Here the pairing $\langle , \rangle$ is given by the cup-product

$$H^k(X, \mathbb{R}) \otimes H^n(X, \mathbb{R}) \to H^{k+n}(X, \mathbb{R})$$

composed with the integration

$$H^{k+n}(X, \mathbb{R}) \to \mathbb{R}.$$

Proof With the notation of the preceding proof, we have

$$\langle [Z], \alpha \rangle_X = \int_X \omega \wedge \tilde{\alpha} = \int_U \omega \wedge \tilde{\alpha}.$$

But as U retracts onto Z, there exists a form β of degree $n-1$ on U such that

$$\tilde{\alpha}_{|U} = \pi^*(\tilde{\alpha}_{|Z}) + d\beta.$$

Stokes' formula and the fact that ω is closed with support disjoint from the boundary of U then show that

$$\int_U \omega \wedge \tilde{\alpha} = \int_U \omega \wedge \pi^*(\tilde{\alpha}_{|Z}).$$

Finally, as the integral of ω over each V_z is equal to 1, Fubini's theorem implies that $\int_U \omega \wedge \pi^*(\tilde{\alpha}_{|Z}) = \int_Z \tilde{\alpha}_{|Z}$. □

Remark 11.16 *A way of rephrasing corollary 11.15 is to say that when Z and X are smooth and compact, the image of* $[Z] \in H^k(X, \mathbb{Z})$ *in* $H_n(X, \mathbb{Z})/\text{torsion} = \text{Hom}\,(H^n(X, \mathbb{Z}), \mathbb{Z})$ *is the homology class of the oriented submanifold Z. In fact, when X is compact, knowing Poincaré duality more precisely:*

$$\text{PD} : H^k(X, \mathbb{Z}) \cong H_n(X, \mathbb{Z}), \quad \dim_{\mathbb{R}} X = n + k$$

shows that PD([Z]) *is the homology class of Z.*

It is useful to introduce the notion of an analytic cycle (or an algebraic cycle, if X is algebraic).

Definition 11.17 *Let X be a complex manifold. An analytic cycle* $Z \subset X$ *of codimension k is a finite combination with integral coefficients* $Z = \sum_i n_i Z_i$, *where each* Z_i *is a closed irreducible analytic subset of codimension k of X.*

Remark 11.18 *If X is a projective algebraic variety, the algebraic subsets of X (defined locally in the Zariski topology by polynomial equations) are exactly the analytic subsets of X (defined locally in the usual topology by holomorphic equations). Thus, we do not need to distinguish between algebraic cycles and analytic cycles, and we generally use the first terminology. This result, due to Chow, was extended by Serre (1956) to all coherent sheaves over a projective manifold ("GAGA" principle).*

The map which associates a cohomology class to a closed analytic subset of codimension k extends by $\mathbb{Z}$-linearity, so that we can define the class $[Z] \in H^{2k}(X, \mathbb{Z})$ of any cycle Z of codimension k.

Remark 11.19 *Consider a sheaf of ideals $\mathcal{I} \subset \mathcal{O}_X$. It allows us to define an analytic subscheme (see Hartshorne 1977) of X, by considering the analytic subset $\tilde{Z}$ defined by $\mathcal{I}$, equipped with the sheaf of functions $\mathcal{O}_X/\mathcal{I}$. We can also associate to it an analytic cycle Z of codimension $k = \operatorname{codim} \tilde{Z}$, by setting $Z = \sum_i n_i Z_i$, where the Z_i are the irreducible components of codimension k of $\tilde{Z}$ and the n_i are the local multiplicities of $\mathcal{I}$ along the Z_i: if $z \in Z_i$ is a smooth generic point, consider a holomorphic smooth k-dimensional submanifold Y of X defined in the neighbourhood of z which cuts Z_i transversally at z. The ideal $\mathcal{I}_{|Y}$ admits as an isolated zero the point z, and we define n_i as the rank of the $\mathbb{C}$-vector space $\mathcal{O}_{Y,z}/\mathcal{I}_{|Y,z}$ (see Fulton 1984).*

11.1.3 The Kähler case

Now let X be a compact Kähler manifold of dimension $n + r$, and let $Z \subset X$ be an analytic subset of codimension r. The cohomology $H^{2r}(X, \mathbb{C})$ admits the Hodge decomposition

$$H^{2r}(X, \mathbb{C}) = \bigoplus_{p+q=2r} H^{p,q}(X).$$

Proposition 11.20 *The image in $H^{2r}(X, \mathbb{C})$ of the class $[Z] \in H^{2r}(X, \mathbb{Z})$ lies in $H^{r,r}(X)$.*

Proof By lemma 7.30, it suffices to show that

$$\langle [Z], \alpha \rangle_X = 0, \ \forall \alpha \text{ of type } (p, q),\ p + q = 2n,\ (p, q) \neq (n, n). \quad (11.6)$$

Now, we have the following result, due to Lelong (1957).

Theorem 11.21 *Let $Z \subset X$ be an analytic subset of a smooth complex manifold. For every differential form with compact support ω on X, the integral $\int_{Z_{\text{smooth}}} \omega$ is convergent, and the current (i.e. the linear form on the space of forms with compact support)*

$$\omega \mapsto \int_{Z_{\text{smooth}}} \omega$$

thus defined is closed (i.e. zero on the exact forms). If X is compact, the element of $H^{2n}(X, \mathbb{C})^*$ thus defined is equal to the image of $[Z]$ under the morphism

$$H^{2r}(X, \mathbb{Z}) \to H^{2r}(X, \mathbb{C}) \to H^{2n}(X, \mathbb{C})^*,$$

where the last map is given by the intersection form (5.14).

This theorem immediately implies the equality (11.6), since $H^{p,q}(X)$ is the set of classes representable by a closed form of type (p,q), and such a form vanishes on Z_{smooth} for $p+q=2n$, $(p,q)\neq(n,n)$. □

Proof of theorem 11.21 It suffices to prove the first two assertions. Indeed, the fact that $\int_{Z_{\text{smooth}}}$ is a closed current shows that

$$\int_{Z_{\text{smooth}}}\alpha=\int_{Z_{\text{smooth}}}\alpha' \text{ for } \alpha'=\alpha+d\beta.$$

Now, lemma 11.13 shows that if α is closed of degree $2n$, there exists such a form α' with compact support in $X-Z_{\text{sing}}$. But then, by corollary 11.15, which can be applied to the case where Z is not compact but where α' has compact support, we have

$$\langle[Z]_{|X-Z_{\text{sing}}},[\alpha']\rangle_{X-Z_{\text{sing}}}=\int_{Z_{\text{smooth}}}\alpha',$$

where the first pairing is the pairing between the cohomology and the cohomology with compact support for an open manifold (Spanier 1966). Finally, it is obvious that

$$\langle[Z]_{|X-Z_{\text{sing}}},[\alpha']\rangle_{X-Z_{\text{sing}}}=\langle[Z],[\alpha']\rangle=\langle[Z],[\alpha]\rangle,$$

where in the first term, $[\alpha']$ is considered as a class with compact support in $X-Z_{\text{sing}}$, and in the second term, it is considered as the induced cohomology class on X. We thus have

$$\langle[Z],[\alpha]\rangle=\int_{Z_{\text{smooth}}}\alpha'=\int_{Z_{\text{smooth}}}\alpha,$$

which proves the final assertion.

The first assertion follows from proposition 11.8. Indeed, a differential form ω on X can be written in holomorphic local coordinates

$$\omega=\sum_{I,J}\omega_{I,J}dz_I\wedge d\overline{z}_J,$$

where the $\omega_{I,J}$ are bounded. Now, there exists a constant $C>0$ such that for every n-dimensional complex subspace $V\subset\mathbb{C}^{n+r}$, we have

$$|dz_I\wedge d\overline{z}_J(v_1\wedge\cdots\wedge v_n\wedge\overline{v}_1\wedge\cdots\wedge\overline{v}_n)|\leq C\,|\Omega^n(v_1\wedge\cdots\wedge v_n\wedge\overline{v}_1\wedge\cdots\wedge\overline{v}_n)|,$$

where Ω is the Kähler form $i\sum_{1\leq l\leq n+r}dz_l\wedge d\overline{z}_l$ and $v_1,\ldots,v_n$ is a basis of V. Indeed, this follows from the compactness of the Grassmannian (cf. section 10.1.1) and the fact that the right-hand term is strictly positive, which implies that the quotient of the first term by the second is a continuous function on the Grassmannian.

It follows that it suffices to show that in local coordinates as above on a relatively compact open set U of X, the integrals $\int_{Z_{\text{smooth}}\cap U} dz_I \wedge d\overline{z}_I$ are convergent, since $\Omega^n = C' \sum_I dz_I \wedge d\overline{z}_I$, where C' is a constant.

Now, by proposition 11.8, we may assume that the coordinates z_i are chosen so that each linear projection ϕ_I given by a choice of n coordinates $(z_i)_{i\in I}$ is a finite map of degree less than or equal to K. We then have

$$\int_{Z_{\text{smooth}}} |dz_I \wedge d\overline{z}_I| = \left|\int_{Z_{\text{smooth}}} dz_I \wedge d\overline{z}_I\right| \leq K \left|\int_{\phi_I(Z\cap U)} dz_I \wedge d\overline{z}_I\right| < \infty.$$

The second assertion is proved similarly. □

11.1.4 Other approaches

There exist other constructions of the class of an analytic set, which are certainly more elegant than the one above, but also less elementary. The first of these, which is valid in the compact case, consists in applying the desingularisation theorem of Hironaka (1964):

Theorem 11.22 *Let $Z \subset X$ be an n-dimensional compact analytic subset. Then there exists a complex compact manifold Z' and a holomorphic morphism $\tau : Z' \to Z$ which is an isomorphism over Z_{smooth}.*

We can then define the class $[Z] \in H^{2r}(X, \mathbb{Z})$, where $\dim X = n+r$, as $\tau_*(1_{Z''})$ and Z'' is the union of the n-dimensional connected components of Z'. The properties proved above are obtained as consequences of general properties of Gysin morphisms (cf. section 7.3.2).

Another approach (Bloch 1972) consists in studying the local hypercohomology of X with support in Z and values in the holomorphic de Rham complex $\Omega_X^{\geq r}$ (this is the generalisation of the relative cohomology of the pair $(X, X - Z)$). This allows one to construct a class in

$$\mathbb{H}^{2r}\left(X, \Omega_X^{\geq r}\right) = F^r\mathbb{H}^{2r}(X, \Omega_X^{\cdot}) = F^r H^{2r}(X, \mathbb{C}).$$

This construction does not provide an integral class, but has the advantage of being purely algebraic. (On the left-hand side, we can put the hypercohomology of the algebraic de Rham complex.)

11.2 Chern classes

11.2.1 Construction

Let X be a topological or differentiable manifold, and let $E \to X$ be a (topological or differentiable) complex vector bundle of rank r.

We will define the Chern classes $c_i(E) \in H^{2i}(X, \mathbb{Z})$, $1 \leq i \leq r$, which, in the differentiable case, depend only on the underlying topological vector bundle. By convention, we set $c_0(E) = 1$ and $c_i(E) = 0$ for $i > r$, and we introduce the Chern polynomial

$$c(E) = \sum_i c_i t^i \in H^*(X, \mathbb{Z})[t].$$

If $r = 1$, the Chern class $c_1(E)$ will be defined as in section 7.1.3, with the holomorphic framework replaced by the topological framework: we consider the exponential exact sequence

$$0 \to \mathbb{Z} \stackrel{2i\pi}{\to} \underline{\mathcal{C}}^0 \stackrel{\exp}{\to} (\underline{\mathcal{C}}^0)^* \to 0,$$

where $\underline{\mathcal{C}}^0$ is the sheaf of continuous complex functions and $(\underline{\mathcal{C}}^0)^*$ is the sheaf of everywhere non-zero functions. By the associated long exact sequence, it gives a morphism

$$c_1 : H^1(X, (\underline{\mathcal{C}}^0)^*) \to H^2(X, \mathbb{Z})$$

which is in fact an isomorphism, since $H^1(X, \underline{\mathcal{C}}^0) = H^2(X, \underline{\mathcal{C}}^0) = 0$ by proposition 4.36 and because a topological manifold admits partitions of unity subordinate to every open cover.

Now, by theorem 4.49, the group $H^1(X, (\underline{\mathcal{C}}^0)^*)$ is the group of isomorphism classes of complex line bundles over X, where the group structure is given by the tensor product.

Theorem 11.23 *There exists a unique "Chern class" map c, which associates to a complex vector bundle E over X an element*

$$c(E) \in H^*(X, \mathbb{Z})[t] \quad \text{with} \quad d^0 c_i = 2i$$

satisfying the following conditions:

(i) *If* $\operatorname{rank} E = 1$, *then* $c(E) = 1 + t c_1(E)$.

(ii) *The "Chern class" map satisfies the following functoriality condition: if $\phi : Y \to X$ is a continuous (or differentiable) map, then $c(\phi^* E) = \phi^*(c(E))$, where $\phi^* : H^{2i}(X, \mathbb{Z}) \to H^{2i}(Y, \mathbb{Z})$ is the pullback introduced in section 7.3.2.*

(iii) (Whitney's formula) *If E is the direct sum of two complex bundles F and G, then*

$$c(E) = c(F)c(G), \tag{11.7}$$

where we use the ring structure on $H^*(X, \mathbb{Z})[t]$.

Proof The uniqueness is proved using the following splitting principle.

Lemma 11.24 *Let $E \to X$ be a complex vector bundle. Then there exists a continuous map $\phi : Y \to X$ satisfying:*
- *The pullback maps $\phi^* : H^l(X, \mathbb{Z}) \to H^l(Y, \mathbb{Z})$ are injective.*
- *The pullback $\phi^* E$ is a direct sum of line bundles.*

Remark 11.25 *If E is differentiable, we can choose ϕ differentiable. If E is holomorphic on a complex manifold, we have a holomorphic analogue of the splitting principle, which consists in requiring that the bundle $\phi^* E$ admits a filtration by holomorphic subbundles whose successive quotients are line bundles (see volume II of this book).*

Lemma 11.24 and conditions(i)–(iii) immediately imply uniqueness, since for E a bundle over X and $\phi : Y \to X$ such that $\phi^* E \cong \bigoplus_i L_i$, we must have the relation

$$\phi^* c(E) = \prod_i (1 + tc_1(L_i)).$$

As the map ϕ^* is injective on $H^*(X, \mathbb{Z})$, this relation determines $c(E)$. □

Proof of lemma 11.24 Consider the projective bundle $\phi_1 : \mathbb{P}(E) \to X$. We have shown (cf. lemma 7.32) that $H^*(\mathbb{P}(E), \mathbb{Z})$ is a free module over the ring $H^*(X, \mathbb{Z})$. In particular, the pullback

$$\phi_1^* : H^*(X, \mathbb{Z}) \to H^*(\mathbb{P}(E), \mathbb{Z})$$

is injective. Moreover, ϕ_1^* contains the tautological line subbundle $S \subset \phi_1^* E$, whose fibre at $\Delta \in \mathbb{P}(E_x)$ is the line $\Delta \subset E_x$. In the differentiable or continuous category, we then have an isomorphism

$$\phi_1^* E \cong S \oplus Q, \quad Q \cong \phi_1^*(E)/S.$$

Indeed, it suffices to put a Hermitian metric on $\phi_1^* E$ and define Q to be the orthogonal complement of S for this metric.

To conclude, it now suffices to use induction on the rank of E, and apply the induction hypothesis to the bundle Q over $\mathbb{P}(E)$. □

It remains to prove existence. For this, we use the projective bundle $\mathbb{P}(E)$ introduced above and the tautological subbundle $S \subset \phi_1^*(E)$. Let $h := c_1(S^*)$. By lemma 7.32, we know that $1, h, \ldots, h^{r-1}$, $r = \operatorname{rank}(E)$ is a basis of $H^*(\mathbb{P}(E), \mathbb{Z})$ over the ring $H^*(X, \mathbb{Z})$. Thus we have a relation

$$h^r + \sum_{0<i\leq r} \phi_1^* c_i(E) h^{r-i} = 0, \tag{11.8}$$

in $H^*(\mathbb{P}(E), \mathbb{Z})$, which defines $c_i(E) \in H^{2i}(X, \mathbb{Z})$ (setting $c_0 = 1$).

The polynomial $c(E)$ is thus characterised by the fact that its reciprocal polynomial $\hat{c} = t^r c(\frac{1}{t})$ is normalised of degree r and annihilates the class h_E. It remains to see that these classes satisfy conditions (i)–(iii) of theorem 11.23.

Condition (i) is obvious; indeed if E is of rank 1, then $\mathbb{P}(E)$ is isomorphic to X and S is isomorphic to E, hence $h = -c_1(S) = -c_1(E)$. The relation (11.8) is then given by

$$h + c_1(E) = 0,$$

in this case, which shows that the two definitions of c_1 coincide for a rank 1 bundle.

The functoriality (ii) is obviously satisfied.

It remains to show that the $c_i(E)$ defined by (11.8) satisfy Whitney's formula (iii).

Let $E = F \oplus G$. Then $\mathbb{P}(E)$ contains the two disjoint projective subbundles $\mathbb{P}(F)$ and $\mathbb{P}(G)$. We have differentiable maps induced by the projections of E onto F and G:

$$\pi_F : Z_G := \mathbb{P}(E) - \mathbb{P}(G) \to \mathbb{P}(F),$$

$$\pi_G : Z_F := \mathbb{P}(E) - \mathbb{P}(F) \to \mathbb{P}(G),$$

and it is clear (thanks to the corresponding linear projection) that $S_{E|Z_G}$ is isomorphic to $\pi_F^*(S_F)$, and $S_{E|Z_F}$ is isomorphic to $\pi_G^*(S_G)$. As we have $\hat{c}(F)(h_F) = 0$ in $H^{2k}(\mathbb{P}(F), \mathbb{Z})$, $k = \operatorname{rank} F$, we deduce that

$$\hat{c}(F)(h_E)_{|Z_G} = 0 \text{ in } H^{2k}(Z_G, \mathbb{Z}) \text{ with } k = \operatorname{rank} F,$$

and similarly,

$$\hat{c}(G)(h_E)_{|Z_F} = 0 \text{ in } H^{2l}(Z_F, \mathbb{Z}) \text{ with } l = \operatorname{rank} G.$$

As $\mathbb{P}(E) = Z_F \cup Z_G$, it follows that the cup-product

$$\hat{c}(F)(h_E) \cup \hat{c}(G)(h_E)$$

vanishes in $H^{2r}(\mathbb{P}(E), \mathbb{Z})$, since these two relations show that in singular or Čech cohomology, we can represent the classes $\hat{c}(F)(h_E)$ and $\hat{c}(G)(h_E)$ by cochains supported in disjoint open sets, namely neighbourhoods of $\mathbb{P}(G)$ and $\mathbb{P}(F)$ respectively.

As the polynomial $\hat{c}(F)\hat{c}(G)$ is normalised of degree r, it must be equal to $\hat{c}(E)$. Passing to the reciprocal polynomials, we find that $c(E) = c(F)c(G)$. □

Remark 11.26 *By construction, the Chern classes $c_i(E)$ of a bundle of rank r are zero for $i > r$. Whitney's formula then shows that if E is a bundle of rank r admitting an everywhere non-zero section, we have $c_r(E) = 0$, since then E is the direct sum of the trivial bundle of rank 1 and a bundle of rank $r - 1$. The relation (11.8) can be interpreted geometrically by noting that the vector bundle $\phi_1^* E \otimes S^*$ admits an everywhere non-zero canonical section given by the inclusion $S \subset \phi_1^* E$. Indeed, relation (11.8) is equivalent to the equation $c_r(\phi_1^* E \otimes S^*) = 0$.*

11.2.2 The Kähler case

Assume now that X is a compact Kähler manifold, and that E is a holomorphic vector bundle over X. Proposition 3.18 says that $\mathbb{P}(E)$ is a Kähler manifold, and lemma 7.32 gives a canonical isomorphism

$$H^k(\mathbb{P}(E), \mathbb{Z}) = \bigoplus\nolimits_{0 \leq l \leq r-1} h_E^l H^{k-2l}(X, \mathbb{Z}) \qquad (11.9)$$

between the integral cohomology groups. As h_E is a class of type $(1, 1)$, the morphism

$$h_E^l : H^{k-2l}(X, \mathbb{Z}) \to H^k(\mathbb{P}(E), \mathbb{Z})$$

is a morphism of Hodge structures, of bidegree (l, l). It follows that if α is a form of type (p, q) on $\mathbb{P}(E)$, its components α_l in the decomposition (11.9) are of type $(p - l, q - l)$. In particular, the components of h^r, which by definition are equal to $-c_{r-l}(E)$, are of type $(r - l, r - l)$. Thus we have the following proposition.

Proposition 11.27 *The Chern classes $c_i(E)$ of a holomorphic vector bundle over a Kähler compact manifold X are integral classes of type (i, i).*

11.3 Hodge classes

11.3.1 Definitions and examples

Let $(V_{\mathbb{Z}}, F^{\cdot} V_{\mathbb{C}})$ be a Hodge structure of even weight $2p$.

Definition 11.28 *The Hodge classes of V are the integral classes of type (p, p) of V.*

We write

$$\mathrm{Hdg}(V) = V_{\mathbb{Z}} \cap V^{p,p}$$

for the set of Hodge classes.

Remark 11.29 *We have* $V^{p,p} \cap V_{\mathbb{R}} = F^p V_{\mathbb{C}} \cap V_{\mathbb{R}}$. *Thus, we also have*

$$\mathrm{Hdg}(V) = V_{\mathbb{Z}} \cap F^p V_{\mathbb{C}}.$$

If X is a Kähler manifold and $V = H^{2p}(X, \mathbb{Z})/\mathrm{torsion}$, we write $\mathrm{Hdg}^{2p}(X) := \mathrm{Hdg}(V)$. We also use the notation $\mathrm{Hdg}^{2p}(X, \mathbb{Z})$ for the set of integral classes whose image in $H^{2p}(X, \mathbb{Z})/\mathrm{torsion}$ is Hodge.

Consider the case $n = 1$. In section 7.1.3, we showed the following result, known as "Lefschetz' theorem on (1, 1) classes".

Theorem 11.30 *Let X be a Kähler manifold. Then* $\mathrm{Hdg}^2(X, \mathbb{Z})$ *is equal to the image of the map*

$$c_1 : \mathrm{Pic}\, X \to H^2(X, \mathbb{Z}),$$

where Pic X *is the group of isomorphism classes of holomorphic line bundles over X.*

Moreover, the results of the preceding sections give ways of constructing Hodge classes on a Kähler manifold. Propositions 11.20 and 11.27 can be summarised as follows.

Theorem 11.31 *Let X be a compact Kähler manifold. Then the cohomology classes of the analytic subsets of X and the Chern classes of the holomorphic vector bundles over X are Hodge classes.*

The relation between the subgroups of $\mathrm{Hdg}^{2k}(X)$ generated by the classes of analytic subsets and the Chern classes of vector bundles is clear only when X is an algebraic variety.

Theorem 11.32 *If X is an algebraic variety, these subgroups of* $\mathrm{Hdg}^{2k}(X)$ *coincide.*

Proof Let us first prove the statement for $k = 1$. If L is a holomorphic line bundle and σ is a holomorphic non-zero section of L, consider the divisor $D \subset X$ of σ. It is the cycle of codimension 1 associated to the subscheme whose sheaf of ideals is locally generated by σ, which we can consider locally as a function by trivialising the bundle L. We have $D = \sum_i n_i D_i$, and we can show that this decomposition is given locally by the decomposition into prime elements of the function corresponding to σ in a local trivialisation of L.

Theorem 11.33 (Lelong) *Let L be a holomorphic bundle of rank* 1, *and let σ be a non-zero holomorphic section of L. Then the cohomology class of the divisor D of σ and the first Chern class $c_1(L)$ are equal in $H^2(X, \mathbb{Z})$.*

Proof Each component D_i of D allows one to define a holomorphic line bundle $L_i := \mathcal{O}_X(D_i) := \mathcal{I}_{D_i}^*$, where $\mathcal{I}_{D_i}$ is the sheaf of free $\mathcal{O}_X$-modules of rank 1 given by the holomorphic functions vanishing on D_i. Indeed, one can show (see Narasimhan 1966) that the sheaf of ideals of D_i is locally generated by an equation f_i. Clearly there exists a section σ_i of L_i whose divisor is equal to D_i: indeed, the inclusion

$$j : \mathcal{I}_{D_i} \to \mathcal{O}_X$$

dualises to an inclusion

$${}^t j : \mathcal{O}_X \to L_i$$

which gives the section $\sigma_i = {}^t j(1)$ of L_i. Now, the equation f_i gives a local generator of $\mathcal{I}_{D_i}$, i.e. a trivialisation

$$\mathcal{I}_{D_i} \cong \mathcal{O}_X,$$

and in this trivialisation, the map j is multiplication by f_i. This also holds for ${}^t j$, which shows that the divisor of $\sigma_i = {}^t j(1)$ is equal to D_i.

We then have $L = \bigotimes_i L_i^{\otimes n_i}$, since the section σ of L gives a morphism

$$\mathcal{O}_X \to L$$

whose dual

$$L^* =: L^{-1} \to \mathcal{O}_X$$

identifies L^{-1} with the free $\mathcal{O}_X$-submodule of rank 1

$$\mathcal{I}_D = \prod_i \mathcal{I}_{D_i}^{n_i} \cong \bigotimes_i L_i^{\otimes - n_i}.$$

Since the first Chern class and the "cycle class" map are both additive, it suffices to prove the result for the section σ_i of L_i, whose divisor D_i is generically smooth.

In fact, we are then immediately reduced to the case where D_i is smooth, thanks to lemma 11.13 and theorem 11.11, which show that

$$H^2(X, \mathbb{Z}) = H^2(X - \operatorname{sing} D_i, \mathbb{Z}).$$

Now consider the exact sequence of relative cohomology

$$\cdots \to H^2(X, X - D_i, \mathbb{Z}) \to H^2(X, \mathbb{Z}) \to H^2(X - D_i, \mathbb{Z}) \to \cdots,$$

and the Thom isomorphism

$$H^2(X, X - D_i, \mathbb{Z}) = H^0(D_i, \mathbb{Z}) = \mathbb{Z}.$$

As the Chern class $c_1(L_i)$ vanishes on $X - D_i$, since L_i is trivial on $X - D_i$, it comes from $H^0(D_i, \mathbb{Z})$, and thus it is an integral multiple of the class of D_i, which by definition (cf. section 11.1.2) is the image of $1 \in H^0(D_i, \mathbb{Z})$. To see that we have $c_1(L_i) = [D_i]$, i.e. that the ratio is equal to 1, it now suffices to show that for every closed form ω with compact support in X of degree $2n - 2 = \dim_{\mathbb{R}} D$, we have

$$\int_{D_i} \omega = \int_X \omega \wedge \alpha, \tag{11.10}$$

where α is a representative of $c_1(L)$ in de Rham cohomology. Recall that in section 7.1.3 we showed that such a representative was given by the form

$$\alpha = \frac{1}{2i\pi} \partial\overline{\partial} \log h,$$

where h is a Hermitian metric on L_i. As

$$\partial\overline{\partial} \log h = \partial\overline{\partial} \log h(\sigma),$$

where the function $\log h(\sigma)$ is singular, formula (11.10) becomes Lelong's formula

$$\int_{D_i} \omega = \frac{1}{2i\pi} \int_X \omega \wedge \partial\overline{\partial} \log h(\sigma).$$

This is obtained by applying Stokes' formula to the right-hand integral over the complement of a tubular neighbourhood T_ϵ of radius ϵ of D_i, and noting that

$$\lim_{\epsilon \to 0} \frac{1}{2i\pi} \int_{\partial T_\epsilon} \omega \wedge \partial \log h(\sigma) = \int_{D_i} \omega,$$

which holds because locally, in suitable local coordinates $z_1, \ldots, z_n$, we have $\sigma = z_1$, $D_i = \{z \mid z_1 = 0\}$ and $\partial \log h(\sigma) = \frac{dz_1}{z_1}$ modulo a C^∞ form. □

The following result is a consequence of Lelong's theorem.

Corollary 11.34 *If X is a complex projective manifold, the Chern classes of holomorphic line bundles on X are the classes of divisors (i.e. cycles of codimension 1).*

Proof If $D = \sum_i n_i D_i$ is a divisor, we have the associated line bundle $\mathcal{O}_X(D) := \bigotimes_i \mathcal{I}_{D_i}^{\otimes -n_i}$ (where we set $L^{-1} = L^*$ for L a line bundle), and theorem 11.33 shows that $c_1(\mathcal{O}_X(D)) = [D]$.

Conversely, it suffices to note that every holomorphic line bundle L over a projective manifold admits a meromorphic section. This follows from the arguments used in the proof of Kodaira's embedding theorem 7.11. Indeed, let h be a metric on L, and let H be an ample line bundle over X equipped with a metric with positive curvature. Then for sufficiently large N, $L \otimes H^{\otimes N}$ and $H^{\otimes N}$ admit non-zero holomorphic sections σ_1, σ_2. L then admits the meromorphic section $\sigma = \frac{\sigma_1}{\sigma_2}$. □

To conclude the proof of theorem 11.32, we first prove the following generalisation of corollary 11.34.

Proposition 11.35 *Let Z be an algebraic cycle of codimension k, and L a line bundle over an algebraic variety X. Then $[Z] \cup c_1(L)$ is the class of an algebraic cycle Z' of codimension $k+1$ of X.*

Proof We can take Z' to be the intersection of Z and the divisor of a sufficiently generic meromorphic section of L. We reduce to the case where Z' is smooth using lemma 11.13. The conclusion is then easy. □

Next, if E is a holomorphic vector bundle over an algebraic variety equipped with an ample line bundle H, then $E' = E \otimes H^{\otimes N}$ is generated by its global sections for sufficiently large N (this follows, as in the proof of theorem 7.11, from Kodaira's vanishing theorem or Serre (1955) applied to the blowups of X at each of its points). The global sections of E' then give a holomorphic map ϕ from X to the Grassmannian $G(N-r, N)$, where $N = \dim H^0(X, \mathcal{E}')$, which to $x \in X$ associates the subspace of the sections of E' which vanish at the point x.

The bundle E' can then be naturally identified with the pullback $\phi^* Q$, where Q is the quotient tautological bundle over the Grassmannian. It is not difficult (see Fulton 1984) to show that the cohomology of the Grassmannian is generated by classes of smooth algebraic cycles, which we may furthermore assume transverse to ϕ (i.e. smooth cycles Z of codimension k such that the scheme

$\phi^{-1}(Z)$ is also smooth of codimension k). We then have

$$\phi^*([Z]) = [\phi^{-1}(Z)],$$

and we deduce that the Chern classes of E' are classes of algebraic cycles.

The Chern classes of E can be computed using the Chern classes of $E' = E \otimes H^N$ and their cup-products with powers of $c_1(H)$. Indeed, the projective bundles $\mathbb{P}(E)$ and $\mathbb{P}(E')$ are isomorphic, but their tautological subbundles are different:

$$S_{E'} = S_E \otimes \pi^* H^{\otimes N}.$$

Thus, we have $h_E - \pi^*(Nc_1(H)) = h_{E'}$, and h_E satisfies the normalised polynomial equation of degree r with coefficients in H^* given by

$$\hat{c}(E')(h_E - \pi^*(Nc_1(H))) = 0.$$

Thus, $\hat{c}(E)(t) = \hat{c}(E')(t - \pi^*(Nc_1(H)))$. Finally, by proposition 11.35, the Chern classes of E are also classes of algebraic cycles. □

11.3.2 The Hodge conjecture

The Hodge conjecture concerns rational Hodge classes on projective manifolds.

Conjecture 11.36 *Let X be a projective manifold, and $\alpha \in \mathrm{Hdg}^{2k}(X)$. Then a multiple $N\alpha$ of α with $N \neq 0$ is the class of an algebraic cycle.*

The case $k = 1$ of this conjecture holds by theorem 11.30 and corollary 11.34. The conjecture was initially formulated for integral classes. A first counterexample concerning torsion classes was given by Atiyah & Hirzebruch (1962). Their counterexample was recently reinterpreted by Totaro (1997), using the complex cobordism ring, through which the cycle class map factors.

Kollár (1992) recently gave a counterexample showing the necessity of considering multiples of the integral classes. His counterexample is given by hypersurfaces X of degree d in the projective space $\mathbb{P}^4$. Such a hypersurface satisfies $H_2(X, \mathbb{Z}) = \mathbb{Z}$, and the class of a plane curve $\mathbb{P}^2 \cap X$ is equal to d times the generator of $H_2(X, \mathbb{Z})$. Kollár shows that this generator is not, however, in general, the class of an algebraic cycle.

The initial Hodge conjecture was much more ambitious. Suppose that $\phi : Y \to X$ is a morphism of projective (or Kähler) algebraic varieties with $\dim X = \dim Y + k$. Then the image of $\phi_* : H^l(Y, \mathbb{Z}) \to H^{l+2k}(X)$ consists of classes

$\alpha = \sum_{p+q=l+2k} \alpha^{p,q}$ whose components $\alpha^{p,q}$ are zero for $p < k$, and thus belong to $F^k H^{2k+l}(X)$. This still holds if Y is singular, in which case the morphism ϕ_* is the morphism

$$\phi_* : H_{2n-l}(Y, \mathbb{Z}) \to H_{2n-l}(X, \mathbb{Z}) \to H^{l+2k}(X, \mathbb{Z}),$$

where $n = \dim Y$ and the last map is given by Poincaré duality.

Hodge conjectured that every integral class contained in $F^k H^*(X)$ on an algebraic variety X is an integral combination of classes coming from a morphism ϕ_* as above. This was contradicted by Grothendieck (1969). In fact, one can construct explicit counterexamples by noting that via the theory of mixed Hodge structures, the image of such a morphism ϕ_* is necessarily a sub-Hodge structure of $H^{l+2k}(X, \mathbb{Z})$ contained in $F^k H^{l+2k}(X)$. Now, it is quite easy to construct examples of integral classes, for example of degree 3 on an algebraic variety X, whose components of type $(3, 0)$ (and $(0, 3)$) vanish, even though $H^3(X)$ has no non-trivial sub-Hodge structure contained in $F^2 H^3(X)$.

The generalised Hodge conjecture was corrected by Grothendieck as follows.

Conjecture 11.37 *Let X be a smooth algebraic variety, and $L \subset H^{2k+l}(X, \mathbb{Q})$ a rational sub-Hodge structure contained in $F^k H^{2k+l}(X)$. Then there exist (not necessarily smooth) algebraic subvarieties $Y_i \overset{j_i}{\hookrightarrow} X$ of codimension k such that L is contained in $\sum_i j_{i*} H_{2n-l}(Y_i, \mathbb{Q})$, $\dim X = n + k$.*

We refer to Lewis (1990) for a detailed discussion of the current state of knowledge about the Hodge conjecture.

11.3.3 Correspondences

Let X and Y be two locally contractible topological spaces such that the cohomology $H^*(X, \mathbb{Z})$ is finitely generated. We then have the following formula of Künneth, which is a special case of the Leray–Hirsch theorem 7.33.

Theorem 11.38 *The cup-products $H^p(X, \mathbb{Z}) \otimes H^q(Y, \mathbb{Z}) \to H^{p+q}(X \times Y, \mathbb{Z})$ induce an isomorphism modulo torsion*

$$H^n(X \times Y, \mathbb{Z}) = \bigoplus_{p+q=n} H^p(X, \mathbb{Z}) \otimes H^q(Y, \mathbb{Z}).$$

Assume now that X and Y are compact Kähler manifolds. Then the cup-product $\mathrm{pr}_1^*\alpha \cup \mathrm{pr}_2^*\beta$ of a class of type (p, q) on X and a class of type (p', q')

on Y is a class of type $(p + p', q + q')$. We thus have

$$H^{r,s}(X \times Y) = \bigoplus_{p+p'=r,q+q'=s} H^{p,q}(X) \otimes H^{p',q'}(Y).$$

Definition 11.39 *The tensor product* $(K, K^{r,s})$ *of two Hodge structures*

$$(V_{\mathbb{Z}}, V^{p,q}) \text{ and } (W_{\mathbb{Z}}, W^{p,q})$$

of weight v *and* w *respectively is the Hodge structure of weight* $v + w$ *defined by*

$$K_{\mathbb{Z}} = V_{\mathbb{Z}} \otimes_{\mathbb{Z}} W_{\mathbb{Z}}, \quad K^{r,s} = \bigoplus_{p+p'=r,q+q'=s} V^{p,q} \otimes_{\mathbb{C}} W^{p',q'}.$$

The following theorem is a consequence of the above.

Theorem 11.40 *Let* X *and* Y *be compact Kähler manifolds. Then* $H^k(X) \otimes H^l(Y)$ *is a sub-Hodge structure of* $H^{k+l}(X \times Y)$*, isomorphic to the tensor product of the Hodge structures on* $H^k(X)$ *and* $H^l(Y)$*.*

Let n be the complex dimension of X. For the cohomology modulo torsion, we have the isomorphism given by the Poincaré duality

$$H^k(X, \mathbb{Z}) \cong (H^{2n-k}(X, \mathbb{Z}))^*,$$

which thus gives an isomorphism (of cohomology groups modulo torsion)

$$H^k(X, \mathbb{Z}) \otimes H^l(Y, \mathbb{Z}) \cong \mathrm{Hom}_{\mathbb{Z}}(H^{2n-k}(X, \mathbb{Z}), H^l(Y, \mathbb{Z})).$$

Lemma 11.41 *Assume that* $k + l$ *is even. Then a class*

$$\alpha \in H^k(X, \mathbb{Z}) \otimes H^l(Y, \mathbb{Z}) \subset H^{k+l}(X \times Y, \mathbb{Z})$$

is a Hodge class if and only if the corresponding morphism

$$\tilde{\alpha} : H^{2n-k}(X, \mathbb{Z}) \to H^l(Y, \mathbb{Z})$$

is a morphism of Hodge structures (of bidegree $(r - n, r - n)$, $k + l = 2r$*).*

Proof The class α is a Hodge class if and only if α is of type (r, r), and this is equivalent to

$$\alpha \in \bigoplus_{\substack{p+p'=r,q+q'=r \\ p+q=k,p'+q'=l}} H^{p,q}(X) \otimes H^{p',q'}(Y).$$

More generally, it thus suffices to show that for a class

$$\beta = \beta^{p,q} \otimes \gamma^{p',q'}, \; p + p' = t, \; q + q' = s, \; p + q = k, \; p' + q' = l,$$

the corresponding element $\tilde{\beta}$ of $\mathrm{Hom}_{\mathbb{C}}(H^{2n-k}(X, \mathbb{C}), H^l(Y, \mathbb{C}))$ sends $H^{u,v}(X)$ to $H^{u+t-n,v+s-n}(Y)$, where $u + v = 2n - k$. Now, we have

$$\tilde{\beta}(\eta) = \langle \eta, \beta^{p,q} \rangle_X \gamma^{p',q'}, \tag{11.11}$$

and the statement thus follows from lemma 7.30, which says that $\beta^{p,q}$ is orthogonal to $H^{u,v}(X)$ for the Poincaré duality if $(u, v) \neq (n - p, n - q)$. If $\eta \in H^{u,v}(X)$, it follows from this lemma and (11.11) that $\tilde{\beta}(\eta) = 0$ if $(u, v) \neq (n - p, n - q)$, and $\tilde{\beta}(\eta)$ is proportional to $\gamma^{p',q'}$ if $(u, v) = (n - p, n - q)$. As $p' = t - p$, $q' = s - q$, we have $\tilde{\beta}(\eta) \in H^{u+t-n,v+s-n}(Y)$. □

Assume that X and Y are algebraic. The Hodge conjecture 11.36 then predicts that a Hodge class on $X \times Y$ is the class of an algebraic cycle with rational coefficients on $X \times Y$. Such a cycle is called a correspondence. The Künneth components of such a class are still Hodge classes, and by the above, they give morphisms of Hodge structures between the cohomology groups of X and those of Y. If we take $Y = X$, the morphisms of Hodge structures $\mathrm{Id}_k : H^k(X, \mathbb{Z}) \to H^k(X, \mathbb{Z})$ give Hodge classes on $X \times X$. The sum $\sum_k \mathrm{Id}_k$ is equal to the cohomology class of the diagonal $\Delta \subset X \times X$, and thus it is the class of an algebraic cycle. However, it is not proven in general that the Künneth components Id_k of the class of the diagonal are classes of algebraic cycles. Another example of a morphism of (rational) Hodge structures, for which we do not know whether the corresponding Hodge class satisfies the Hodge conjecture, is given by the inverse $(L^{n-k})^{-1}$ of the Lefschetz isomorphisms

$$L^{n-k} : H^k(X, \mathbb{Q}) \to H^{2n-k}(X, \mathbb{Q}), \; n = \dim X.$$

Exercises

1. Let X be an oriented differentiable compact manifold, and $Z \overset{j}{\hookrightarrow} X$ an oriented submanifold of (real) codimension r.
   (a) Show that

   $$j_* \circ j^* : H^k(X, \mathbb{R}) \to H^{k+r}(X, \mathbb{R})$$

   is equal to $\cup [Z]$ (the cup-product with the class of Z).
   (b) Show that

   $$j^* \circ j_* : H^k(Z, \mathbb{R}) \to H^{k+r}(Z, \mathbb{R})$$

is equal to $\cup [Z]_{|Z}$ (the cup-product with the restriction to Z of the class of Z).

Let us now assume that X is a compact Kähler manifold and that Z is a complex hypersurface which is ample. (That is, the class $[Z]$ is a Kähler class.)

(c) Deduce from the hard Lefschetz theorem that the restriction map

$$j^* : H^k(X, \mathbb{Q}) \to H^k(Z, \mathbb{Q})$$

is injective for $k \leq \dim_{\mathbb{C}} Z$.

(d) Deduce from the hard Lefschetz theorem that the Gysin map

$$j_* : H^k(Z, \mathbb{Q}) \to H^{k+2}(X, \mathbb{Q})$$

is injective for $k < \dim_{\mathbb{C}} Z$.

These statements are part of the Lefschetz theorem on hyperplane sections, which will be proved in the second volume of this book.

2. *Curves on surfaces and the Hodge index theorem.* Let S be a smooth projective surface and let $\phi : S \to \mathbb{P}^n$ be a holomorphic map satisfying the following properties:
   (i) There exists a curve $C \subset S$ such that $\phi(C)$ is a finite set of points. (Here C is a not necessarily connected finite union of irreducible hypersurfaces C_i which are not necessarily smooth.)
   (ii) There exists an open set of S where ϕ is an embedding.
   (a) Show that the class $h = \phi^* c_1(\mathcal{O}_{\mathbb{P}^n}(1))$ satisfies $\langle h, h \rangle > 0$, where $\langle , \rangle$ is the intersection form on $H^2(S, \mathbb{Z})$.
   (b) Show that $\langle [C_i], h \rangle = 0$ for any i. Deduce from the Hodge index theorem that the intersection form $\langle , \rangle$ is negative definite on the subgroup of $H^2(S, \mathbb{Z})$ generated by the $[C_i]$'s.
   (c) Show that $\langle C_i, C_j \rangle \geq 0$ for $i \neq j$. Deduce then from the above that the $[C_i]$'s are independent in $H^2(S, \mathbb{Z})$.
3. *Chern classes and Euler classes.* Let E be a real oriented vector bundle of rank r on a differentiable manifold X. We define its Euler class $e(E)$ in the following way: the section 0 of E provides a submanifold

$$X \overset{0}{\hookrightarrow} E$$

of E. This submanifold has an oriented normal bundle and hence possesses a class $[X] \in H^r(E, \mathbb{Z})$. Then we define

$$e(E) := 0^*[X].$$

(a) Let L be a holomorphic line bundle on a complex manifold X. We can see L as a real differentiable vector bundle $L_{\mathbb{R}}$ of rank 2. Show that $c_1(L) = e(L_{\mathbb{R}})$ in $H^2(X, \mathbb{Z})$. (Use the Lelong formula on L.)

(b) Let E be a oriented real differentiable vector bundle of rank $2k$ which is a direct sum of k complex line bundles L_i, or more precisely of their underlying real vector bundles $L_{i,\mathbb{R}}$. Show that $e(E) = e(L_{1,\mathbb{R}}) \cup \cdots \cup e(L_{k,\mathbb{R}})$.

(c) Deduce from (a), (b) and the splitting principle that for any holomorphic vector E of rank k on X, with underlying real vector bundle $E_{\mathbb{R}}$ of rank $2k$, we have

$$c_k(E) = e(E_{\mathbb{R}}) \text{ in } H^{2k}(X, \mathbb{Z}).$$

(Note that this remains true in the differentiable case.)

(d) Let $Z \subset X$ be a complex submanifold of codimension r of a complex manifold X. Show with the help of the construction of $[Z]$ in a tubular neighbourhood of Z in X and of (c) that

$$[Z]_{|Z} = c_r(N_{Z/X}) \text{ in } H^{2r}(Z, \mathbb{Z}).$$

12

Deligne–Beilinson Cohomology and the Abel–Jacobi map

In this chapter, we define a refined invariant of an analytic cycle homologous to 0 on a compact Kähler manifold, namely its Abel–Jacobi invariant, which generalises the Abel–Jacobi invariant for the 0-cycles on curves (see Arbarello *et al.* 1985). In the last section, following Deligne, we will show that we can even construct a Deligne class, which determines the Hodge class, and which is equal to the Abel–Jacobi invariant for a cycle homologous to 0.

The intermediate Jacobians $J^{2k-1}(X)$ of such a manifold X are the complex tori defined by

$$J^{2k-1}(X) = H^{2k-1}(X, \mathbb{C})/(F^k H^{2k-1}(X) \oplus H^{2k-1}(X, \mathbb{Z})).$$

When a cycle Z is homologous to 0, if we interpret its homology class in terms of singular homology, we can find a differentiable chain γ whose boundary is equal to Z. Hodge theory then shows that the integration current $\int_\gamma$ gives a well-defined functional on $F^{n-k} H^{2n-2k+1}(X)$, where $n = \dim X$. As γ is defined up to a cycle, we can then define

$$\Phi_X^k(Z) = \int_\gamma \in F^{n-k} H^{2n-2k+1}(X)^* / H_{2n-2k+1}(X, \mathbb{Z}),$$

where the left-hand term is equal to $J^{2k-1}(X)$ by Poincaré duality.

We will establish some properties of the Abel–Jacobi map. For example, we show that the Abel–Jacobi invariant varies holomorphically with the cycle Z. We begin by studying the case $k = n$, where the Abel–Jacobi map is identified with the Albanese map. In particular, in the case $k = 1$, we prove Abel's theorem, which states that the Abel–Jacobi map is given by the identification

$$\mathrm{Pic}^0(X) = H^{0,1}(X)/H^1(X, \mathbb{Z})$$

coming from the exponential exact sequence.

In arbitrary codimension, if $\mathcal{Z} \subset C \times X$ is a cycle of codimension k, which we can view as a family of cycles $(Z_t)_{t\in C}$ parametrised by a complete smooth curve C, then by the universal property of the Albanese map, the Abel–Jacobi map

$$\phi : C \to J^{2k-1}(X), \quad c \mapsto \Phi_X^k(Z_c - Z_0)$$

induces a morphism of complex tori

$$\psi : J(C) \to J^{2k-1}(X).$$

We will prove the following theorem.

Theorem 12.1 *The morphism ψ is the morphism of complex tori induced by the morphism of Hodge structures*

$$[\mathcal{Z}]_* : H^1(C, \mathbb{Z}) \to H^{2k-1}(X, \mathbb{Z})$$

given by the Hodge class of the cycle $\mathcal{Z} \subset C \times X$.

We conclude this chapter by defining the Deligne cohomology. We show that we can identify the Deligne cohomology $H_D^{2k}(X, \mathbb{Z}(k))$ with a quotient of a subgroup $\Xi_{\mathrm{diff}}^{2k-1}(X)^{k,k}$ of the group $\Xi_{\mathrm{diff}}^{2k-1}(X)$ of the differential characters introduced by Cheeger and Simons. We then define the Deligne class of a cycle of codimension k, using a Green current, which allows us to associate a differential character to the cycle.

12.1 The Abel–Jacobi map

12.1.1 Intermediate Jacobians

Let X be a compact Kähler manifold. For every integer $k > 0$, the Hodge filtration on $H^{2k-1}(X, \mathbb{C})$ gives the decomposition as a direct sum

$$H^{2k-1}(X, \mathbb{C}) = F^k H^{2k-1}(X) \oplus \overline{F^k H^{2k-1}(X)}.$$

Thus, $F^k H^{2k-1}(X) \cap H^{2k-1}(X, \mathbb{R}) = \{0\}$, and the composition map

$$H^{2k-1}(X, \mathbb{R}) \to H^{2k-1}(X, \mathbb{C})/F^k H^{2k-1}(X)$$

is an isomorphism of $\mathbb{R}$-vector spaces. Therefore, the lattice

$$H^{2k-1}(X, \mathbb{Z}) \subset H^{2k-1}(X, \mathbb{R})$$

gives a lattice in the $\mathbb{C}$-vector space $H^{2k-1}(X, \mathbb{C})/F^k H^{2k-1}(X)$.

Definition 12.2 *The kth intermediate Jacobian $J^{2k-1}(X)$ is the complex torus*

$$J^{2k-1}(X) = H^{2k-1}(X, \mathbb{C})/(F^k H^{2k-1}(X) \oplus H^{2k-1}(X, \mathbb{Z})).$$

For $k = 1$, it is essentially the complex torus

$$\mathrm{Pic}^0(X) = H^1(X, \mathcal{O}_X)/2i\pi H^1(X, \mathbb{Z})$$

introduced in section 7.2.2, via the identification

$$H^1(X, \mathbb{C})/F^1 H^1(X, \mathbb{C}) \stackrel{2i\pi}{=} H^1(X, \mathcal{O}_X).$$

In general, it is a transcendental object, whose nature is much more difficult to understand than $\mathrm{Pic}^0(X)$. For example, even when X is algebraic, $J^{2k-1}(X)$ is not in general an abelian variety, in contrast with proposition 7.16, which shows that this is the case for $J^1(X) = \mathrm{Pic}^0(X)$.

Remark 12.3 *We can define a complex torus*

$$J^{2p-1}(V) := V_{\mathbb{C}}/(F^p V \oplus V_{\mathbb{Z}})$$

for every Hodge structure of weight $2p - 1$. This construction is functorial, in the sense that every morphism of Hodge structures

$$(V_{\mathbb{Z}}, F^{\cdot} V) \to (W_{\mathbb{Z}}, F^{\cdot+r} W)$$

of bidegree (r, r) induces a morphism of complex tori

$$J^{2p-1}(V) \to J^{2(p+r)-1}(W).$$

12.1.2 The Abel–Jacobi map

The Abel–Jacobi map Φ_X^k was defined by Griffiths, as a generalisation of the Jacobi map for curves. It is a morphism from the group $\mathcal{Z}^k(X)_{\mathrm{hom}}$ of cycles of codimension k cohomologous to 0 to $J^{2k-1}(X)$, defined as follows. Let $Z \in \mathcal{Z}^k(X)_{\mathrm{hom}}$. The cohomology class of Z is equal to $\sum_i n_i[Z_i]$, where $[Z_i]$ is dual in the sense of Poincaré, i.e. $[Z_i]$ is the image under the Poincaré duality isomorphism

$$H_{2n-2k}(X, \mathbb{Z}) = H^{2k}(X, \mathbb{Z}), \quad n = \dim X$$

of the fundamental homology class of the irreducible compact complex analytic subset Z_i. This homology class can be computed in singular homology using a triangulation by differentiable chains of Z_i (cf. section 4.3.2). This fact follows from the equality (cf. section 11.1.4)

$$[Z_i] = \tau_*([\tilde{Z}_i]_{\text{fund}}),$$

where $[\tilde{Z}_i]_{\text{fund}}$ is the fundamental class of the compact oriented manifold $\tilde{Z}_i$, and $\tau : \tilde{Z}_i \to Z_i \to X$ is a desingularisation of Z_i. The condition $[Z] = 0$ in $H^{2k}(X, \mathbb{Z})$ translates into the fact that the differentiable chain associated to $\sum_i n_i Z_i$ by a triangulation of each Z_i is homologous to 0, i.e. by the existence of a differentiable chain $\Gamma \subset X$ of dimension $2n - 2k + 1$ such that $\partial\Gamma = Z$.

The Abel–Jacobi invariant $\Phi_X^k(Z) \in J^{2k-1}(X)$ is then constructed as follows. Poincaré duality gives a commutative diagram

$$\begin{array}{ccc} H^{2k-1}(X, \mathbb{Z})/\text{torsion} & = & H_{2n-2k+1}(X, \mathbb{Z})/\text{torsion} \\ \downarrow & & \downarrow \\ H^{2k-1}(X, \mathbb{C}) & = & H^{2n-2k+1}(X, \mathbb{C})^*, \end{array}$$

where the second vertical map is given by integrating forms over chains. As

$$F^k H^{2k-1}(X) = F^{n-k+1} H^{2n-2k+1}(X)^{\perp},$$

the lower horizontal map induces an isomorphism

$$H^{2k-1}(X, \mathbb{C})/F^k H^{2k-1}(X) = F^{n-k+1} H^{2n-2k+1}(X)^*,$$

and thus we have an isomorphism

$$J^{2k-1}(X) = F^{n-k+1} H^{2n-2k+1}(X)^*/H_{2n-2k+1}(X, \mathbb{Z}),$$

where we write $H_{2n-2k+1}(X, \mathbb{Z})$ for the homology modulo torsion of X, and the map

$$H_{2n-2k+1}(X, \mathbb{Z}) \to F^{n-k+1} H^{2n-2k+1}(X)^* \qquad (12.1)$$

is given by integrating forms over chains. (These integrals and particularly those of the holomorphic forms $\omega \in H^{2k-1,0}(X)$ are called the *periods of* X.)

Recall now (proposition 7.5) the isomorphism

$$F^{n-k+1} H^{2n-2k+1}(X, \mathbb{C}) = \frac{F^{n-k+1} A^{2n-2k+1}(X) \cap \operatorname{Ker} d}{d F^{n-k+1} A^{2n-2k}(X)}. \qquad (12.2)$$

Let $\phi \in F^{n-k+1} H^{2n-2k+1}(X, \mathbb{C})$, and let η be a representative of ϕ lying in $F^{n-k+1} A^{2n-2k+1}(X) \cap \operatorname{Ker} d$. Set

$$\int_\Gamma \phi = \int_\Gamma \eta.$$

The result does not depend on the choice of η, since by (12.2), η is defined up to addition of a form $d\psi$, $\psi \in F^{n-k+1}A^{2n-2k}(X)$; Stokes' formula then says that

$$\int_\Gamma d\psi = \int_Z \psi,$$

and this is zero since $\psi_{|Z} = 0$ for reasons of type. Thus,

$$\int_\Gamma \in F^{n-k+1}H^{2n-2k+1}(X, \mathbb{C})^*$$

is well-defined. Moreover, if Γ' is another chain such that $\partial\Gamma' = Z$, then $\partial(\Gamma - \Gamma') = 0$ and thus $\int_\Gamma - \int_{\Gamma'}$ lies in the image of (12.1). Therefore, $\int_\Gamma$ is well-defined independently of the choice of Γ such that $\partial\Gamma = Z$, in the quotient $F^{n-k+1}H^{2n-2k+1}(X)^*/H_{2n-2k+1}(X, \mathbb{Z})$. We set

$$\Phi_X^k(Z) = \int_\Gamma \in F^{n-k+1}H^{2n-2k+1}(X)^*/H_{2n-2k+1}(X, \mathbb{Z}) = J^{2k-1}(X).$$

We have the following important result.

Theorem 12.4 (Griffiths 1968) *Let Y be a connected complex manifold, $y_0 \in Y$ a reference point, and $Z \subset Y \times X$ a cycle of codimension k. We will assume that $Z = \sum_i n_i Z_i$ where each Z_i is smooth and such that $\mathrm{pr}_1 : Z_i \to Y$ is a submersion. Then the fibres $Z_y = \sum_i n_i Z_{i,y}$, $Z_{i,y} := \mathrm{pr}_1^{-1}(y) \subset X$, are all homologous in X, and the map ϕ*

$$y \mapsto \Phi_X^k(Z_y - Z_{y_0})$$

from Y to the complex torus $J^{2k-1}(X)$ is holomorphic.

Remark 12.5 *The hypotheses are actually too strong. Instead of working with smooth cycles over Y, we should in greatest generality work with cycles which are flat (see Hartshorne 1977) over Y, a condition which ensures that the fibres Z_y vary continuously, and under which theorem 12.4 remains valid (Griffiths 1968, II).*

Proof of theorem 12.4 We can obviously assume that $Z = Z_i$ is itself a smooth manifold, since the map ϕ is equal to $\sum_i n_i\phi_i$, where the sum is taken in the torus $J^{2k-1}(X)$ and $\phi_i : Y \to J^{2k-1}(X)$ is the Abel–Jacobi map associated to the cycle Z_i: $\phi_i(y) = \Phi_X^k(Z_{i,y} - Z_{i,y_0})$. Let $y_1 \in Y$, and let $U \subset Y$ be a neighbourhood of y_1 in Y isomorphic to a ball, via a choice of (real) coordinates x_i. For every $y \in U$, let

$$\Gamma_y = p_Y^{-1}([y_1, y]) \subset Z,$$

where p_Y is the restriction to Z of the first projection $\mathrm{pr}_1 : Y \times X \to Y$.

Γ_y is a manifold with boundary contained in Z, which is sent differentiably to X by the second projection p_X, and its boundary, which is equal to $y \times Z_y - y_1 \times Z_{y_1}$, is sent by p_X to $Z_y - Z_{y_1}$. The choice of a triangulation of Γ_y thus gives a differentiable chain whose boundary is equal to $Z_y - Z_{y_1}$ (i.e. is a differentiable chain obtained by triangulation of Z_y and Z_{y_1}).

As we have $\phi(y) = \phi(y_1) + \Phi_X^k(Z_y - Z_{y_1})$, it suffices to show that the map $\phi' : y \mapsto \Phi_X^k(Z_y - Z_{y_1})$ is holomorphic in U. By definition of Φ_X^k, the map ϕ' admits a continuous lifting to a map ψ with values in the universal cover $F^{n-k+1}H^{2n-2k+1}(X)^*$ of $J^{2k-1}(X)$, given by

$$\psi(y)(\eta) = \int_{\Gamma_y} p_X^*(\tilde{\eta}),$$

where η lies in $F^{n-k+1}H^{2n-2k+1}(X)$ and $\tilde{\eta}$ is a representative of η lying in $F^{n-k+1}A^{2n-2k+1}(X) \cap \operatorname{Ker} d$. To see that ϕ' is holomorphic, it suffices to see that ψ is holomorphic (as a function with values in a complex vector space), and this is equivalent to the fact that the function $\psi_\eta(y) := \psi(y)(\eta)$ is holomorphic on U for every $\eta \in F^{n-k+1}H^{2n-2k+1}(X)$.

Lemma 12.6 *The differential of the function ψ_η at the point y is given by the formula*

$$d\psi_\eta(y)(v) = \int_{Z_y} \operatorname{int}(\chi_v)(p_X^*(\tilde{\eta})), \tag{12.3}$$

where χ_v is a vector field normal to Z_y in Z such that $p_{Y*}(\chi_v) = v$ at every point of Z_y.

Note that the differential form $\operatorname{int}(\chi_v)(p_X^*(\tilde{\eta}))_{|Z_y}$ does not depend on the choice of a normal vector field, since such a field is defined up to a tangent field to Z_y, and the degree of $p_X^*(\tilde{\eta})$ is equal to $\dim_{\mathbb{R}} Z_y + 1$.

Proof The form $p_X^*(\tilde{\eta})$, which is closed and of degree $2n-2k+1 = \dim_{\mathbb{R}} Z_{y_1} + 1$, has restriction to Z_{y_1} equal to zero, and is thus exact on $Z_U = p_Y^{-1}(U)$ since Z_U is diffeomorphic to $Z_{y_1} \times U$ and U is contractible, so that the restriction map $H^*(Z_U) \to H^*(Z_{y_1})$ is an isomorphism. Let us write $p_X^*(\tilde{\eta}) = d\beta$. Then the function ψ_η becomes

$$\psi_\eta(y) = \int_{Z_y} \beta - \int_{Z_{y_1}} \beta$$

by Stokes' formula. It remains only to apply formula (9.8), which gives

$$d_v\left(\int_{Z_y} \beta\right) = \int_{Z_y} \operatorname{int}(\chi_v) d\beta = \int_{Z_y} \operatorname{int}(\chi_v)(p_X^*(\tilde{\eta})). \qquad \square$$

It is now easy to show that the differential of ψ_η is of type (1, 0). Indeed, the differential form $p_X^*(\tilde{\eta})$ lies in $F^{n-k+1}A^{2n-2k+1}(Z)$. If v is a tangent vector to Y at y of type (0, 1), we can take a lifting χ_v of v to a vector field normal to Z_y and of type (0, 1). The differential form $\mathrm{int}(\chi_v)(p_X^*(\tilde{\eta}))_{|Z_y}$ is then zero for reasons of type. By formula (12.3), we thus conclude that $d\psi_\eta(v) = 0$. □

12.1.3 Picard and Albanese varieties

Let us first consider the case $k = 1$. Let Z be a cycle of codimension 1 in X, i.e. a divisor. To Z corresponds the holomorphic line bundle $\mathcal{L} := \mathcal{O}_X(Z)$ considered in the proof of corollary 11.34, which is equipped with a meromorphic section σ whose divisor is equal to Z. By theorem 11.33, Z is cohomologous to 0 if and only if $c_1(\mathcal{L}) = 0$. By the exponential exact sequence, the isomorphism class of $\mathcal{L}$ is thus naturally an element α_Z of $\mathrm{Pic}^0(X) = J^1(X)$.

Proposition 12.7 *α_Z is equal to $\phi_X^1(Z)$.*

Proof Assume for simplicity that Z satisfies the property that every irreducible hypersurface appears in Z with multiplicity 1. (Note that if X is projective, then every divisor homologous to 0 can be written as a sum of divisors satisfying this property.)

The fact that $c_1(\mathcal{L}) = 0$ can be interpreted as saying that we can find holomorphic trivialisations of $\mathcal{L}$ over the open sets U_i of a covering of X, such that the corresponding transition functions g_{ij} are of the form $\exp(2i\pi f_{ij})$, with

$$f_{ij} + f_{jk} + f_{ki} = 0$$

on U_{ijk}. Indeed, this follows from the definition of $c_1(\mathcal{L})$, using the exponential exact sequence

$$0 \to \mathbb{Z} \to \mathcal{O}_X \to \mathcal{O}_X^* \to 1.$$

Take a covering U_i of X such that $\mathcal{L}_{|U_i}$ is trivial, and such that the transition functions g_{ij} on $U_i \cap U_j$ associated to these trivialisations are of the form $\exp(2i\pi f_{ij})$. Then we can change the f_{ij} to $f_{ij} - b_{ij}$ with $b_{ij} \in \mathbb{Z}$, and by definition, $c_1(\mathcal{L})$ is represented by the Čech cocycle with integral coefficients $(a_{ijk}) = \delta(f_{ij})$. The fact that the class of this cocycle is zero shows that if the covering also satisfies the conditions of theorem 4.41, then there exists a cochain (b_{ij}) with integral coefficients such that $(a_{ijk}) = \delta(b_{ij})$. Modifying the f_{ij} to $f_{ij} - b_{ij}$, we obtain the desired result.

We can now write $f_{ij} = f_i - f_j$ on U_{ij}, where the f_i are C^∞ functions on U_i. The forms of type (0, 1) on U_i given by $\alpha_i = -\overline{\partial} f_i$ coincide on U_{ij}. The form α of type (0, 1) thus constructed on X is $\overline{\partial}$-closed, and it is clear by the exponential exact sequence that its class in $H^1(X, \mathcal{O}_X) = H^1(X, \mathbb{C})/F^1H^1(X)$ is $\frac{1}{2i\pi}$ times a lifting of $\alpha_Z \in H^1(X, \mathcal{O}_X)/2i\pi H^1(X, \mathbb{Z})$.

Note that the functions $g_i = \exp(2i\pi f_i)$ satisfy the condition $g_i = g_{ij}g_j$ on U_{ij}, and thus give an everywhere non-zero $\mathcal{C}^\infty$ section τ of the vector bundle L of rank 1 corresponding to $\mathcal{L}$. The quotient $\chi = \frac{\sigma}{\tau}$ is then a singular function along the negative part of the divisor Z (i.e. the union of the hypersurfaces of X appearing in Z with a negative coefficient), and on $X - Z$ we have

$$\alpha = \frac{1}{2i\pi}\overline{\partial}\log\chi,$$

where the function $\log\chi$ is multivalued. Suppose for simplicity that the locus of the poles and the locus of the zeros of σ are disjoint. We can then consider the function χ as a function with values in $\mathbb{P}^1$. For a smooth generic path γ on $\mathbb{P}^1$ such that $\partial\gamma = 0 - \infty$, $\Gamma = \chi^{-1}(\gamma)$ can be seen as a differentiable chain on X of boundary Z, using a triangulation of Γ. Now, by the definition of Φ_X^1, it suffices to show that if η is a closed form of type $(n, n-1)$ on X, then

$$\int_\Gamma \eta = \int_X \alpha\wedge\eta. \qquad (12.4)$$

But the function $\log\chi$ is single-valued on $X - \Gamma$, so we can apply Stokes' theorem to the exact form

$$\alpha\wedge\eta = \frac{1}{2i\pi}\overline{\partial}(\log\chi)\wedge\eta = \frac{1}{2i\pi}d((\log\chi)\eta) \qquad (12.5)$$

on $X - \Gamma_\epsilon$, where $\Gamma_\epsilon = \chi^{-1}(\gamma_\epsilon)$ and γ_ϵ is an arbitrarily small neighbourhood of the segment γ. (In the equality (12.5), we used the fact that η is closed of type $(n, n-1)$.)

We obtain

$$\int_X \alpha\wedge\eta = \lim_{\epsilon\to 0}\int_{X-\Gamma_\epsilon}\alpha\wedge\eta = \lim_{\epsilon\to 0}\int_{\partial\Gamma_\epsilon}\frac{-1}{2i\pi}(\log\chi)\eta.$$

But the boundary $\partial\Gamma_\epsilon$ can be cut into two pieces $\Gamma_\epsilon^+ := \Gamma_\epsilon \cap \{\Im z \geq 0\}$, and $\Gamma_\epsilon^- := \Gamma_\epsilon \cap \{\Im z \leq 0\}$. Each of these pieces tends to Γ (in the sense of currents, or as differentiable chains), but the function $\log\chi$ on Γ_ϵ^+ tends to $\log\chi^+$ on Γ and the function $\log\chi$ on Γ_ϵ^- tends to $\log\chi^-$ on Γ, with $\log\chi^- = \log\chi^+ + 2i\pi$. We thus have

$$\lim_{\epsilon\to 0}\int_{\partial\Gamma_\epsilon}\frac{1}{2i\pi}(\log\chi)\eta = \int_\Gamma \eta,$$

which proves formula (12.4). □

Corollary 12.8 (Abel's theorem) *Let D be a divisor homologous to 0 on X. Then $\Phi^1_X(D) = 0$ if and only if the bundle $\mathcal{O}_X(D)$ is trivial.*

Note that this condition is equivalent to the fact that D is the divisor of a meromorphic function on X. Indeed, by construction, $\mathcal{O}_X(D)$ has a meromorphic section σ_D of divisor D. If $\mathcal{O}_X(D)$ is trivial, then by this trivialisation, we can see σ_D as a meromorphic function on X. Conversely, if ϕ is a meromorphic function of divisor equal to D, then ϕ gives an everywhere non-zero section of the line bundle $\mathcal{O}_X(D)$. This leads to the following definition.

Definition 12.9 *We say that a divisor D of X is rationally equivalent to 0, or is principal, if it is the divisor of a meromorphic function on X.*

By Abel's theorem, the divisors rationally equivalent to 0 are exactly those which are annihilated by the Abel–Jacobi map.

We next consider the case where $k = n = \dim X$. The cycles of codimension n of X are then the combinations $\sum_i n_i p_i$ of points p_i of X. If X is connected, such a cycle is homologous to 0 if and only if it is of degree 0, i.e $\sum_i n_i = 0$.

Applying theorem 12.4 to the diagonal $\Delta \subset X \times X$, and choosing a reference point $x_0 \in X$, we obtain a holomorphic map alb_X, $x \mapsto \Phi^{2n-1}_X(x - x_0)$ from X to the intermediate Jacobian $J^{2n-1}(X)$.

Definition 12.10 *The complex torus* $\mathrm{Alb}\, X := J^{2n-1}(X)$ *is called the Albanese variety of X, and the map* alb_X *is called the Albanese map.*

The Albanese morphism with values in the complex torus $\mathrm{Alb}X$ satisfies the following property.

Lemma 12.11 *The image* $\mathrm{alb}_X(X)$ *generates the torus* $\mathrm{Alb}X$ *as a group. More precisely, for sufficiently large k, the morphism*

$$\mathrm{alb}^k_X : X^k \to \mathrm{Alb}X, \quad (x_1, \ldots, x_k) \mapsto \sum_i \mathrm{alb}_X(x_i),$$

where the right-hand sum is taken using the group structure of the torus $\mathrm{Alb}X$*, is surjective.*

Proof As X^k is compact, it suffices to show that alb^k_X is a submersion at at least one point $x = (x_1, \ldots, x_k)$ of X^k for sufficiently large k. As the cotangent bundle of $\mathrm{Alb}X$ is trivial, this comes down to showing that the composition map

$$\alpha : \left(\mathrm{alb}^k_X\right)^* : H^0(\mathrm{Alb}X, \Omega_{\mathrm{Alb}\, X}) \to H^0(X^k, \Omega_{X^k}) \to \Omega_{X^k,x}$$

is injective. By the definition of $\mathrm{Alb}X = H^0(X, \Omega_X)^*/H_1(X, \mathbb{Z})$, we have

$$H^0(\mathrm{Alb}X, \Omega_{\mathrm{Alb}X}) = H^0(X, \Omega_X),$$

and formula (12.3) shows that the map α can be identified with the restriction map

$$H^0(X, \Omega_X) \to \Omega_{X,x_1} \oplus \cdots \oplus \Omega_{X,x_k}.$$

For sufficiently large k, there obviously exists at least one $x \in X^k$ for which this restriction is injective. □

Corollary 12.12 *If X is a projective variety, then* $\mathrm{Alb}X$ *is an abelian variety.*

Proof If X is projective, $\mathrm{Alb}X$ is a Moishezon variety (i.e. the transcendence degree of its field of meromorphic functions is equal to its dimension). Indeed, we have a surjective morphism $\phi : X^k \to \mathrm{Alb}X$, and X^k is projective. This implies that we can find a subvariety Y of X^k of dimension equal to $\dim \mathrm{Alb}\, X$ and such that $\phi_{|Y}$ remains surjective. The morphism $\phi_{|Y}$ is thus generically finite. For every meromorphic function f on Y, we can define the trace $\mathrm{Tr}\, f$, which is a meromorphic function on $\mathrm{Alb}X$, and this enables us to show that the inclusion of the field of meromorphic functions $\mathbb{C}(\mathrm{Alb}X) \subset \mathbb{C}(Y)$ is an algebraic extension of degree equal to $\deg \phi_{|Y}$.

As $\mathrm{Alb}X$ is Kähler, $\mathrm{Alb}X$ is then projective by theorem 12.13 below. □

Theorem 12.13 (Moishezon 1967a, b) *A Moishezon variety is projective if and only if it is Kähler.*

Remark 12.14 *We can also give a transcendental proof of corollary 12.12, by constructing a polarisation on* $\mathrm{Alb}\, X$ *as in section 7.2.2.*

We have the following characterisation of the Albanese morphism.

Theorem 12.15 *The Albanese morphism satisfies the following universal property:*
(*) *For every holomorphic map $f : X \to T$ with values in a complex torus and satisfying $f(x_0) = 0$, there exists a unique morphism of complex tori*

$$g : \mathrm{Alb}\, X \to T$$

such that $f = g \circ \mathrm{alb}_X$.

Proof The uniqueness follows from lemma 12.11. The existence of g is proved by noting that having a morphism

$$g : \mathrm{Alb}\, X = H^0(X, \Omega_X)^*/H_1(X, \mathbb{Z}) \to T = H^0(T, \Omega_T)^*/H_1(T, \mathbb{Z})$$

is equivalent to having the morphism of groups

$$g_* : H_1(X, \mathbb{Z}) \to H_1(T, \mathbb{Z}),$$

subject to the condition that its $\mathbb{R}$-linear extension to $H^0(\Omega_X)^*$ is $\mathbb{C}$-linear, so as to ensure that the morphism of real tori induced by g_* is holomorphic. It then suffices to set

$$g_* = f_* : H_1(X, \mathbb{Z}) \to H_1(T, \mathbb{Z}).$$

The equality $f = g \circ \mathrm{alb}_X$ comes from the fact that these two morphisms have the same induced pullback map

$$H^0(T, \Omega_T) \to H^0(X, \Omega_X),$$

i.e. the same differential, and that moreover, they coincide at x_0. □

Remark 12.16 *In Serre (1958/1959), we find a purely algebraic construction of the Albanese variety, as a solution of the universal problem (*), formulated uniquely for morphisms with values in abelian varieties.*

12.2 Properties

12.2.1 Correspondences

Let X and Y be two smooth compact Kähler manifolds, with Y connected, and let $Z \subset Y \times X$ be a cycle of codimension k flat (see Hartshorne 1977) over Y. Let $y_0 \in Y$ be a reference point. Theorem 12.4 (or rather, its generalisation to the flat case) gives a holomorphic map

$$\phi : Y \to J^{2k-1}(X), \quad y \mapsto \Phi_X^k(Z_y - Z_{y_0}).$$

Theorem 12.15 then implies the existence of a morphism of complex tori

$$\psi : \mathrm{Alb} Y \to J^{2k-1}(X)$$

such that $\phi = \psi \circ \mathrm{alb}_Y$.

Theorem 12.17 *ψ is the morphism of complex tori $\widetilde{[Z]}$ induced by the morphism of Hodge structures*

$$[Z] : H^{2m-1}(Y, \mathbb{Z}) \to H^{2k-1}(X, \mathbb{Z}),$$

where $m = \dim Y$, given by the Künneth component $[Z]^{1,2k-1} \in H^1(Y,\mathbb{Z}) \otimes H^{2k-1}(X,\mathbb{Z})$ of $[Z]$ (cf. section 11.3.3).

Proof The morphism ψ is determined by the morphism

$$\phi_* : H_1(Y,\mathbb{Z}) \to H_1(J^{2k-1}(X),\mathbb{Z}) = H^{2k-1}(X,\mathbb{Z}),$$

and this is also true for the morphism $\widetilde{[Z]}$, of which the associated morphism $\widetilde{[Z]}_*$ is by definition the morphism of groups given by

$$[Z] : H^{2m-1}(Y,\mathbb{Z}) = H_1(Y,\mathbb{Z}) \to H^{2k-1}(X,\mathbb{Z}).$$

Thus, it suffices to show that $\phi_* = [Z]$. Let $\gamma : [0,1] \to Y$ be a loop in Y based at 0. By definition of the Abel–Jacobi map, we have a continuous lifting $\phi \circ \gamma$ to a map $\widetilde{\phi \circ \gamma}$ with values in the universal cover

$$H^{2k-1}(X,\mathbb{C})/F^k H^{2k-1}(X) = F^{n-k+1}H^{2n-2k+1}(X)^*,\ n = \dim X,$$

of $J^{2k-1}(X)$, given by

$$\widetilde{\phi \circ \gamma}(t)(\eta) = \int_{\Gamma_t} (p_X^*\eta), \quad \forall \eta \in F^{n-k+1}H^{2n-2k+1}(X)$$

where $\Gamma_t = p_Y^{-1}(\gamma([0,t])) \subset Z$.

Now, the image of $\phi_*([\gamma]) \in H_{2n-2k+1}(X,\mathbb{Z})$ in $F^{n-k+1}H^{2n-2k+1}(X)^*$, where $[\gamma]$ is the class of γ in $H_1(Y,\mathbb{Z})$, is by definition equal to

$$\widetilde{\phi \circ \gamma}(1) - \widetilde{\phi \circ \gamma}(0) \in F^{n-k+1}H^{2n-2k+1}(X)^*,$$

which is equal to $\eta \mapsto \int_\Gamma p_X^*\eta$, where $\Gamma = p_Y^{-1}(\gamma)$. We have thus shown that

$$\phi_*([\gamma]) = (p_X)_*([\Gamma]) \in H_{2n-2k+1}(X,\mathbb{Z}),\ \Gamma = p_Y^{-1}(\gamma).$$

The proof of theorem 12.17 is completed by the following lemma. □

Lemma 12.18 *By Poincaré duality, the map*

$$[\gamma] \mapsto \left[p_X\left(p_Y^{-1}(\gamma)\right)\right], \quad H_1(Y,\mathbb{Z}) \to H_{2n-2k+1}(X,\mathbb{Z})$$

can be identified with

$$[Z] : H^{2m-1}(Y,\mathbb{Z}) \to H^{2k-1}(X,\mathbb{Z}).$$

Proof In fact, we only need a version of this lemma modulo torsion. To prove it, it suffices to test the equality on closed differential forms, which is not difficult. □

Corollary 12.19 *The image of ψ is a complex subtorus of $J^{2k-1}(X)$ having the property that its tangent space at 0 is contained in $H^{k-1,k}(X)$.*

Here, we identify the tangent space at 0 of $J^{2k-1}(X)$ with the complex vector space $H^{2k-1}(X,\mathbb{C})/F^kH^{2k-1}(X)$, which naturally contains $H^{k-1,k}(X)$.

Proof The morphism of Hodge structures $[Z]$ is of bidegree $(k-m, k-m)$, and as the Hodge structure on $H^{2m-1}(Y)$ is of type $(m, m-1)+(m-1, m)$, the image of $[Z]$ is thus contained in $H^{k,k-1}(X)\oplus H^{k-1,k}(X)$. □

Remark 12.20 *Corollary 12.19 can also be obtained by means of formula (12.3) for the differential of the map ϕ. Indeed, this shows that*

$$d\phi(T_{Y,y})\subset F^{n-k+1}H^{2n-2k+1}(X)^*$$

annihilates $F^{n-k+2}H^{2n-2k+1}(X)$. Dually, this exactly means that

$$d\phi(T_{Y,y})\subset H^{2k-1}(X,\mathbb{C})/F^kH^{2k-1}(X)$$

is contained in $H^{k-1,k}(X)$.

12.2.2 Some results

If X is a projective manifold equipped with an ample divisor H, we can show that there exists a finite number of projective schemes M_i (see Hartshorne 1977) and flat subschemes $Z_i\subset M_i\times X$ of codimension k over M_i, such that every subscheme of X of pure codimension k and of degree less than or equal to d for some fixed integer d is the fibre of one of these families at a point. (The degree is the integral of a suitable power of the Chern form of H for any metric.) The schemes M_i are called Hilbert schemes. It follows that the group $\mathrm{Griff}^k(X)$ of the cycles of codimension k homologous to 0, modulo the subgroup $\mathcal{Z}^k(X)_{\mathrm{alg}}$ of cycles algebraically equivalent to 0, is countable; here $\mathcal{Z}^k(X)_{\mathrm{alg}}$ is the group generated by the cycles $Z_c - Z_{c'}$ for every flat family

$$\begin{array}{ccc}\mathcal{Z} & \stackrel{p}{\to} & X\\ \downarrow^{q} & & \\ C, & & \end{array}$$

where C is a smooth projective connected curve, c, c' are two points of C, and $Z_c = p(q^{-1}(c))$.

Now let

$$J^{2k-1}(X)_{\rm alg} \subset J^{2k-1}(X)$$

be the largest complex subtorus of $J^{2k-1}(X)$ whose tangent space is contained in $H^{k-1,k}(X)$. Corollary 12.19 shows that $\Phi_X^k(\mathcal{Z}^k(X)_{\rm alg})$ is a subtorus of $J^{2k-1}(X)$ contained in $J^{2k-1}(X)_{\rm alg}$.

We can thus define the transcendental part of the Abel–Jacobi map

$$\Phi_{X,tr}^k : {\rm Griff}^k(X) \to J^{2k-1}(X)_{tr} := J^{2k-1}(X)/J^{2k-1}(X)_{\rm alg}$$

as the factorisation of Φ_X^k. The image of $\Phi_{X,{\rm tr}}^k$ is a countable group.

When $k = 1$, we have shown that the Abel–Jacobi map is surjective. Moreover, up to rational equivalence, a divisor homologous to 0 is parametrised by a point of ${\rm Pic}^0(X)$. Furthermore, two rationally equivalent divisors are clearly algebraically equivalent (i.e. their difference lies in $\mathcal{Z}^k(X)_{\rm alg}$). Indeed, the divisor $D - D'$ of the function σ deforms to the zero divisor by the family

$$Z_t = {\rm div}\,(\alpha\sigma - \beta), \quad t = (\alpha, \beta) \in \mathbb{P}^1.$$

As ${\rm Pic}^0(X)$ is a connected algebraic variety (cf. section 7.2.2), the divisors homologous to 0 are algebraically equivalent to 0.

In codimension $k > 1$, Griffiths showed that this result no longer holds, even up to torsion.

Theorem 12.21 (Griffiths 1969) *Let X be a general quintic hypersurface of $\mathbb{P}^4$. Then X contains a finite number > 1 of lines, and the difference Z of two such lines is a cycle homologous to 0 whose image under the Abel–Jacobi map is not a torsion point. We have $J^3(X)_{\rm alg} = 0$, and thus Z is not a torsion point in* ${\rm Griff}^2(X)$.

Here, "general" means that the polynomial of degree 5 defining X must be chosen outside a countable union of algebraic subsets of $H^0(\mathbb{P}^4, \mathcal{O}_{\mathbb{P}^4}(5))$.

Clemens (1983a) even showed that the group ${\rm Griff}^2(X)$ for X as above is not necessarily finitely generated, although it is countable. This result was extended by the author (2000) to all families of Calabi–Yau manifolds of dimension 3.

As regards the continuous part of Φ_X^k, i.e.

$$\Phi_X^k : \mathcal{Z}^k(X)_{\rm alg} \to J^{2k-1}(X)_{\rm alg},$$

we note that it is conjecturally surjective. Indeed, this would follow from the Hodge conjecture, since when X is projective, $J^{2k-1}(X)_{\rm alg}$ is in fact an abelian variety A by theorem 6.32 (cf. section 7.2.2 for polarisations of tori). We have an isomorphism of Hodge structures

$$\alpha : H^{2N-1}(A,\mathbb{Z}) = H^{2k-1}(X,\mathbb{Z})_{\text{alg}}, \quad N = \dim A,$$

where $H^{2k-1}(X,\mathbb{Z})_{\text{alg}} \subset H^{2k-1}(X,\mathbb{Z})$ is the sub-Hodge structure of type (k,k − 1)+$(k-1,k)$ corresponding to the subtorus $J^{2k-1}(X)_{\text{alg}} \subset J^{2k-1}(X)$. By lemma 11.41, α corresponds to a Hodge class $[\alpha]$ of codimension k on $A \times X$, and the Hodge conjecture predicts the existence of a cycle Z of codimension k and of class $N[\alpha]$, $N \neq 0$ on $A \times X$. By theorem 12.17, the Abel–Jacobi map $A = \text{Alb}\, A \to J^{2k-1}(X)$ induced by Z is equal to N times the initial isomorphism $A = J^{2k-1}(X)_{\text{alg}}$, and thus it is surjective.

We have very few actual results on this consequence of the Hodge conjecture. Let us mention the following result, which generalises that of Clemens & Griffiths (1972).

Theorem 12.22 (Exercise 2) *Let X be a 3-dimensional manifold covered by rational curves (i.e. curves whose normalisation is isomorphic to $\mathbb{P}^1$). Then the Abel–Jacobi map*

$$\Phi_X^2 : \mathcal{Z}^2(X)_{\text{alg}} \to J^3(X)$$

is surjective.

Remark 12.23 *Such a manifold X satisfies $H^{3,0}(X) = 0$, and thus $J^3(X) = J^3(X)_{\text{alg}}$. This follows from the fact that the canonical bundle K_X has no non-zero holomorphic sections on the rational curves covering X. Indeed, for such a generic curve C, the normal bundle of C in X must be generically generated by its global sections. If u, v are two sections generically generating $N_{C/X}$, the interior product by $u \wedge v$ gives an injection $K_{X|C} \subset K_C$. But the canonical bundle of $\mathbb{P}^1$ has no non-zero holomorphic sections.*

12.3 Deligne cohomology

12.3.1 The Deligne complex

Let X be a complex manifold. Let $p \geq 1$ be an integer. The Deligne complex $\mathbb{Z}_D(p)$ is by definition the complex

$$0 \to \mathbb{Z} \stackrel{(2i\pi)^p}{\to} \mathcal{O}_X \stackrel{d}{\to} \Omega_X \to \cdots \to \Omega_X^{p-1} \to 0,$$

where $\mathbb{Z}$ is placed in degree 0, and Ω_X^k in degree $k+1$.

Definition 12.24 *The Deligne cohomology groups $H^k_{\mathrm{D}}(X, \mathbb{Z}(p))$ are the hypercohomology groups $\mathbb{H}^k(X, \mathbb{Z}_{\mathrm{D}}(p))$.*

Example 12.25 *For $p = 1$, $\mathbb{Z}_D(1)$ is quasi-isomorphic to the sheaf $\mathcal{O}_X^*$, placed in degree 1 by the exponential exact sequence. Thus, $H^2_{\mathrm{D}}(X, \mathbb{Z}(1)) = H^1(X, \mathcal{O}_X^*)$.*

Proposition 12.26 *If X is a compact Kähler manifold, we have a long exact sequence*

$$\cdots \to H^k_{\mathrm{D}}(X, \mathbb{Z}(p)) \to H^k(X, \mathbb{Z})$$
$$\to H^k(X, \mathbb{C})/F^p H^k(X, \mathbb{C}) \to H^{k+1}_{\mathrm{D}}(X, \mathbb{Z}(p)) \to \cdots,$$

where $F^p H^k(X, \mathbb{C})$ is the Hodge filtration on $H^k(X, \mathbb{C})$.

Proof We have the exact sequence of complexes

$$0 \to \Omega_X^{\leq p-1}[1] \to \mathbb{Z}_{\mathrm{D}}(p) \to \mathbb{Z} \to 0,$$

where in the left-hand complex, the notation [1] means that Ω_X^k is placed in degree $k+1$. This exact sequence induces a long exact sequence of hypercohomology

$$\cdots H^k_{\mathrm{D}}(X, \mathbb{Z}(p)) \to H^k(X, \mathbb{Z}) \to \mathbb{H}^k\left(X, \Omega_X^{\leq p-1}\right) \to H^{k+1}_{\mathrm{D}}(X, \mathbb{Z}(p)) \cdots,$$

and thus it suffices to show that

$$\mathbb{H}^k\left(X, \Omega_X^{\leq p-1}\right) = H^k(X, \mathbb{C})/F^p H^k(X, \mathbb{C}).$$

But this follows from proposition 7.5, which says exactly that the hypercohomology of the truncated de Rham complex $\mathbb{H}^k(X, \Omega_X^{\geq p})$ is isomorphic to $F^p H^k(X, \mathbb{C})$, and from the short exact sequence of complexes

$$0 \to \Omega_X^{\geq p} \to \Omega_X^{\cdot} \to \Omega_X^{\leq p-1} \to 0.$$

□

In particular, for $k = 2p$ we obtain the following result.

Corollary 12.27 *The Deligne cohomology group $H^{2p}(X, \mathbb{Z}_{\mathrm{D}}(p))$ is an extension of the group $\mathrm{Hdg}^{2p}(X, \mathbb{Z})$ consisting of the integral classes whose image in $H^{2p}(X, \mathbb{C})$ is of type (p, p) by the intermediate Jacobian $J^{2p-1}(X)$:*

$$0 \to J^{2p-1}(X) \to H^{2p}_{\mathrm{D}}(X, \mathbb{Z}(p)) \to \mathrm{Hdg}^{2p}(X, \mathbb{Z}) \to 0. \qquad (12.6)$$

Proof The group $\mathrm{Hdg}^{2p}(X, \mathbb{Z})$ is equal to

$$\mathrm{Ker}\,(H^{2p}(X, \mathbb{Z}) \to H^{2p}(X, \mathbb{C})/F^p H^{2p}(X)).$$

□

12.3.2 Differential characters

Let X be a differentiable manifold. Let $C_l^{\mathrm{diff}}(X)$ be the group of singular differentiable chains of dimension l, and let Z_l^{diff} be the subgroup of closed chains for the differential ∂ (cf. section 4.3.2). The group of differential characters $\Xi_{\mathrm{diff}}(X)$ was introduced by Cheeger & Simons (1985):

Definition 12.28 $\Xi^l_{\mathrm{diff}}(X)$ *is the subgroup of* $\mathrm{Hom}\,(Z_l^{diff}, \mathbb{R}/\mathbb{Z})$ *consisting of the* $\chi : Z_l^{\mathrm{diff}} \to \mathbb{R}/\mathbb{Z}$ *such that there exists a real differential form (obviously uniquely determined by* χ*)* $\omega \in A^{l+1}(X)$ *satisfying*

$$\chi(\partial\phi) = \int_{\Delta_{l+1}} \phi^*\omega \bmod \mathbb{Z}, \quad \forall \phi \in C^{\mathrm{diff}}_{l+1}(X). \tag{12.7}$$

For every differentiable map $\phi : \Delta_{l+2} \to X$, the form $d\omega$ then satisfies

$$\int_{\Delta_{l+2}} \phi^*(d\omega) = \int_{\partial\Delta_{l+2}} \phi^*\omega \bmod \mathbb{Z} = \chi(\partial \circ \partial(\phi)) = 0 \text{ in } \mathbb{R}/\mathbb{Z}.$$

Thus, for every differentiable map $\phi : \Delta_{l+2} \to X$, we have $\int_{\Delta_{l+2}} \phi^*(d\omega) \in \mathbb{Z}$, and this obviously implies that $d\omega = 0$.

Furthermore, by equation (12.7), we have $\int_{\Delta_{l+1}} \phi^*\omega = 0$ in $\mathbb{R}/\mathbb{Z}$ when $\partial\phi = 0$. Thus, the periods (i.e. the integrals over the integral homology classes) of ω are integral, and the de Rham class $[\omega]$ of ω lies in the image of the map $\alpha : H^{l+1}(X, \mathbb{Z}) \to H^{l+1}(X, \mathbb{R})$.

Suppose that X is a complex manifold, and let $\mu \in A^{l-1}(X)$ be a real form. Then for every closed chain $\phi = \sum_i n_i\phi_i$, $\phi_i : \Delta_l \to X$, $\int_{\Delta_l} \phi^*(i\partial\mu) := \sum_i n_i \int_{\Delta_l} \phi_i^*(i\partial\mu) \in \mathbb{R}$, since by Stokes' formula we have

$$\overline{\int_{\Delta_l} \phi^*(i\partial\mu)} = \int_{\Delta_l} \phi^*(-i\bar\partial\mu) = \int_{\Delta_l} \phi^*(-id\mu + i\partial\mu) = \int_{\Delta_l} \phi^*(i\partial\mu).$$

Thus, we can associate to μ the differential character $\phi \mapsto \int_{\Delta_l} \phi^*(i\partial\mu) \bmod \mathbb{Z}$, which we write $\int i\partial\mu$. We then have the following result.

Proposition 12.29 *Assume that X is compact Kähler. Consider the subgroup*

$$\Xi^{2p-1}_{\mathrm{diff}}(X)^{p,p} \subset \Xi^{2p-1}_{\mathrm{diff}}(X)$$

consisting of the differential characters whose associated form ω is of type (p, p). Then $H_D^{2p}(X, \mathbb{Z}(p))$ is naturally isomorphic to the quotient $K_{\text{diff}}^{2p-1}(X)$ of the group $\Xi_{\text{diff}}^{2p-1}(X)^{p,p}$ by the subgroup generated by the $\int i\partial\mu$, $\mu \in A_{\mathbb{R}}^{p-1,p-1}(X)$.

Proof Note that the Deligne complex is quasi-isomorphic to the cone (cf. section 8.1.2) of the morphism of complexes of sheaves over X:

$$\mathbb{Z} \oplus \Omega_X^{\geq p} \stackrel{(\alpha_1,\alpha_2)}{\rightarrow} \Omega_X^{\cdot},$$

where α_1 is equal to $(2i\pi)^p$ times the natural inclusion of $\mathbb{Z}$ into $\mathcal{O}_X$, and α_2 is the natural inclusion up to sign.

We will drop the coefficient $(2i\pi)^p$ from now on, since the complex obtained by taking the natural inclusion of $\mathbb{Z}$ in $\Omega_X^{\cdot}$ instead of of α_1 is obviously isomorphic to the Deligne complex. The sheaf $\mathcal{C}_{\text{sing}}^{\cdot}(X, \mathbb{Z})$ associated to the presheaf $U \mapsto C_{\text{sing}}^{\cdot}(U, \mathbb{Z})$ of differentiable cochains (cf. proof of theorem 4.47) is a Γ-acyclic resolution of $\mathbb{Z}$, and similarly, the sheaf $\mathcal{C}_{\text{sing}}^{\cdot}(X, \mathbb{C})$ is a Γ-acyclic resolution of $\mathbb{C}$ and is thus quasi-isomorphic to $\Omega_X^{\cdot}$. Moreover, the complex $F^p\mathcal{A}^{\cdot}$ of $\mathcal{C}^\infty$ forms of type $(p, \cdot - p) + \cdots + (\cdot, 0)$ is a Γ-acyclic complex of sheaves quasi-isomorphic to $F^p\Omega_X^{\cdot}$ by lemma 8.5. Thus, we obtain a Γ-acyclic complex of sheaves quasi-isomorphic to $\mathbb{Z}_D(p)$ by taking the cone of the morphism

$$\mathcal{C}_{\text{sing}}^{\cdot}(X, \mathbb{Z}) \oplus F^p\mathcal{A}^{\cdot} \stackrel{(\alpha_1,\alpha_2)}{\rightarrow} \mathcal{C}_{\text{sing}}^{\cdot}(X, \mathbb{C}),$$

where α_2 is, up to sign, the map given by integrating forms over the singular differentiable chains.

By proposition 8.12, the Deligne cohomology is thus equal to the cohomology of the cone of the map (α_1, α_2) induced on the level of the global sections:

$$\Gamma(\mathcal{C}_{\text{sing}}^{\cdot}(X, \mathbb{Z})) \oplus F^pA^{\cdot}(X) \stackrel{(\alpha_1,\alpha_2)}{\rightarrow} \Gamma(\mathcal{C}_{sing}^{\cdot}(X, \mathbb{C})).$$

Finally, we noted in the proof of theorem 4.47 that the complex of global sections of $\mathcal{C}_{\text{sing}}^{\cdot}$ is quasi-isomorphic to the complex of singular cochains, and this still holds if we restrict ourselves to the complex of differentiable singular cochains. In conclusion, we have an isomorphism

$$H_D^k(X, \mathbb{Z}(p)) = H^k(C^{\cdot}(\alpha_1, \alpha_2)),$$

where $C^{\cdot}(\alpha_1, \alpha_2)$ denotes the cone of the morphism of complexes (α_1, α_2), and (α_1, α_2) is the morphism of complexes given by

$$C_{\text{sing,diff}}^{\cdot}(X, \mathbb{Z}) \oplus F^pA^{\cdot}(X) \rightarrow C_{\text{sing,diff}}^{\cdot}(X, \mathbb{C}),$$

where α_1 is the natural inclusion, and α_2 is up to sign the morphism given by integrating forms over differentiable chains. By definition, an element of $C^k(\alpha_1, \alpha_2)$ is thus a triple $(a^k_{\mathbb{Z}}, b^k_F, c^{k-1}_{\mathbb{C}})$, where $a^k_{\mathbb{Z}}$ is an integral singular cochain of degree k, b^k_F is a form in $F^pA^k(X)$ and $c^{k-1}_{\mathbb{C}}$ is a complex singular cochain of degree $k-1$. Moreover, the differential of the cone is given by

$$d\big(a^k_{\mathbb{Z}}, b^k_F, c^{k-1}_{\mathbb{C}}\big) = \big(da^k_{\mathbb{Z}}, db^k_F, dc^{k-1}_{\mathbb{C}} + a^k_{\mathbb{Z}} - b^k_F\big). \tag{12.8}$$

Thus, a Deligne cohomology class $\eta \in H^k_D(X, \mathbb{Z}(p))$ is represented by a triple $(a^k_{\mathbb{Z}}, b^k_F, c^{k-1}_{\mathbb{C}})$ satisfying

$$da^k_{\mathbb{Z}} = 0, \quad db^k_F = 0, \quad dc^{k-1}_{\mathbb{C}} = -a^k_{\mathbb{Z}} + b^k_F. \tag{12.9}$$

Moreover, η is defined modulo the triples satisfying

$$a^k_{\mathbb{Z}} = da^{k-1}_{\mathbb{Z}}, \quad b^k_F = db^{k-1}_F, \tag{12.10}$$

$$c^{k-1}_{\mathbb{C}} = dc^{k-2}_{\mathbb{C}} + a^{k-1}_{\mathbb{Z}} - b^{k-1}_F. \tag{12.11}$$

The conditions (12.9) show that $a^k_{\mathbb{Z}}$ and b^k_F are closed and that their classes in $H^k(X, \mathbb{Z})$ and $F^pH^k(X, \mathbb{C})$ have equal images under the natural maps

$$H^k(X, \mathbb{Z}) \to H^k(X, \mathbb{C}), \quad F^pH^k(X) \to H^k(X, \mathbb{C}).$$

The conditions (12.10) show that these classes do not depend on the choice of the representative η.

Assume now that $k = 2p$. Then the form b^{2p}_F is of real cohomology class, and lies in $F^pH^{2p}(X)$. The uniqueness of the Hodge decomposition and the Hodge symmetry thus shows that it is representable by a real closed form of type (p, p). By the relations (12.10), we can thus represent η by a triple such that the form b^{2p}_F is real and closed of type (p, p). Let us consider the cochain $c^{2p-1}_{\mathbb{C}}$. By condition (12.9), its imaginary part is closed of degree $2p-1$, and thus admits a class in $H^{2p-1}(X, \mathbb{R})$. But by the condition (12.11) applied with $a^{2p-1}_{\mathbb{Z}} = 0$, $db^{2p-1}_F = 0$, this class is defined modulo the image of the map

$$\Im : F^pH^{2p-1}(X, \mathbb{C}) \to H^{2p-1}(X, \mathbb{R}),$$

which to a given class associates its imaginary part. Now, as X is Kähler, this map is surjective. We may thus assume that η is represented by a triple $(a^{2p}_{\mathbb{Z}}, b^{2p}_F, c^{2p-1}_{\mathbb{C}})$ such that the form b^{2p}_F is real of type (p, p) and the class of the imaginary part of $c^{2p-1}_{\mathbb{C}}$ is zero. Applying conditions (12.11) with $a^{2p-1}_{\mathbb{Z}} = 0$, $b^{2p-1}_F = 0$, we finally see that we can represent η by a triple $(a^{2p}_{\mathbb{Z}}, b^{2p}_F, c^{2p-1}_{\mathbb{C}})$ satisfying equation (12.9) and such that:

(i) The form b^{2p}_F is real of type (p, p).
(ii) The cochain $c^{2p-1}_{\mathbb{C}}$ is real.

Let us then associate to such an η the differential character of degree $2p-1$ given by

$$\chi = c^{2p-1} \bmod \mathbb{Z},$$

where c^{2p-1} is the restriction of $c_{\mathbb{C}}^{2p-1}$ to closed chains.

Obviously, by Stokes' formula, this differential character has associated form ω given by b_F^{2p}, which is of type (p,p). To conclude the construction of the desired morphism

$$\alpha : H_{\mathrm{D}}^{2p}(X, \mathbb{Z}(p)) = \Xi_{\mathrm{diff}}^{2p-1}(X)^{p,p} \Big/ \left\{ \int i\partial\mu \mid \mu \in A_{\mathbb{R}}^{p-1,p-1}(X) \right\} = K_{\mathrm{diff}}^{2p-1}(X),$$

it remains only to see that the triples $(a_{\mathbb{Z}}^{2p}, b_F^{2p}, c_{\mathbb{C}}^{2p-1})$ which are exact and satisfy conditions (i) and (ii) are sent to differential characters of the form $\int i\partial\mu$, $\mu \in A^{p-1,p-1}(X) \bmod \mathbb{Z}$. But let

$$a_{\mathbb{Z}}^{2p} = da_{\mathbb{Z}}^{2p-1}, \quad b_F^{2p} = db_F^{2p-1},$$

$$c_{\mathbb{C}}^{2p-1} = dc_{\mathbb{C}}^{2p-2} + a_{\mathbb{Z}}^{2p-1} - b_F^{2p-1}$$

be such a coboundary. The associated character

$$\chi = c^{2p-1} = \int c_{\mathbb{C}}^{2p-1} \bmod \mathbb{Z}$$

does not depend on $c_{\mathbb{C}}^{2p-2}$ or $a_{\mathbb{Z}}^{2p-1}$ by Stokes' formula, and because we are only considering cochains modulo $\mathbb{Z}$. Thus, we have

$$\chi = -\int b_F^{2p-1} \bmod \mathbb{Z},$$

with the condition that $b_F^{2p-1} \in F^p A^{2p-1}(X)$ and $dc_{\mathbb{C}}^{2p-2} - b_F^{2p-1}$ is a real cochain. It then suffices to show the following result.

Lemma 12.30 *Let X be a complex manifold, and let $\phi \in F^p A^{2p-1}(X)$ be such that $\int \phi$ is the sum of a real cochain and a coboundary. Then up to an exact form, ϕ is equal to a form $i\partial\psi$, with ψ real of type $(p-1, p-1)$.*

Temporarily admitting this lemma, we have thus constructed the map α, and it remains to see that it is an isomorphism.

We only prove the injectivity here. Let $\eta \in H_{\mathrm{D}}^{2p}(X, \mathbb{Z}(p))$ be such that $\alpha(\eta) = 0$ in $K_{\mathrm{diff}}^{2p-1}(X)$. We can represent η by a triple $(a_{\mathbb{Z}}^{2p}, b_F^{2p}, c_{\mathbb{C}}^{2p-1})$ such that:

(i) The form b_F^{2p} is real of type (p,p).

(ii) The cochain $c_{\mathbb{C}}^{2p-1}$ is real.
(iii) $da_{\mathbb{Z}}^{2p}=0$, $db_F^{2p}=0$, $dc_{\mathbb{C}}^{2p-1}=b_F^{2p}-a_{\mathbb{Z}}^{2p}$.
The condition $\alpha(\eta)=0$ then says that the corresponding differential character $c_{\mathbb{C}}^{2p-1}$ mod $\mathbb{Z}$ is equal to $i\int\partial\mu$ mod $\mathbb{Z}$. In other words, there exist cochains

$$a_{\mathbb{Z}}^{2p-1}\in C^{2p-1}(X,\mathbb{Z})_{\mathrm{sing}},\quad c_{\mathbb{C}}^{2p-2}\in C^{2p-2}(X,\mathbb{C})_{\mathrm{sing}}$$

and a form $i\partial\mu\in A^{p,p-1}(X)$ such that

$$c_{\mathbb{C}}^{2p-1}=dc_{\mathbb{C}}^{2p-2}-i\partial\mu+a_{\mathbb{Z}}^{2p-1}.$$

One can then modify the initial triple to

$$\left(a'^{2p}_{\mathbb{Z}},b'^{2p}_{F},c'^{2p-1}_{\mathbb{C}}\right)=\left(a_{\mathbb{Z}}^{2p},b_F^{2p},c_{\mathbb{C}}^{2p-1}\right)-d\left(a_{\mathbb{Z}}^{2p-1},i\partial\mu,c_{\mathbb{C}}^{2p-2}\right),$$

which also represents η, and satisfies the condition $c'^{2p-1}_{\mathbb{C}}=0$. As we have

$$d\left(a'^{2p}_{\mathbb{Z}},b'^{2p}_{F},c'^{2p-1}_{\mathbb{C}}\right)=0,\quad c'^{2p-1}_{\mathbb{C}}=0,$$

we thus obtain by (12.8)

$$a'^{2p}_{\mathbb{Z}}=b'^{2p}_{F}\ \text{in}\ C^{2p}_{\mathrm{sing}}(X),$$

and this immediately implies that $a'^{2p}_{\mathbb{Z}}=b'^{2p}_{F}=0$. □

Proof of lemma 12.30. The hypothesis implies that the imaginary part of ϕ is an exact form. Let us write $\phi-\overline{\phi}=d(i\beta)$, with $\beta\in A^{2p-2}(X)$. We can of course assume β real. Decompose $\beta=\beta_1+\beta_2+\overline{\beta_1}$, with $\beta_1\in F^pA^{2p-2}(X)$, $\beta_2\in A^{p-1,p-1}(X)$, β_2 real. As $\phi\in F^pA^{2p-1}(X)$, we obtain

$$\phi=d(i\beta_1)+i\partial\beta_2.$$

□

12.3.3 Cycle class

Let us now explain how to associate to a smooth cycle Z of codimension p in X a Deligne class $[Z]_{\mathrm{D}}$ which lifts the cohomology class $[Z]\in H^{2p}(X,\mathbb{Z})$ by the exact sequence (12.6). Furthermore, for $Z=\sum_i n_iZ_i$ homologous to 0, we will have

$$[Z]_{\mathrm{D}}:=\sum_i n_i[Z_i]_{\mathrm{D}}=\Phi_X^p(Z)\in J^{2p-1}(X).$$

In fact, the smoothness hypothesis is not necessary. Assuming X algebraic, we can show that the class constructed in this way depends only on the rational equivalence class of Z (see Fulton 1984), and satisfies the compatibility with the product (see Fulton (1984) for intersection theory on cycles modulo rational equivalence and Esnault & Viehweg (1988) for the product on the Deligne cohomology).

In fact, we will construct a class $[Z]_D$ in $K_{\text{diff}}^{2p-1}(X)$, and use proposition 12.29 to view it as an element of $H_D^{2p}(X, \mathbb{Z}(p))$. For this, we need the following result, whose proof can be found in Soulé (1992, th. 3, p. 44).

Theorem 12.31 *Let $Z \subset X$ be a closed smooth algebraic subvariety of codimension p of a complex algebraic variety. Then there exists a real form ψ of type $(p-1, p-1)$ on $X - Z$ satisfying the following conditions:*
(i) ψ is integrable.
(ii) We have the equality of currents

$$i\partial\overline{\partial}\psi = Z - \omega,$$

where Z is the current of integration on Z and ω is a C^∞ real closed form of type (p, p) representating the cohomology class of Z. In particular, $\omega = -id\partial\psi$ on $X - Z$.
(iii) For Γ a differentiable submanifold of X defined in the neighbourhood of a point z of Z and meeting Z transversally at z, equipped with the complex orientation ($T_{\Gamma,z} = N_{Z/X,z}$), we have

$$\lim_{\epsilon\to 0}\int_{S_\epsilon^{2p-1}} i\partial\psi = 1,$$

where $S_\epsilon^{2p-1} \subset \Gamma$ is a ball of radius ϵ centred at z (for suitable coordinates on Γ).

Recall here that currents are linear forms on the space of the differential forms with compact support. The differential of a current T is defined by $\partial T(\eta) = T(d\eta)$. An integrable differential form ψ (i.e. one whose coefficients are locally integrable functions) gives the current $\eta \mapsto \int_X \psi \wedge \eta$.

We use the singular form ψ given by theorem 12.31 to define the class $[Z]_D \in K_{\text{diff}}^{2p-1}(X)$ of the differential character $\chi_{Z,\psi}$, which is itself defined as follows: for a closed differentiable chain γ of dimension $2p-1$, there exists a closed differentiable chain γ' of dimension $2p-1$ which does not meet Z, and a differentiable chain Γ of dimension $2p$, such that $\gamma = \gamma' + \partial\Gamma$. (This is an elementary result. For instance, it suffices to deform γ using a generic vector field on X defined in the neighbourhood of γ.)

We then set

$$\chi_{Z,\psi}(\gamma) = \int_{\gamma'} i\partial\psi + \int_{\Gamma} \omega \bmod \mathbb{Z}.$$

Let us first show that $\chi_{Z,\psi}$ is well-defined. Let γ'' and Γ' be differentiable chains as above. We have

$$\gamma' - \gamma'' = \partial(\Gamma' - \Gamma).$$

The integral $\int_{\partial\phi} \omega$ is zero for an exact chain $\partial\phi$, since ω is closed. As the boundary of $\Gamma' - \Gamma$ does not meet Z, we can write

$$\Gamma' - \Gamma = \Gamma'' + \partial\phi,$$

where Γ'' meets Z transversally. (Again, this is an elementary approximation result.) Let us now apply Stokes' theorem to the form ω on Γ''_ϵ, where the index ϵ means that we have removed a ball of radius ϵ in the neighbourhood of each of the points of intersection with Z. Thanks to property (ii), we obtain

$$\int_{\Gamma''} \omega = \lim_{\epsilon \to 0} \int_{\Gamma''_\epsilon} \omega = \lim_{\epsilon \to 0} \sum_z \varepsilon(z) \int_{S_\epsilon(z)} i\partial\psi + i \int_{\gamma' - \gamma''} \partial\psi,$$

where z runs through the set of the intersection points of Γ and Γ' with Z, and $\varepsilon(z)$ is a sign which depends on the compatibility of the orientation of Γ (or Γ') at z with the complex orientation. Finally, by property (iii), we find that each contribution $\lim_{\epsilon \to 0} \varepsilon(z) \int_{S_\epsilon(z)} i\partial\psi$ is equal to ± 1, so

$$\int_{\Gamma' - \Gamma} \omega = \int_{\Gamma''} \omega = i \int_{\gamma' - \gamma''} \partial\psi \bmod \mathbb{Z},$$

which shows that

$$\int_{\gamma'} i\partial\psi + \int_{\Gamma} \omega = \int_{\gamma''} i\partial\psi + \int_{\Gamma'} \omega \bmod \mathbb{Z}.$$

It remains to see how the differential character $\chi_{Z,\psi}$ depends on ψ or on ω. For fixed ω, let ψ' be another form satisfying conditions (i) and (ii) above. Then we have the equality of currents

$$\overline{\partial}\partial(\psi' - \psi) = 0.$$

The current $\partial(\psi' - \psi)$ on X is thus both ∂-closed and $\overline{\partial}$ exact. Proposition 6.17, which remains valid for currents, then shows that the current $\partial(\psi' - \psi)$ is exact, and thus that the form $\partial(\psi' - \psi)$ is exact on $X - Z$. Thus, by Stokes' formula, if ω is fixed, $\chi_{Z,\psi}$ does not depend on ψ. Finally, let ω' be another real closed representative of type (p, p) of the class of Z. Proposition 6.17 shows that $\omega-\omega' = i\overline{\partial}\partial\mu$, where μ is a real form of type $(p-1, p-1)$. Thus, if ψ is a current satisfying conditions (i)–(iii) of theorem 12.31 for ω, then $\psi' = \psi + \mu$ is a current satisfying conditions (i),–(iii) of lemma 12.31 for ω'. We therefore find that

$$\chi_{Z,\psi'} = \chi_{Z,\psi} + \int i\partial\mu,$$

and $\chi_{Z,\psi} = \chi_{Z,\psi'}$ in $K_{\mathrm{diff}}^{2p-1}(X)$.

Exercises

1. *Abel–Jacobi map and blowup.* Let X be a compact Kähler manifold of dimension 3 and let $C \subset X$ be a smooth curve. We denote by $\tau : \tilde{X} \to X$ the blowup of X along C (cf. section 3.3.3). Let $j : E \hookrightarrow \tilde{X}$ be the exceptional divisor of the blowup τ. We denote by τ_E the restriction of τ to E. The points c of C parametrize the curves $E_c := j(\tau_E^{-1}(c))$ (which are isomorphic to $\mathbb{P}^1$) of $\tilde{X}$.
   (a) Using theorem 7.31 and theorem 12.17, show that the induced Abel–Jacobi map
   $$J(C) \to J^3(\tilde{X})$$
   identifies $J(C)$ with the direct factor of $J^3(X)$ provided by the sub-Hodge structure
   $$j_*\tau_E^*(H^1(C)) \subset H^3(\tilde{X}).$$
   (b) Deduce from this that if the Abel–Jacobi map
   $$\Phi_X : \mathcal{Z}^2_{\mathrm{hom}}(X) \to J^3(X)$$
   is surjective, then the Abel–Jacobi map of $\tilde{X}$ is also surjective.
2. *The Abel–Jacobi map for uniruled threefolds.* Let S be a smooth projective surface and E be a vector bundle of rank 2 on S. Let $X = \mathbb{P}(E)$, which is a projective variety of dimension 3 admitting a morphism $p : X \to S$ and equiped with a holomorphic line bundle $H = \mathcal{O}_X(1)$.
   (a) Show that $J^3(X)$ is isomorphic to $\mathrm{Alb}\, S \oplus Pic^0(S)$ by the map induced by the morphism of Hodge structures
   $$p^* + c_1(H) \circ p^* : H^3(S) + H^1(S) \to H^3(X).$$

(b) Deduce from theorem 12.17 and proposition 12.7 that the Abel–Jacobi map

$$\Phi_X : \mathcal{Z}^2_{\mathrm{hom}}(X) \to J^3(X)$$

is surjective.

(c) Let $\phi : Y \to X$ be a surjective morphism between two compact Kähler threefolds. Deduce from lemma 7.28 and theorem 12.17 that if the Abel–Jacobi map of Y is surjective, that of X is also surjective.

Hence we have shown the following result:

Let X be a Kähler threefold which is uniruled, that is, which is covered by rational curves. (This is equivalent to the fact that X is dominated by a threefold Y which is obtained by blowing up curves in a projective bundle over a surface.) Then the Abel–Jacobi map

$$\Phi_X : \mathcal{Z}^2_{\mathrm{hom}}(X) \to J^3(X)$$

is surjective.

Bibliography

A. Andreotti, A. Mayer (1967). On period relations for abelian integrals on algebraic curves, *Ann. Scuola Norm. Sup. Pisa* **21**, 189–238.

E. Arbarello, M. Cornalba, P. Griffiths, J. Harris (1985). *Geometry of Algebraic Curves* vol. I, Grundlehren der Math. Wiss. **267**, Springer.

V. Arnold (1984). *Ordinary Differential Equations*, third edition, MIT Press.

M. Atiyah, F. Hirzebruch (1962). Analytic cycles on complex manifolds, *Topology* **1**, 25–45.

W. Barth, C. Peters, A. Van de Ven (1984). *Compact Complex Surfaces*, Ergebnisse der Mathematik und ihrer Grenzgebiete 3. Folge, Band 4, Springer-Verlag.

A. Beauville (1978). *Surfaces Algébriques Complexes, Astérisque* **54**, Société Mathématique de France. English translation *Complex Algebraic Surfaces*, Cambridge University Press, (1983); second edition (1996).

A. Beauville, J.-P. Bourguignon (ed.) (1985) Géométrie des surfaces $K3$: modules et périodes, Séminaire Palaiseau, *Astérisque* **126**, Société mathématique de France.

S. Bloch (1972). Semi-regularity and de Rham cohomology, *Inventiones Math.* **17**, 51–66.

S. Bloch, A. Ogus (1974) Gersten's conjecture and the homology of schemes, *Ann. Scient. Éc. Norm. Sup.*, 4ème série, **7**, 181–202.

A. Borel, A. Haefliger (1961). La classe d'homologie fondamentale d'un espace analytique, *Bull. Soc. Math. France* **89**, 461–513.

A. Borel, J.-P (1959). Serre. Le théorème de Riemann-Roch (d'après Grothendieck), *Bull. Soc. Math. France* **86**, 97–136.

R. Bott, L. Tu (1982). *Differential Forms in Algebraic Topology*, Graduate Texts in Math. **82**, Springer.

J. Carlson, M. Green, P. Griffiths, J. Harris (1983). Infinitesimal variations of Hodge structure (I), *Compositio Mathematica* **50** 109–205.

J. Cheeger, J. Simons (1985). *Differential Characters and Geometric Invariants*, Lecture Notes in Math. **1167**, Springer 50–80.

H. Clemens (1983a). Homological equivalence, modulo algebraic equivalence, is not finitely generated, *Publ. Math. IHES* **58**, 19–38.

H. Clemens (1983b). Double solids, *Advances in Math.* **47**, 107–230.

H. Clemens, P. Griffiths (1972). The intermediate Jacobian of a cubic threefold, *Annals of Math.* **95**, 281–356.

P. Deligne (1971). Théorie de Hodge II, *Publ. Math. IHES* **40** 5–57.

P. Deligne (1975). Théorie de Hodge III, *Publ. Math. IHES* **44**, 5–77.
P. Deligne, L. Illusie (1987). Relèvements modulo p^2 et décomposition du complexe de de Rham, *Inv. Math.* **89**, 247–270.
P. Deligne, D. Mumford (1969). The irreducibility of the space of curves of a given genus, *Publ. Math. IHES* **36**, 75–109.
J.- P. Demailly (1996). Théorie de Hodge L^2 et théorèmes d'annulation. In *Introduction à la Théorie de Hodge*, Panoramas et synthèses **3**, Société mathématique de France, 3–111.
A. Dieudonné (1982). *Éléments d'Analyse 9*, Cahiers scientifiques XLII, Gauthier-Villars.
F. El Zein, S. Zucker (1984). Extendability of normal functions associated to algebraic cycles. In *Topics in Transcendental Algebraic Geometry* (ed. P. Griffiths) *Ann. Math. Studies* **106**, 269–288.
H. Esnault (1988). Characteristic classes of flat bundles, *Topology* **27** (**3**), 323–352.
H. Esnault, E. Viehweg (1988). Deligne-Beilinson cohomology. In *Beilinson's Conjectures on Special Values of L-functions*, (ed. Rapoport, Schappacher and Schneider), Perspect. Math. **4**, Academic Press 43–91.
H. Esnault, E. Viehweg (1992). *Lectures on Vanishing Theorems*, DMV Seminar band **20**, Birkhäuser.
R. Friedman (1991). On threefolds with trivial canonical bundle. In *Complex Geometry and Lie Theory*, Proc. Symp. Pure. Math. **53**, 103–134.
W. Fulton (1984). *Intersection Theory*, Ergebnisse der Math. und ihrer grenzgebiete 3. Folge, Band **2**, Springer.
R. Godement (1958). *Topologie algébrique et Théorie des Faisceaux*, Hermann, Paris.
A. Gramain (1971). *Topologie des Surfaces*, Collection SUP, Presses Universitaires de France.
H. Grauert, K. Fritzsche (1976). *Several Complex Variables*, Graduate texts in Mathematics **38**, Springer-Verlag.
M. Green (1989). Griffiths' infinitesimal invariant and the Abel-Jacobi map, *J. Diff. Geom.* **29**, 545–555.
M. Green, J. P. Murre, C. Voisin (1993). *Algebraic Cycles and Hodge Theory*, CIME course, Lecture Notes in Math. **1594**, Springer.
P. Griffiths (1968). Periods of integrals on algebraic manifolds, I, II, *Amer. J. Math.* **90**, 568–626, 805–865.
P. Griffiths (1969). On the periods of certain rational integrals I, II, *Ann. of Math.* **90** 460–541.
P. Griffiths, J. Harris (1978). *Principles of Algebraic Geometry*, Wiley, New York.
A. Grothendieck (1966). On the de Rham cohomology of algebraic varieties, *Publ. Math. IHES* **29**, 95–103.
A. Grothendieck (1969). Hodge's general conjecture is false for trivial reasons, *Topology* **8**, 299–303.
B. Harris (1989). Differential characters and the Abel-Jacobi map. In *Algebraic K-theory: Connexions with Geometry and Topology*, edited by J. F. Jardine and V. P. Snaith, 69–86.
R. Hartshorne (1970). *Ample Subvarieties of Algebraic Varieties*, Lecture Notes in Math. **156**, Springer.
R. Hartshorne (1975). Equivalence relations on on algebraic cycles and subvarieties of small codimension, in *Algebraic Geometry*, Proc. Symp. Pure Math. **29** 129–164.
R. Hartshorne (1977). *Algebraic Geometry*, Graduate Texts in Math. **52**, Springer.
H. Hironaka (1964). Resolution of singularities of an algebraic variety over a field of characteristic zero I, II, *Ann. Math.* **79** 109–326.

W. Hodge (second edition 1952). *The Theory and Applications of Harmonic Integrals*, Cambridge University Press, Cambridge.

L. Hörmander (second edition 1979). *An Introduction to Complex Analysis in Several Variables*, North-Holland.

L. Illusie (1996). Frobenius et dégénérescence de Hodge. In *Introduction à la Théorie de Hodge*, Panoramas et synthèses **3**, Société mathématique de France, 113–168.

M. Karoubi (1990). Théorie générale des classes caractéristiques secondaires, *K-Theory* **4**, 55–87.

N. Katz, T. Oda (1968). On the differentiation of de Rham cohomology classes with respect to parameters, *J. Math. Kyoto Univ.* **8**, 199–213.

S. Kleiman (1968). Algebraic cycles and the Weil conjectures. In *Dix Exposés sur la Théorie des Schémas*, Advanced Studies in Pure Math. **3**, North-Holland, Amsterdam 359–386.

K. Kodaira (1954). On Kähler varieties of restricted type; (an intrinsic characterization of algebraic varieties.) *Ann. Math.* **60** 28–48.

K. Kodaira (1986). *Complex Manifolds and Deformations of Complex Structures* Grundlehren der Math. Wiss. **283**, Springer.

J. Kollár (1992). Lemma p. 134 in *Classification of Irregular Varieties*, edited by E. Ballico, F. Catanese, C. Ciliberto, Lecture Notes in Math. **1515**, Springer.

P. Lelong (1957). Intégration sur un ensemble analytique complexe, *Bull. Soc. Math. France*, **85** 239–262.

J.D. Lewis (1990). *A Survey of the Hodge Conjecture*, Publication du Centre de Recherche Mathématique de Montréal.

Yu. Manin (1968). Correspondences, motifs, and monoidal transformations, *Math. USSR Sbornik* **6**, 439–470.

J. Milnor (1963). *Morse Theory*, *Ann. Math. Studies* **51**, Princeton.

J. Milnor (1965). *Topology from the Differentiable Viewpoint*, The University Press of Virginia.

B. Moishezon (1967a). On n-dimensional compact varieties with n algebraically independent meromorphic functions. *Amer. Math. Soc. Translations* **63**, 51–177.

B. Moishezon (1967b). A criterion for projectivity of complete algebraic abstract varieties, *Amer. Math. Soc. Translations* **63**, 1–50.

J. Moser (1965). On the volume element of a manifold, *Trans. Amer. Math. Soc.* **120**, 286–294.

D. Mumford (1970). *Abelian Varieties*, Oxford University Press, Oxford.

D. Mumford (1988). *The Red Book of Varieties and Schemes*, Lecture Notes in Mathematics **1358**, Springer-Verlag.

R. Narasimhan (1966). *Introduction to the Theory of Analytic Spaces*, Lecture Notes in Mathematics **25**, Springer-Verlag.

J.-P. Serre (1955). Faisceaux algébriques cohérents, *Ann. Math.* **61**, 197–278.

J.-P. Serre (1956). Géométrie algébrique et géométrie analytique, *Ann. Inst. Fourier* **6**, 1–42.

J.-P. Serre (1958/1959). In Séminaire Chevalley (1958/1959). *Variétés de Picard*, 3ème année.

C. Soulé (1989). Connexions et classes caractéristiques de Beilinson, *Contemp. Math.* **83**, 349–376.

C. Soulé et al. (1992). *Lectures on Arakelov Geometry*, Cambridge Studies in Advanced Mathematics 33, Cambridge University Press.

E. Spanier (1966). *Algebraic Topology*, Tata McGraw-Hill publishing company, New Delhi.

G. Tian (1987). Smoothness of the universal deformation space of compact Calabi-Yau manifolds and its Peterson-Weyl metric, in *Mathematical Aspects of String Theory* (ed. S. T. Yau), World Scientific Press, 629–646.

B. Totaro (1997). Torsion algebraic cycles and complex cobordism, *J. Amer. Math. Soc.* **10** (2), 467–493.

C. Voisin (2000). The Griffiths group of a general Calabi-Yau threefold is not finitely generated, *Duke Math. J.* **102** (1), 151–186.

C. Voisin (2002). A counterexample to the Hodge conjecture extended to Kähler varieties, *IMRN 2002*, No 20, 1057–1075.

C. Voisin. *Hodge theory and Complex Algebraic Geometry* II, to appear Cambridge University Press, Cambridge.

A. Weil (1957). *Variétés Kählériennes*, Hermann, Paris.

Index

Printed in the United States
By Bookmasters